LCA Compendium – The Complete World of Life Cycle Assessment

Series editors
Walter Klöpffer, LCA Consult & Review, Frankfurt am Main, Germany
Mary Ann Curran, BAMAC, Ltd., LCA & Sustainability Consultant, Rock Hill, SC, USA

Aims and Scope

Life Cycle Assessment (LCA) has become the recognized instrument to assess the ecological burdens and human health impacts connected with the complete life cycle (creation, use, end-of-life) of products, processes and activities, enabling the assessor to model the entire system from which products are derived or in which processes and activities operate. Due to the steady, world-wide growth of the field of LCA, the wealth of information produced in journals, reports, books and electronic media has made it difficult for readers to stay abreast of activity and recent developments in the field. This led to the realization of the need for a comprehensive and authoritative publication.

The *LCA Compendium* Book Series will discuss the main drivers in LCA (SETAC, UNEP/SETAC Life Cycle Initiative, etc.), the strengths and limitations of LCA, the LCA phases as defined by ISO standards, specific applications of LCA, Life Cycle Management (LCM) and Life Cycle Sustainability Assessment (LCSA). Further volumes, which are closely related to these themes will cover examples of exemplary LCA studies ordered according to the importance of the fields of application. They will also present new insights and new developments and will keep the whole work current. The aim of the series is to provide a well-structured treatise of the field of LCA to give orientation and guidance through detailed descriptions on all steps necessary to conduct an LCA study according to the state of the art and in full agreement with the standards.

The *LCA Compendium* Book Series anticipates publishing volumes on the following themes:

- Background and Future Prospects in Life Cycle Assessment (published in March 2014)
- Goal and Scope Definition in Life Cycle Assessment
- Life Cycle Inventory Analysis (LCI)
- Life Cycle Impact Assessment (LCIA) (published in March 2015)
- Interpretation, Critical Review and Reporting in Life Cycle Assessment
- Applications of Life Cycle Assessment
- Special Types of Life Cycle Assessment
- Life Cycle Management (LCM) (to be published in August 2015)
- Life Cycle Sustainability Assessment (LCSA)
- Life Cycle Assessment Worldwide

More information about this series at http://www.springer.com/series/11776

Guido Sonnemann • Manuele Margni
Editors

Life Cycle Management

Springer Open

Editors
Guido Sonnemann
University of Bordeaux, ISM, UMR 5255
Talence Cedex, France

Manuele Margni
CIRAIG, Polytechnique Montréal
Montréal, QC, Canada

ISSN 2214-3505 ISSN 2214-3513 (electronic)
LCA Compendium – The Complete World of Life Cycle Assessment
ISBN 978-94-024-0443-2 ISBN 978-94-017-7221-1 (eBook)
DOI 10.1007/978-94-017-7221-1

Springer Dordrecht Heidelberg New York London

Softcover reprint of the hardcover 1st edition 2015

Printed on acid-free paper

Springer Science+Business Media B.V. Dordrecht is part of Springer Science+Business Media (www.springer.com)

Preface

Life Cycle Assessment (LCA) is a science-based technique to assess resource consumption and potential environmental impacts associated with a product or service throughout its whole life cycle, from extraction via manufacturing and use to end-of-life by compiling an inventory of relevant energy, material, water and land inputs, and releases to the environment.

Life Cycle Management (LCM) is a management concept applied in industrial and service sectors to improve products and services while enhancing the overall sustainability performance of business and its value chains. In this regard, Life Cycle Management is an opportunity to differentiate through sustainability performance on the market place, working with all departments of a company such as research and development, procurement, and marketing, and enhance the collaboration with stakeholders along a company's value chain. LCM is used beyond short-term business success and aims at long-term achievements minimizing environmental and socioeconomic burden while maximizing economic and social value.

What was our reason to prepare this LCM book? We believe that Life Cycle Management is a key concept within the life cycle community that allows operationalizing sustainability within organizations by putting life cycle thinking into business practice. In the LCM context, Life Cycle Assessment is one important technique among others, hence, the need for providing with this book a space for further explaining what LCM is about and its relationship to Life Cycle Assessment.

Moreover, there is confusion with different similar terms such as product life cycle management and application life cycle management, which are not linked to sustainability, so that we identified a need for clarification. Finally, there is progress in implementing LCM, and there are challenges in mainstreaming LCM in business practice and beyond in public policy, which we felt was worth reporting on.

This volume of the *LCA Compendium* aims to give the reader a thorough insight into Life Cycle Management presenting its origin, evolution, and the state of practice, including progress made, current challenges, the way forward to its operationalization in more and more organizations, and its linkage to business value creation.

With this open-access book, which we were able to prepare thanks to our role as cochairs of the LCM 2015 Conference, we target not only the scientific community but in particular also the life cycle professionals in business and industry as well as administration. We expect the readers to find inspiration on how to implement Life Cycle Management in organizations throughout multiple value chains.

The book is structured in five parts:

Part I Introducing Life Cycle Management)

The first part defines what Life Cycle Management is within the realm of sustainability and what are the opportunities and challenges to implement it into business practice.

Part II Advancing the Implementation of Life Cycle Management in Business Practice

The second part continues focusing on progress made with regard to implementation processes of life cycle approaches and its linkage to business value creation.

Part III Life Cycle Management as Part of Sustainable Consumption and Production Strategies and Policies

The third part broadens the scope of Life Cycle Management and presents it as part of sustainable consumption and production discussing strategic opportunities for policy action and related responsibilities of consumers and policy makers, among other stakeholders, along the value chain.

Part IV Mainstreaming and Capacity Building on Life Cycle Management

The fourth part provides a series of chapters addressing the challenges of mainstreaming Life Cycle Management. It discusses opportunities to build operational capability and the potential for mainstreaming LCM in emerging economies through capacity building, concluding on the need to enhance communication and collaboration within the global LCA community.

Part V Implementation and Case Studies of Life Cycle Management in Different Business and Industry Sectors

Finally this book concludes by providing a few practice examples of Life Cycle Management in different economic sectors.

All the chapters of this book have been elaborated by recognized and experienced experts in the LCM domain to provide the reader a qualified and comprehensive insight into the dynamic and increasingly relevant field of Life Cycle Management. Each chapter functions as a self-containing unit within each part of the book, simultaneously playing its individual role in the overall concept of this volume of the *LCA Compendium*.

Our vision is a sustainable global society where Life Cycle Management approaches are well established and fully integrated into regular decision-making processes. We will be grateful if this book helps the reader to make a step forward in this direction.

Bordeaux, France
Montréal, QC, Canada

Guido Sonnemann
Manuele Margni
Editors – Life Cycle Management

Acknowledgments

The *LCA Compendium* Book Series complements *The International Journal of Life Cycle Assessment*, which has been published by Springer since 2008. This is the volume on Life Cycle Management of the *LCA Compendium*. We first acknowledge and appreciate that Springer accepted to publish a volume on Life Cycle Management (LCM) jointly with LCM 2015, the 7th International Conference on Life Cycle Management, in Bordeaux, 30 August–2 September 2015. In particular we thank Fritz Schmuhl (Environmental Sciences) for recognizing the window of opportunity to publish such a comprehensive volume thanks to the momentum generated by the LCM 2015 Conference.

Second, we would like to thank Walter Kloepffer and Mary Ann Curran, the series editors of the *LCA Compendium*, for their support in making the idea of publishing the volume jointly with the LCM 2015 Conference coming up. Without their experience as editors of journal articles, it would have been difficult to get this collection of chapters reviewed and edited in the short time frame we had available between the first submission of the draft chapters to us and the final submission to Springer.

Our third thank you is to Almut B. Heinrich, the former managing editor of *The International Journal of Life Cycle Assessment*, who is acting as the managing editor for this book series and backed us throughout the process.

Fourth, we would like to thank our numerous chapter authors and coauthors who did not know if their chapter would be accepted or not. Without their willingness to cooperate on this volume, devotion of time, and sharing of expertise and experience, it would not have been possible to produce this book. In this context, our special thanks go to the authors from the industry and business who provided first hand insight on how LCM is implemented.

A special thank you goes also to Eskinder Gemechu, a postdoc at the Life Cycle Group CyVi of the University of Bordeaux, who helped to manage the status and to edit the chapters.

Finally, we would like to express our gratitude to our families who have had us less with them during these months of preparing the LCM 2015 Conference and this LCM volume in particular.

Guido Sonnemann and Manuele Margni
Editors – Life Cycle Management

Contents

Contributors

Angela Adams Quantis-International, Lausanne, Switzerland

Naeem Adibi [avniR] Platform – cd2e, Rue de Bourgogne, Loos en Gohelle, France

Mourad Ben Amor Department of Civil Engineering, Université de Sherbrooke, Sherbrooke, QC, Canada

Anne-Claire Asselin Division of Technology, Industry and Economics, UNEP, Paris, France

Miguel Fernandez Astudillo Department of Civil Engineering, Université de Sherbrooke, Quebec, Canada

Martin Baitz thinkstep AG, Leinfelden-Echterdingen, Germany

Sanjeevan Bajaj Federation of Indian Chambers of Commerce & Industry, New Delhi, India

Fritz Balkau Sustainable Solutions, Paris, France

Russ T. Balzer WorldAutoSteel, Springboro, OH, USA

Mark Barthel Waste & Resources Action Programme (WRAP), Banbury, Oxon, UK

Christian Bauer Laboratory for Energy Systems Analysis, Paul Scherrer Institut (PSI), Villigen, Switzerland

Catherine Beutin Aquimer competitivity cluster, Boulogne-sur-Mer, France

Christophe Bogaert ADEME Nord-Pas de Calais, Centre tertiaire de l'Arsenal, Douai, France

Henk Bosch DSM Nutritional Products, Kaiseraugst, Switzerland

Joanne Boudehenn Nord Picardie Bois, Villeneuve-d'Ascq, France

Stefan Brattig Product Sustainability, BMW Group, Munich, Germany

Jodie Bricout [avniR] Platform – cd2e, Rue de Bourgogne, Loos en Gohelle, France

Clare Broadbent World Steel Association, Brussels, Belgium

Florian Broch Volkswagen AG, Environmental Affairs Product, Wolfsburg, Germany

Sophie Cabaret CCI Grand Lille, Lille, France

François Xavier Callens cd2e excellence cluster, Rue de Bourgogne, Loos en Gohelle, France

Llorenç Milà i Canals Division of Technology, Industry and Economics, UNEP, Paris, France

Kirana Chomkhamsri P.P.P. Intertrader, Nonthaburi, Thailand

Soledad Contreras PRé Consultants, Amersfoort, The Netherlands

Marie Darul cd2e excellence cluster, Rue de Bourgogne, Loos en Gohelle, France

Archana Datta Federation of Indian Chambers of Commerce and Industry (FICCI), New Delhi, India

Christelle Demaretz Nord-Pas de Calais Regional Council, Lille, France

Jon Dettling Quantis-International, Boston, MA, USA

Felipe Duarte Vale, Rio de Janeiro, Brazil

James A. Fava thinkstep, West Chester, PA, USA

Matthias Finkbeiner Technische Universität Berlin, Berlin, Germany

Angela Fisher General Electric, Ecoassessment Center of Excellence, Niskayuna, NY, USA

William Flanagan General Electric, Ecoassessment Center of Excellence, Niskayuna, NY, USA

Jeppe Frydendal Ecolabelling Denmark, Stockholm, Sweden

Danish Standards Foundation, Charlottenlund, Denmark

Anne Gaasbeek PRé Consultants, Amersfoort, The Netherlands

Xavier Gabarrell Sostenipra (ICTA-IRTA-Inèdit; 2014 SGR 1412)-Institut de Ciència i Tecnologia Ambientals (ICTA), Universitat Autònoma de Barcelona (UAB), Bellaterra, Spain

Department of Chemical Engineering, Universitat Autònoma de Barcelona, Bellaterra, Spain

Eskinder Demisse Gemechu University of Bordeaux, ISM, UMR 5255, Talence, France

Mark Goedkoop PRé Consultants, Amersfoort, The Netherlands

Kinga Grenda University of Coimbra, Coimbra, Portugal

Marc Haquette cd2e excellence cluster, Rue de Bourgogne, Loos en Gohelle, France

Sarra Harbi Quantis-International, EPFL Innovation Park, Bât D, Lausanne, Switzerland

Curtis A. Harnanan thinkstep, Ottawa, ON, Canada

David Hunkeler aquaTECH, Geneva, Switzerland

Allan Astrup Jensen Nordic Institute of Product Sustainability, Environmental Chemistry and Toxicology (NIPSECT), Copenhagen/Frederiksberg, Denmark

Alejandro Josa Department of Geotechnical Engineering and Geosciences, School of Civil Engineering, Universitat Politècnica de Catalunya-Barcelona Tech (UPC), Barcelona, Spain

Institute of Sustainability, UPC, Barcelona, Spain

Yuki Hamilton Onda Kabe Brakem, São Paulo, Brazil

Sofia Khan UNEP, Albany, NY, USA

Patrick Kim Product Sustainability, BMW Group, Munich, Germany

Stephan Krinke Volkswagen AG, Environmental Affairs Product, Wolfsburg, Germany

Annekatrin Lehmann Technische Universität Berlin, Berlin, Germany

Yves Loerincik Quantis-International, Lausanne, Switzerland

B.M. Krishna Manda Copernicus Institute of Sustainable Development, Utrecht University, Utrecht, The Netherlands

Manuele Margni Quantis-International, EPFL Innovation Park, Bât D, Lausanne, Switzerland

CIRAIG, Polytechnique Montréal, Montréal, Canada

Bernard Mazijn Ghent University and IDO vzw, Bruges, Belgium

Jeanne Meillier UP-tex competitivity cluster, Marcq-en-Baroeul, France

Joan Manuel F. Mendoza Sostenipra (ICTA-IRTA-Inèdit; 2014 SGR 1412)-Institut de Ciència i Tecnologia Ambientals (ICTA), Universitat Autònoma de Barcelona (UAB), Bellaterra, Spain

Sustainable Industrial Systems (SIS) Group, School of Chemical Engineering and Analytical Science, University of Manchester, Manchester, UK

Eric Mieras PRé Consultants, Amersfoort, The Netherlands

Simon Miller 3keel LLP, Oxford, UK

Stephane Morel Renault, Alliance Technology Development, Guyancourt, France

Oscar Alberto Vargas Moreno Secretaría Distrial de Ambiente (SDA), Bogotá, Colombia

Maria Luisa Nerys Vale, Rio de Janeiro, Brazil

Jordi Oliver-Solà Sostenipra (ICTA-IRTA-Inèdit; 2014 SGR 1412)-Institut de Ciència i Tecnologia Ambientals (ICTA), Universitat Autònoma de Barcelona (UAB), Bellaterra, Spain

Inèdit. Inèdit Innovació, S.L. UAB Research Park, Cabrils, Spain

Patrick Orlans CETIM Technical Centre for the Mechanical Industry, Marcq-en-Baroeul, France

Luiz Gustavo Ortega Brakem, São Paulo, Brazil

Vanessa Pasquet [avniR] Platform – cd2e, Rue de Bourgogne, Loos en Gohelle, France

Anna Petit-Boix Sostenipra (ICTA-IRTA-Inèdit; 2014 SGR 1412)-Institut de Ciència i Tecnologia Ambientals (ICTA), Universitat Autònoma de Barcelona (UAB), Bellaterra, Spain

Maria da Graça C.B. Popi Oxiteno, São Paulo, Brazil

Ana Quiros Association for Life Cycle Assessment in Latin America (ALCALA), San Jose, Costa Rica

M. Graça Rasteiro University of Coimbra, Coimbra, Portugal

Gerald Rebitzer Amcor, Zurich, Switzerland

Arne Remmen Aalborg University, Aalborg, Denmark

Quentin Renault Agroe, excellence cluster, Quartier des Trois Parallèles, La Citadelle Avenue du Mémorial des Fusillés, Arras, France

Jean-Pierre Revéret CIRAIG – Université du Québec à Montréal, Montréal, Canada

Sophie Reynaud MATIKEM competitivity cluster, Parc scientifique de la Haute Borne, Villeneuve d'Ascq, France

Joan Rieradevall Sostenipra (ICTA-IRTA-Inèdit; 2014 SGR 1412)-Institut de Ciència i Tecnologia Ambientals (ICTA), Universitat Autònoma de Barcelona (UAB), Bellaterra, Spain

Department of Chemical Engineering, Universitat Autònoma de Barcelona, Bellaterra, Spain

Marina Santa Rosa Rocha Brazilian Business Council for Sustainable Development, Rio de Janeiro, Brazil

Aubin Roy [avniR] Platform – cd2e, Rue de Bourgogne, Loos en Gohelle, France

Frieder Rubik Institute for Ecological Economy Research (IÖW), Heidelberg, Germany

Alice Salamon [avniR] Platform – cd2e, Rue de Bourgogne, Loos en Gohelle, France

Peter Saling BASF SE, Sustainability strategy, Ludwigshafen, Germany

David Sanjuan-Delmás Sostenipra (ICTA-IRTA-Inèdit; 2014 SGR 1412)-Institut de Ciència i Tecnologia Ambientals (ICTA), Universitat Autònoma de Barcelona (UAB), Bellaterra, Spain

Esther Sanyé-Mengual Sostenipra (ICTA-IRTA-Inèdit; 2014 SGR 1412)-Institut de Ciència i Tecnologia Ambientals (ICTA), Universitat Autònoma de Barcelona (UAB), Bellaterra, Spain

Urs Schenker Nestlé Research Center, Lausanne 26, Switzerland

Guido Sonnemann University of Bordeaux, ISM, UMR 5255, Talence, France

Philip Strothmann Strothmann Consulting, Berlin, Germany

Thomas E. Swarr Yale University, New Haven, CT, USA

Marzia Traverso Product Sustainability, BMW Group, Munich, Germany

Karin Treyer Laboratory for Energy Systems Analysis, Paul Scherrer Institut (PSI), Villigen, Switzerland

Cassia Maria Lie Ugaya Universidade Tecnológica Federal do Paraná (UTFPR), Curitiba, Brazil

Sonia Valdivia World Resources Forum, St. Gallen, Switzerland

Cornelis Theunis Van Der Lugt Center for Corporate Governance in Africa, Stellenbosch University Business School, Cape Town, South Africa

Volkmar Wagner Product Sustainability, BMW Group, Munich, Germany

Jens Warsen Volkswagen AG, Environmental Affairs Product, Wolfsburg, Germany

Marc-Andree Wolf maki Consulting, Stephanskirchen, Germany

Ernst Worrell Copernicus Institute of Sustainable Development, Utrecht University, Utrecht, The Netherlands

Part I
Introducing Life Cycle Management

This part defines what life cycle management is within the realm of sustainability and what are the opportunities and challenges to implement it into business practice.

Chapter 1
Introduction: Life Cycle Management

Gerald Rebitzer

Abstract Environmental management practices in most business organizations from the early 1990s were characterized by their focus on internal operations, cost savings, and compliance and risk management approaches. Such a traditional view of sustainability management, however, is not sufficient to address current business challenges – to create competitive advantages while contributing to sustainable development. There is a need for a life cycle management practice that expands the scope through including the complete value chain and that links sustainability management and performance of organizations and products to business value and value creation. Being an extremely powerful concept and process, life cycle management can ensure businesses deliver real-world improvements for all stakeholders. In the long term, it can also help to transform the market by making sustainability a differentiator just as quality is today.

Keywords Business organization • Life cycle assessment • Life cycle management • Life cycle sustainability assessment • Life cycle sustainability management • Sustainability • Value creation

1 The Business Context

Paradigm shifts in the world of international business and economics and a shift from a view that focuses purely on profit to one that takes a more balanced and long-term approach to also address environmental, governance, and social factors have been discussed for many years now. There is a growing understanding that businesses cannot only focus on short-term profitability and internal factors such as productivity improvements to be successful in the long run. On the other side,

G. Rebitzer (✉)
Amcor, Affolternstrasse 56, CH-8050 Zurich, Switzerland
e-mail: gerald.rebitzer@amcor.com

G. Sonnemann, M. Margni (eds.), *Life Cycle Management*,
LCA Compendium – The Complete World of Life Cycle Assessment,
DOI 10.1007/978-94-017-7221-1_1

however, for many business sectors, competition is now truly global and fiercer than ever, making it necessary to have the focus on profitable growth and to being able to react very flexible to changing market requirements. The ancient quote "Change is the only constant in life" (generally attributed to Heraclitus of Ephesus, a Greek philosopher, who lived from 535 BC to 475 BC) is today more valid than ever.

How can this fast-paced business reality that inevitably aims at profitability, short and long-term, be married with the need to balance long-term financial, environmental, governance, and social impacts and benefits?

The traditional view in environmental management and later in sustainability management can be characterized by:

- Concentrating on internal operations ("inside the factory walls")
- Targeting cost savings through efficiency improvements and related reductions in material and energy use as well as waste generation
- Assurance of compliance to regulatory and other explicitly stated requirements (international standards, customer requests, etc.)
- Risk management, mainly to avoid liability issues and reputational damage

This perspective, which was shared by most governments and other stakeholders, was prevalent in most business organizations from the early 1990s and well into the new millennium and is still the standard in many organizations today. It is often represented by organizations, where the sustainability function is a sub-function of Environment, Health, and Safety (EHS).

It is obvious that this internally focused cost savings, compliance, and risk management approach can only be a basis, but will never be sufficient to address the aforementioned business challenges and align with the primary profitability goals of any business organization in a market economy.

2 The Role of Life Cycle Management

This is where life cycle management, first discussed in the pioneering 1st International Conference on Life Cycle Management organized by Allan Astrup Jensen (2001), then formally introduced by David Hunkeler (Hunkeler et al. 2004) and later extended by Matthias Finkbeiner (Finkbeiner, ed, 2011) towards life cycle sustainability management, comes in by:

- Expanding the scope to also address upstream (supply chain) and downstream activities (customers and their customers, and products)
- Addressing not only environmental but also social and economic aspects throughout the life cycle of products and services
- Linking sustainability management and performance of organizations and products to business value and value creation

Expanding the scope means to include the complete value chain, both from the product perspective (life cycle thinking), but also in the sense of value chain cooperation between organizations. With such an approach, organizational and product performance become building blocks of the relationships with suppliers and customers and therefore part of the dialogue and performance criteria between partners in the value chain. This way optimal solutions can be found, looking at the complete picture, and trade-offs and inefficient activities can be avoided.

Linking sustainability to value is all about how sustainability can help to create added value. This goes far beyond cost savings and managing risks and compliance. It is an opportunity to leverage sustainability as an element of differentiation on the market and driver of profitable growth. It is an opportunity to make sustainability a key factor in research and development, operations, procurement, sales and marketing, etc.

How can this be achieved? It is extremely challenging or even impossible to get people excited about "doing less bad" or just being aligned with regulations and explicit requests (which are unfortunately often only "tick the box" exercises), but if one leverages the opportunities from an integrated value proposition that takes the sustainability offering into account, one gets to a completely new dimension that can move sustainability out of the "green corner" and into the business mainstream. And this is not limited to end-producers, who sell products to consumers. It is relevant for the complete value chain, since most of the time the contributions of businesses in the supply chain are essential to implement sustainability for a given end-product or service.

It is important to stress that this value proposition can relate to both direct product performance (e.g., a product with improved environmental life cycle performance) and management performance in the supply chain without measurable changes in the product (such as for ethically sourced products).

Leading businesses that are successful in making life cycle management an enabler that helps to make the day-to-day job of the aforementioned functions more efficient and/or better are achieving an edge and are outperforming their competitors. In order to make this happen, sustainability has to be integrated into standard business processes, very similar to the way quality or cost aspects are integrated today (Remmen et al. 2007; UNEP/SETAC 2009).

3 Conclusions and Perspectives

In summary one can conclude that life cycle management is an extremely powerful concept and process and can enable businesses and other organizations to make sustainability part of "business as usual" and deliver real-world improvements for them and their customers. Life cycle sustainability management, if developed and implemented appropriately for a given organization, has the power to move sustainability management from a cost of doing business to a driver of profitability affecting all three elements of the triple bottom line.

In the long term, it is expected that life cycle management can help to transform the market by making sustainability a differentiator just as quality is today. Only if sustainability is a factor of competition on the market, can market forces kick in to drive performance. Life cycle management can be enabler to unleash these market forces and deliver the step-change improvements, e.g., in combating climate change impacts, that are so desperately needed for the long-term prosperity and survival of mankind.

References

Jensen AA (2001) Life cycle management – a bridge to sustainable products. In: 1st international conference of life cycle management (LCM). dk-TEKNIK ENERGY & ENVIRONMENT, Copenhagen/Soeborg, 27–28 Aug 2001

Finkbeiner M (ed) (2011) Towards life cycle sustainability management. Springer, Dordrecht/Heidelberg/London/New York

Hunkeler D, Saur K, Rebitzer G, Schmidt W, Jensen A, Stranddorf H, Christiansen K (2004) Life cycle management. SETAC, Pensacola FL

Remmen A, Jensen AA, Frydendal J (2007) Life cycle management. A business guide to sustainability. UNEP DTIE

UNEP/SETAC (2009) Life cycle management: how business uses it to decrease footprint, create opportunities and make value chains more sustainable. UNEP, Paris

Chapter 2
Life Cycle Management: Implementing Sustainability in Business Practice

Guido Sonnemann, Eskinder Demisse Gemechu, Arne Remmen, Jeppe Frydendal, and Allan Astrup Jensen

Abstract Life cycle management is a business management concept applied in industrial and service sectors to improve products and services, while enhancing the overall sustainability performance of the business and its value chains. Life cycle thinking and product sustainability is operational for businesses that are ambitious and committed to reducing their environmental and socio-economic burden while maximizing economic and social value. In this regard, life cycle management is used beyond short-term business success and aims at long-term achievements. The term "life cycle management" has been confused with other uses in engineering and manufacturing (product life cycle management) and in software development (application life cycle management), in buildings, plants, information management and so on. There is a need to clarify this term and its definition more than a decade since the concept was first introduced. This chapter aims at elaborating the concept and definitions of life cycle management as currently found in literature and as extending it from focusing on implementation of life cycle sustainability assessment into business practice to include it as part of sustainable consumption and production strategies and policies. Methods and tools used and the general framework for life cycle sustainability management covering environmental, social and economic aspects in business practices are discussed in detail.

G. Sonnemann (✉) • E.D. Gemechu
University of Bordeaux, ISM, UMR 5255, 33400 Talence, France

CNRS, ISM, UMR 5255, 33400 Talence, France
e-mail: guido.sonnemann@u-bordeaux.fr

A. Remmen
Aalborg University, Aalborg, Denmark

J. Frydendal
Ecolabelling Denmark

Danish Standards Foundation

A.A. Jensen
Nordic Institute of Product Sustainability, Environmental Chemistry and Toxicology (NIPSECT), Copenhagen/Frederiksberg, Denmark

G. Sonnemann, M. Margni (eds.), *Life Cycle Management*,
LCA Compendium – The Complete World of Life Cycle Assessment,
DOI 10.1007/978-94-017-7221-1_2

Keywords Application life cycle management • Data life cycle management - Information life cycle management • Life cycle assessment • Life cycle management • Life cycle sustainability management • Product life cycle management • Sustainability

1 Life Cycle Management and Life Cycle Sustainability Management: A Clarification of Terms

A web search on life cycle management (LCM) results in a link to the UNEP/SETAC Life Cycle Initiative's website as well as the official UNEP website on life cycle management and the publication of the UNEP/SETAC Life Cycle Initiative on *Life Cycle Management: A Business Guide to Sustainability* (Remmen et al. 2007), to which the authors of this chapter as co-initiator and co-authors of the UNEP publication will refer to. Moreover, links of LCM events from the first held in Copenhagen to the recent conferences in Gothenburg (http://lcm2013.org/) and the upcoming event in Bordeaux (http://lcm2015.org/) are among the top ten search results jointly with references to product life cycle management (PLCM) and information life cycle management (ILCM).

These considerations are the motivation for the authors to use Life Cycle Management: Implementing Sustainability in Business Practice as the title for this chapter.

Looking at some of the existing definitions of LCM that are summarized in Table 2.1, LCM seems to be a concept with a broad variety of approaches and methodological tools. Companies apply it in a number of different ways in order to achieve the desired outcomes, as far as it relates to their sustainability performance. The theoretical background for LCM has been developed by a SETAC Working Group (Hunkeler et al. 2004). Yet there is no universal definition of LCM.[1]

LCM is mainly a business management concept for sustainable products that can be applied in the industrial and service sectors with the aim of improving specific goods and services and enhancing the overall sustainability performance of the business and its value chains in general. It makes life cycle thinking and product sustainability operational for businesses that are ambitious and are committed to reduce their environmental and socio-economic burden, while maximizing economic and social values. In this regard, LCM is used beyond the short-term business success; rather it aims at taking businesses forward towards long-term achievements and sustainable value creation. So LCM requires a holistic view and a full understanding of interdependency of businesses in order to support relevant decisions and actions so as to improve sustainability performance that takes into account both the environmental and social benefits and at the same time offer a number of value creation opportunities to the business.

[1] Seemingly similar, but unrelated, terms include product lifecycle management (PLM), application life cycle management (ALM) for software, and data lifecycle management (DLM).

Table 2.1 Different definitions of life cycle management (further developed based on the work by Seuring 2004)

Reference	LCM definitions
Linnanen (1995)	Life cycle management consists of three views: (1) the management view – integrating environmental issues into the decision making of the company; (2) the engineering view – optimizing the environmental impact caused by the product during its life cycle; and (3) the leadership view – creating a new organizational culture
Fava (1997)	Life cycle management is the linkage between life cycle environmental criteria and an organization's strategies and plans to achieve business benefits
Finkbeiner et al. (1998)	A comprehensive approach towards product and origination related environmental management tools that follow a life cycle perspective
Heiskanen (2002)	LCA-based ideas and tools can be viewed as emerging institutional logics of their own. While LCA makes use of many scientific models and principles, it is more a form of accounting than an empirical, observational science. Thus, the life cycle approach implies a kind of "social planner's view' on environmental issues, rather than the minimization of a company's direct environmental liabilities"
Hunkeler et al. (2004)	Life cycle management (LCM) is an integrated framework of concepts and techniques to address environmental, economic, technological and social aspects of products, services and organizations. LCM, as any other management pattern, is applied on a voluntary basis and can be adapted to the specific needs and characteristics of individual organizations
Baumann and Tillman (2004)	LCM is "the managerial practices and organizational arrangements that apply life cycle thinking. This means that environmental concerns and work are coordinated in the whole life cycle instead of being independent concerns in each company"
Remmen et al. (2007)	LCM is a product management system aiming to minimize environmental and socioeconomic burdens associated with an organization's product or product portfolio during its entire life cycle and value chain
UNEP/SETAC (2009)	"… a business management approach that can be used by all types of businesses (and other organizations) to improve their products and thus the sustainability performance of the companies and associated value chains" "It can be used to target, organize, analyze and manage product-related information and activities towards continuous improvement along the life cycle"
Jensen (2012)	"… a systematic integration of life cycle thinking in modern business practice with the aim to provide the societies with more sustainable goods and services and to manage the total lifecycle's of an organizations product portfolio towards more sustainable production and consumption"

The definitions of LCM are thus wide and its concept needs further development, to which this book aims to contribute with theoretical and practical contributions, in particular from industries and businesses. New aspects include, for instance, activities on mainstreaming and capacity building as well as the use of LCM in the context of emerging economies, SMEs and regional development. The focus on the earlier definitions of LCM was mainly linked to the management view of only the environmental aspect of a product or a company – environmental LCM (Fava 1997;

Finkbeiner et al. 1998; Linnanen 1995). However, recent definitions of LCM cover environmental, social and economic issues (Hunkeler et al. 2004; Remmen et al. 2007) along a product life cycle, which is in line with recent developments in the area of life cycle assessment (LCA) that further expand the context of LCA to include social and economic elements under the life cycle sustainability assessment (LCSA) framework (Finkbeiner et al. 2010; Klöpffer 2008; UNEP 2011) to cover, for instance, new challenges related to the criticality of materials (Sonnemann et al. 2015). Hence, the implementation of LCSA into real world decision-making processes both at product, process or individual organizational level is to be ensured through the application of a broader LCM concept that aims at maximizing the triple bottom line. Finkbeiner (2011) referred to it as life cycle sustainability management (LCSM) for the first time.

2 Life Cycle Management: Concepts and Definition

As indicated in the *Business Guide to Sustainability* by Life Cycle Management (Remmen et al. 2007), which itself is based on Remmen and Münster's (2003) report to the Danish Ministry of Environment and the pioneering SETAC publication on Life Cycle Management by Hunkeler et al. (2004), LCM has been developed on the basis of fundamental concepts related to sustainable development, which are the triple bottom line and life cycle thinking. The most popular definition of sustainable development is the one from the United National World Commission on Environment and Development "Development that meets the needs of the present without compromising the ability of future generations to meet their own needs" (Brundtland Commission 1987). This definition is based on two key concepts "needs" (the essential needs of the world's poor, to which overriding priority should be given) and "limitations" (the restriction imposed by technologies and socio-economic factors on the ability of the environment to meet the needs of present and future generation).

The triple bottom line (3BL) is a framework that integrates the "three dimensions of sustainability: economic, environmental and social" (Fig. 2.1) (Remmen et al. 2007). They are also called the three Ps: people, planet and profit. Businesses traditionally used to account only the economic aspect of their "bottom line" through profit they gain or lose. However, the modern accounting broadens the definition of bottom line to a full cost accounting by including the environmental cost on ecosystem service and a cost on the society. The consideration of 3BL in the conceptualization of LCM allows companies to broaden their focus from only economic aspects to the environmental and social dimensions.

The goal of life cycle thinking is to avoid burden shifting by assessing a product's use of natural resources and its impact on the environment, the economy and society throughout its entire life cycle. The life cycle of a given product involves a number of stages from the extraction of raw materials through processing, manufacturing, distribution, use, recycling, reuse or final disposal (Fig. 2.2). Life cycle thinking

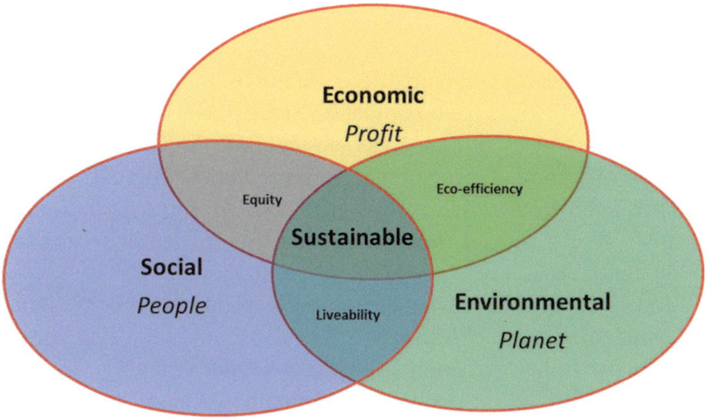

Fig. 2.1 The three dimensions of sustainability

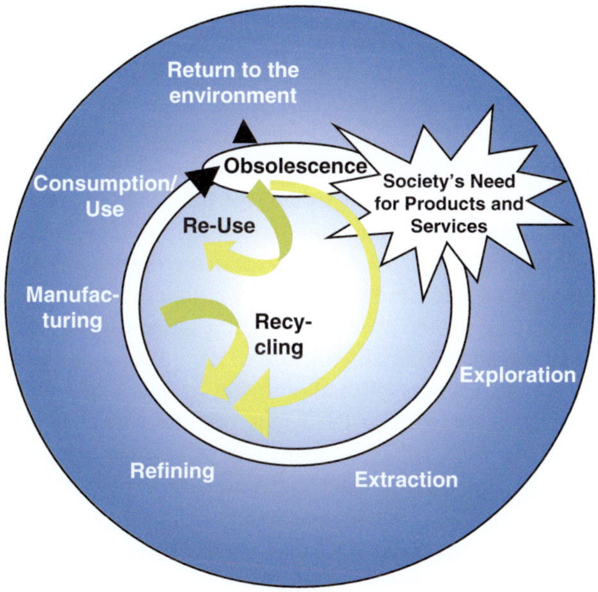

Fig. 2.2 The life cycle of products and services

enables to consider the environmental, social and economic impacts associated with the production or consumption of the product by taking into account all the stages of the product life cycle. It provides a means of ensuring that improvements in one stage are not creating a greater cumulative impact by simply shifting the burden to another stage of the life cycle. Therefore, it also allows companies to see the influence of their choices with regard to sustainability and help them take decisions, so trade-offs can be balanced positively to impact the economy, the environment and society.

In this context, LCA as an ISO standardized analysis is just one decision-making tool used to assess and identify the environmental aspects and potential environmental impacts of a product over its life cycle (ISO 2006a, b). Life cycle sustainability assessment (LCSA) is the combination of LCA, life cycle costing (LCC) and social life cycle assessment (SLCA) to assess the three dimensions of sustainability for products (Finkbeiner et al. 2010; Klöpffer 2008; UNEP 2011). In addition, also other tools such as material flow analysis (MFA), input–output analysis[2] and environmental risk assessment (ERA) are used by organizations to address their sustainability challenges.

LCM allows organizations to put life cycle thinking into modern business practice by using these tools. However, LCM is also about the systematic integration of product sustainability in company strategy and planning, product design and development, purchasing decisions and communication programs (Remmen et al. 2007).

A particular aspect of LCM is the question of who has which responsibilities in the product life cycle with regard to sustainability and who can do what under which circumstances. Evidently a retailer like Wal-Mart has more power to initiate change within the supply chain than for instance a supplier of automotive parts. Furthermore, the existence of sector wide collaborations like the Global e-Sustainability Initiative (GeSI) shows that there is space for joint work of companies of one sectors to address supply chain challenges, which in the case of GeSI have been addressed for instance by E-TASC (Electronics – Tool for Accountable Supply Chains), which a web-based tool utilized by companies to manage their own factories, communicate with their customers and assess their suppliers on corporate responsibility risks.

A related facet of life cycle (sustainability) management is how it is embedded in sustainable consumption and production (SCP) policies. SCP is understood as the "The use of services and related products, which respond to basic needs and bring a better quality of life while minimizing the use of natural resources and toxic materials as well as the emissions of waste and pollutants over the life cycle of the service or product so as not to jeopardize the needs of future generations" (Norwegian Ministry of the Environment 1994). It means that SCP is a holistic approach that has at its core a life cycle perspective, which is the attitude of becoming mindful of how everyday life has an impact on the environment and society.

According to UNEP (2012), SCP focuses on the sustainable and efficient management of resources at all stages of value chains of goods and services encourages the development of processes that use fewer resources and generate less waste, including hazardous substances, while yielding environmental benefits and frequently productivity and economic gains. Such improvements can also increase the competitiveness of enterprises, turning solutions for sustainability challenge into business, employment and export opportunities. SCP also encourages capturing and reusing or recycling valuable resources, thereby turning waste streams into value

[2] See "LCA Compendium", volume "Special Types of Life Cycle Assessment" (editor: Matthias Finkbeiner), chapter 6 "Input–output and Hybrid LCA" by Shinichiro Nakamura and Keisuke Nansai.

Fig. 2.3 SCP policies along the product life cycle (UNEP 2012)

streams. The fundamental objective of SCP is to decouple socio-economic development from environmental degradation.

SCP policies cover all the areas highlighted in Fig. 2.3. A core element linked to SCP is resource efficiency that is about ensuring that natural resources are efficiently produced and processed, and consumed in a more sustainable way, as well as about reducing the environmental impact from the consumption and production of products over their full life cycles. By producing more wellbeing with less material consumption, resource efficiency enhances the means to meet human needs while respecting the ecological carrying capacity of the earth.

That means SCP is based on a life cycle approach but the link to how it is put into business practice using multiple tools mentioned above is not addressed in the same way as in LCM. Therefore, the question on how tools like LCA are actually used in public policy making and hence might influence business operations has been covered under the heading of LCM in the past. Overall there seem to be high expectations of the future use of LCA in SCP policy areas such as sustainable public procurement and eco-design directives as well as consumer information. However, there are still certain challenges to overcome such as the lack of good quality and available data, the lack of valid and internationally recognized calculation principle, more capacity building and resources.

3 Systems and Tools for Life Cycle Management

Being an integrated management framework of concepts, techniques and procedures, LCM connects different operational concepts, policies, systems, methods, tools and data that incorporate environmental, economic and social aspects and looks how they are interconnected and how to best address these throughout the product or process life cycle. As indicated in the previous section, a wide range of methods, tools and concepts can be used in LCM. Analytical tools are life cycle assessment (LCA), life cycle costing (LCC), social life cycle assessment (SLCA), organizational LCA (OLCA), hotspot analysis, different forms of footprinting such as water footprint and carbon footprint, cost benefit analysis (CBA), material flow analysis (MFA), substance flow analysis (SFA), input–output analysis (IOA), environmental risk assessment (ERA), etc. Procedural tools include auditing, checklists, eco-design, eco-labeling, etc. and supportive tools such as weighting, e.g. by Delphi expert panels, uncertainty analysis, sensitivity analysis, etc. could be applied. LCM also includes design concepts such as design for the environment, design for sustainability, design for recycling etc. It also refers to policies and strategies such as circular economy, sustainable consumption and production, integrated product policy (IPP), resource efficiency, eco-efficiency, dematerialization, industrial ecology, etc. as well as organizational systems or programs such as extended product responsibility (EPR), product development process (PDP), certification, environmental communication, value chain management, etc. All these analytical and procedural tools as well as policies, strategies and systems/programs are part of LCM (Nilsson-Lindén et al. 2014; Remmen et al. 2007; Sonnemann and Leeuw 2006). The initial ideas for this integrated approach of using multiple tools and methods stem from the ChainNet project (Wrisberg and Udo de Haes 2002).

The choice of policies, strategies, systems, programs and different types of tools represented in Fig. 2.4 mainly depends on the principal goals and the level of ambition of each company. Companies use LCM to support their goals of providing products that are as sustainable as possible. Companies need to go beyond their organizational boundaries and be willing to expand their scope of collaboration through external communications to all stakeholders of their value chain as it makes them more visible, may improve their public image, improve their relations with stakeholders and may increase their market penetration through mapping their product chains and develop criteria for product enhancement and value creation. Life cycle information may be included in:

- Communication to shareholders and stakeholders in general by, for example, green accounting and annual environmental or sustainability reports
- Communication with customers through such items as life cycle based environmental product declarations, LCA data, product environmental performance indicators or product profiles
- Communication with public authorities via product information schemes and green public procurement guidelines
- Communication with the public, consumers (including professional purchasers in businesses) and retailers using product brochures and various eco-labeling systems and information campaigns

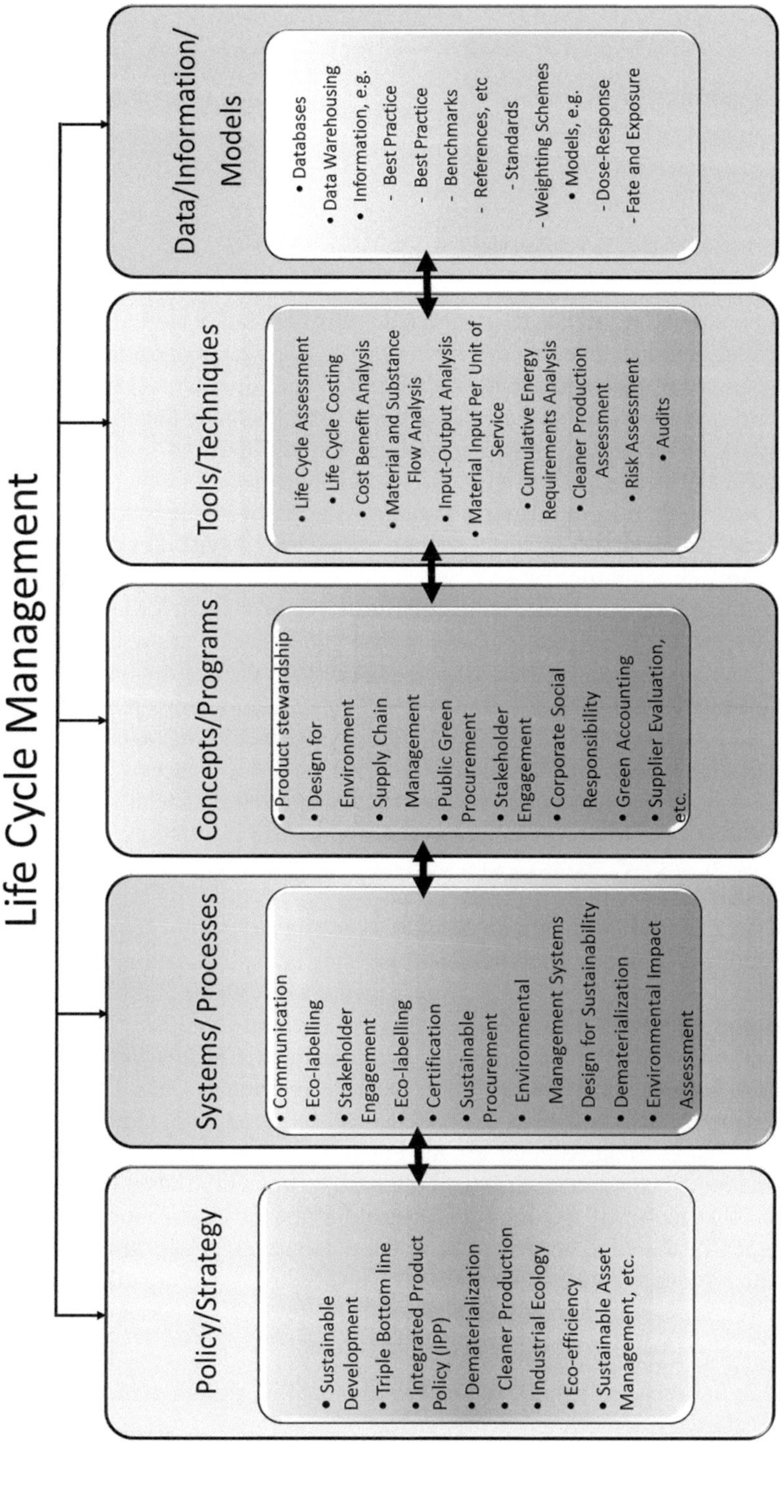

Fig. 2.4 Choice of policies, strategies, systems, programs and different types of tools within the life cycle management context (Based on Remmen et al. 2007)

- Communication with suppliers, including SMEs, using company codes or manuals of conduct, audit or supplier evaluation systems

Larger companies often combine supplier communication with offering training exercises.

4 Organizational Challenges

LCM can be applied in all organizations from a very small-scale local vendor to large and multinational companies. However, the application procedure and its organization may vary in each company. This is mainly due the fact that the relevance of different aspects of sustainability varies from company to company, and it depends on factors such as the type of product system involved in the company, specific social and environmental issues they would like to address, their geographic scope and supply chain complexity and so on. As a life cycle approach, LCM is a dynamic process in which companies may begin applying it with specific goals and objectives depending on the resource they have. They may begin with using LCA as a tool to evaluate their environmental performance of a single product and find an alternative solution to reduce the environmental burden from this product. Through time they can adjust their goal and move forward step-by-step from one project to a more advanced and sophisticated life cycle management practice, with a process in place for multiple products, which require advanced tools and data-intensive programs. Another point of departure for a company could be to benchmark their products or services against ecolabel criteria to determine relevant aspects to consider in a life cycle perspective and to find inspiration for improvements. This, of course, requires that ecolabel criteria exist for the products or services in question.

One of the critical reasons for companies to be engaged with LCM practices is their pursuit for continuous improvements, covering economic, environmental and social aspects. Nowadays, companies are influenced by external and internal factors so that they envisage such improvements and develop strategic policies, apply a number of tools and establish programs that integrate LCM into the core operations of their business. Business strategy, market opportunities and requirements from the finance sector, as well as national legislations, trade block related regulations and international agreements are the key drivers behind the implementation of LCM. Evidently, companies have to apply it based upon high-level management decisions, only then LCM becomes an integral part of the organizations' policies and strategies in the short term and long term. However, it can also be implemented with a pioneer in one of the multiple departments of an organization.

Successful implementation of LCM demands continuous support from top level management such as:

- Providing the required resources for the sustainability initiative including time and educational resources

- Participating actively in setting up the strategic sustainability goals of the organization
- Communicating explicitly throughout the organization regarding the sustainability aims in an effective and clear manner
- Involving actively the employees with regard to ideas and suggestions for the use of life cycle approaches

However, in order for LCM to be accepted and get continuous support from top level management, it needs to highlight the economic benefits the company can profit from its implementation in addition to the social and environmental performance improvements.

A successful implementation of LCM also needs full participation by a range of employees in order to ensure that the initiative will be deeply rooted in the organization and that the focus will be on concrete improvements to a product's sustainability profile, rather than mere talk and data collection. Furthermore, broad participation ensures that the LCM program does not 'die' if a key employee involved leaves the organization.

Leading companies will undertake initiatives to increase market share and enhance the potential for product innovation. A business striving for increased resource efficiency may see a strategy for product sustainability as an opportunity to reduce costs. In more conservatively operating companies, intrinsic factors will include reduced penalties and risks since taking a life cycle approach can help identify important opportunities and risks. Other organizations may seek to gain competitive advantage through innovation, brand value enhancement and strategic positioning in the market.

In the case of product design and development processes, for example, design decisions take place within the broader corporate management structure. An integrated management system – covering quality, environment and health & safety – with policies, goals, performance measures and a strategic plan that supports continuous improvements will be a driver for integration of sustainability performance metrics and measures. In this context, life cycle (sustainability) management offers a framework that allows management to organize and align the various applied concepts and tools in such a way as to exploit the synergies and interrelations between them.

Another key factor for the success of LCM practices in an organization is the involvement of all departments. Such an initiative could impact all functions and departments of a company. For instance, an implementation of a new design idea may need the support from procurement and marketing departments. Any decision that changes the material composition of a product not only affects its quality, price and environmental profile but also raises questions regarding procurement of new material, potential new markets, consequences to the production process, new logistical demands, etc. Therefore, communication and sharing ideas within and across departments in an organization is key to LCM. Communication and interaction helps generate a range of news ideas and helps push ideas into realization. It is important to recognize also the environmental and social initiatives, which already exist in various

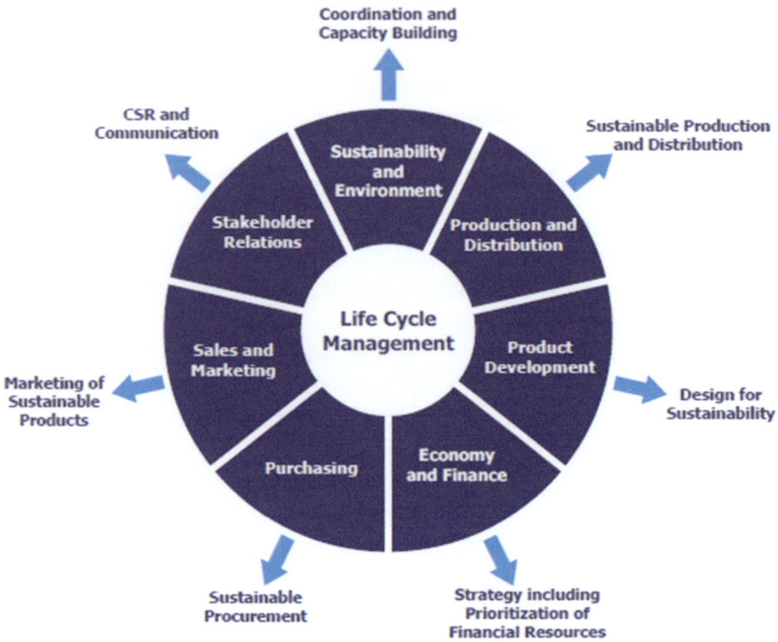

Fig. 2.5 All functions play an important role in life cycle management. The figure shows examples of how different departments in an organization can contribute to an LCM program (Remmen et al. 2007)

departments. Involving all departments means learning from what has done before. A life cycle perspective requires that all departments or functions work together, including product development, purchasing, production, logistics as well as marketing. All functions (illustrated in the following as departments) must therefore participate with ideas for initiatives and solutions, based on their particular expertise.

Figure 2.5 illustrates how different departments in a company can contribute to the establishment and running of LCM initiatives. In most cases, large, multinational companies have environmental, social responsibility and sustainability departments that could coordinate the implementation of LCM. These departments are typically responsible for reporting developments within environmental and sustainable policy and they can provide valuable inputs through training of employees in the other departments. However, it is crucial that the whole company is motivated and 'speaks the same language'.

However, most small and medium-sized enterprises (SMEs) do not usually have such departments. Therefore, the coordination of LCM activities could also be managed by forming a cross-organizational or cross-functional team with a representative from each relevant function, where a motivated employee (the pioneer) can act as coordinator and at the same time make sure that everybody has the necessary tools and materials to inspire and carry out the activities. The relevance for putting LCM into business practice of each department is summarized in Table 2.2.

Table 2.2 Departments in an organization and their relevance for life cycle management in practice (Based on the work from Remmen et al. 2007)

Department	Main activities
Production and Distribution	• Assess the environmental and social impacts associated with production processes and thus suggest an alternative solution to reducing resource consumption and the related impacts • Provide a novel ideas and data for product and process improvements • Identify and suggest a solution to reduce the impacts associated with the energy consumption connected to the transportations of raw materials, intermediate and finished products

Product Development and Design	• Move environmental and social considerations higher up on the design criteria list
	• Develop a new product with the starting point of social, ethical and environmental considerations– for example a new clothes collection based on organic cotton and fair trade
	• Make the existing product more sustainable, for example by replacing an environmentally harmful substance with a less harmful substance
	• Shift from producing a product to supplying a service – the sale of answering machines shifts to an electronic answering service delivered by the phone company
	• Assess the environmental and socioeconomic aspects of a product from two different angles based on a definition of the product system:
	  • A product life cycle perspective with assessment of the environmental and socio-economic impacts of a product system with tools such as Life Cycle Assessment (LCA) or Life Cycle Costing (LCC)
	  • A stakeholder perspective with assessment of impacts based on the stakeholders' view such as legal requirements, market demands, and competitors' products. Quality Function Deployment (QFD), interviews, etc. are commonly applied tools

Economy and Finance	• Provide a good financial performance to allow the company to see its impact on driving the company towards sustainability and LCSM • Assess the life cycle avoided costs due to the implementation of LCSM project, for example, by tracking both the annual cost reduction and commutative savings from prior years
Purchase	• Play an important role in selecting the optimal raw materials, semi-products and products for production, by applying some tools that integrate environmental and social considerations together with other factors such as price, quality, and functionality • Encourage environmental considerations at their suppliers via questions and demands an overview of the supplier's environmental and social initiatives as well as policies; documentation of the impacts from the previous life-cycle stages; overview of working conditions at suppliers and sub suppliers; and/or specific environmental and social data regarding raw materials, secondary materials, etc.
Sales and Marketing	• Ensure a good flow of information to and from the customers such as consumer behavior and preferences, product's eco-friendly use and disposal, etc. • Promote the eco-friendly product e.g. by the use of ecolabels
Stakeholder Relations	• Identify and engage stakeholders (employees, suppliers, customers, etc.) so as to anticipate their opinions on the business, products and services and to identify what really matters to them

5 Conclusion and Outlook

In this chapter, the authors propose to consider the term life cycle sustainability management to clearly differentiate the term from other disciplines using also the term life cycle management. The current literature is summarized, in particular Remmen et al. (2007) and Finkbeiner (2011) on Life Cycle Management and Life Cycle Sustainability Management, focusing on putting life cycle thinking into business practice using relevant tools, including life cycle sustainability assessment (LCSA) to cover the three dimensions of sustainable development. The authors show that different companies have different ways of engaging their departments in LCM practices; they also use different tools and set different priorities.

A particular effort is made to explain life cycle (sustainability) management as part of sustainable consumption and production (SCP) strategies and policies. Strategies to change consumption and production patterns need to take into account the varying responsibilities of different actors in the value chain, including the consumers. Overall, there seem to be high expectations of the future use of LCA in SCP policy areas such as sustainable public procurement and eco-design directives as well as consumer information.

With regard to the future, management science will increasingly be brought into the topic of LCM, as shown for instance through recent work done by, inter alia, Nilsson-Lindén et al. (2014), and that there is an important need for capacity building and a great potential for mainstreaming. With regard to capability development using the LCM Capability Maturity Model (UNEP/SETAC 2013) is a good way to help companies to catch up with leading ones in the area of LCM. The next steps in companies are to move from projects to processes to establish the LCM team as a business partner for the long term. For mainstreaming to happen, professional communication targeting the opinion leaders and collaboration among life cycle experts and networks around the world are important elements for sustainable value creation.

References

Baumann H, Tillman A-M (2004) The hitch hiker's guide to LCA: an orientation in life cycle assessment methodology and application. Studentlitteratur, Lund

Brundtland Commission (1987) Report of the World Commission on Environment and Development

Fava JA (1997) LCA: concept, methodology, or strategy? J Ind Ecol 1(2):8–10. doi:10.1162/jiec.1997.1.2.8

Finkbeiner M, Wiedemann M, Saur K (1998) A comprehensive approach towards product and organisation related environmental management tools. Int J Life Cycle Assess 3(3):169–178. doi:10.1007/BF02978825

Finkbeiner M, Schau EM, Lehmann A, Traverso M (2010) Towards life cycle sustainability assessment. Sustainability 2:3309–3322

Finkbeiner M (ed) (2011) Towards life cycle sustainability management. Springer, Dordrecht/Heidelberg/London/New York

Heiskanen E (2002) The institutional logic of life cycle thinking. J Clean Prod 10(5):427–437. doi:10.1016/S0959-6526(02)00014-8

Hunkeler D, Saur K, Rebitzer G, Schmidt W, Jensen A, Stranddorf H, Christiansen K (2004) Life cycle management. SETAC, Pensacola FL

ISO (2006a) ISO 14040 international standard. In: Environmental management – life cycle assessment – principles and framework. International Organisation, Geneva

ISO (2006b) ISO 14040 international standard. In: Environmental management – life cycle assessment – requirements and guidelines. International Organisation, Geneva

Jensen AA (2012) The business case of life cycle management – how to create a sustainable value chain. Shanghai Second Polytechnic University, Shanghai

Klöpffer W (2008) Life cycle sustainability assessment of products. Int J Life Cycle Assess 13(2):89–95

Linnanen L (1995) Life cycle management: integrated approach towards corporate environmental issues. Bus Strategy Environ 4(3):117–127. doi:10.1002/bse.3280040303

Nilsson-Lindén H, Baumann H, Rosén M, Diedrich A (2014) Organizing life cycle management in practice: challenges of a multinational manufacturing corporation. Int J Life Cycle Assess doi:10.1007/s11367-014-0818-y

Norwegian Ministry of the Environment (1994) Oslo roundtable on sustainable production and consumption. Retrieved 15 Mar 2015. http://www.iisd.ca/consume/oslo000.html

Remmen A, Münster M (2003) An introduction to life-cycle thinking and management. Danish EPA, Copenhagen

Remmen A, Jensen AA, Frydendal J (2007) Life cycle management. A business guide to sustainability. UNEP DTIE

Seuring S (2004) Industrial ecology, life cycles, supply chains: differences and interrelations. Bus Strategy Environ 13(5):306–319. doi:10.1002/bse.418

Sonnemann G, de Leeuw B (2006) Life cycle management in developing countries: state of the art and outlook. Int J Life Cycle Assess 11(1):123–126. doi:10.1065/lca2006.04.020

Sonnemann G, Gemechu ED, Adibi N, Bruille VD, Bulle C (2015) From a critical review to a conceptual framework for integrating the criticality of resources into life cycle sustainability assessment. J Clean Prod. doi:10.1016/j.jclepro.2015.01.082

UNEP (2011) Towards a life cycle sustainability assessment: making informed choices on products. UNEP/SETAC, Paris

UNEP (2012) Global outlook on SCP policies: taking action together. UNEP, Paris

UNEP/SETAC (2009) Life cycle management: how business uses it to decrease footprint, create opportunities and make value chains more sustainable. UNEP, Paris

UNEP/SETAC (2013) LCM training material. Retrieved 7 Feb 2015. http://www.lifecycleinitiative.org/resources/training/

Wrisberg N, Udo de Haes HA (2002) Analytical tools for environmental design and management in a systems perspective: the combined use of analytical tools. Kluwer, Dordrecht & Boston

Chapter 3
Life Cycle Management as a Way to Operationalize Sustainability Within Organizations

Sarra Harbi, Manuele Margni, Yves Loerincik, and Jon Dettling

Abstract This chapter proposes a value creation framework to operationalize sustainability within organizations through an improved link between life cycle management tools and business drivers for value creation. Internal and external stakeholders' need to be first identified and accounted for, and value creation must be clearly identified in order to be acknowledged and communicated. The question "what do we want to achieve?" needs to be answered before thinking how to best achieve the identified business value. We propose to apply "reverse-engineering" to define the value creation path and identify the departments and collaborators to be involved at different level of the organization. LCM offers an essential and flexible integrated management framework of concepts, techniques and procedures to think how to best operationalize sustainable actions to achieve the identified business value. The sustainability action involves a team that should include, at least a sponsor or a pilot from the department that is expecting final value creation (Human Resources – HR for employee engagement, marketing for product positioning, etc.) and a representative from each department involved in the value creation path. Each of them will need an LCM tool adapted to their need and specific objectives. Last but not least, one needs measurable indicators on global goals that are to be monitored by the overall project sponsor, through KPI (key performance indicators) and follow-up.

S. Harbi • Y. Loerincik
Quantis-International, EPFL Innovation Park, Bât D, CH-1015 Lausanne, Switzerland

M. Margni (✉)
Quantis-International, EPFL Innovation Park, Bât D, CH-1015 Lausanne, Switzerland

CIRAIG, Polytechnique Montréal, C.P. 6079, Succ. Centre-Ville, Montréal, QC H3C 3A7, Canada
e-mail: manuele.margni@polymtl.ca

J. Dettling
Quantis-International, 295 Hanover Street, Suite 4A, Boston, MA 02113, USA

G. Sonnemann, M. Margni (eds.), *Life Cycle Management*,
LCA Compendium – The Complete World of Life Cycle Assessment,
DOI 10.1007/978-94-017-7221-1_3

Keywords Life cycle assessment • Life cycle management • Sustainability • Value creation

1 Introduction

The role sustainability plays differs from one organization to another. This is also true across different functions within the same company. Indeed, an organization is not homogeneous, and the needs of departments such as operations, marketing, product development, sales and other stakeholders within the company are often very different. They are also very different in how they are impacted by the risks and opportunities posed by the sustainability topic. A one-fit-for-all solution to make sustainability operational within an organization is therefore not feasible, but rather the right approach needs to be tailored to the unique context, resources and constraints of the company or department in question.

Many companies have started their journey towards sustainability in response to stakeholder or customer requests, or sometimes through a strong personal commitment of key individuals. However, today's corporations are inherently profit-driven by the necessity of competition and so, to be integrated into the company and become part of the company's DNA and strategy, sustainability needs to create value for the corporation itself (Bonini and Schwartz 2014). If not connected to the business it can easily be disregarded in challenging economic conditions. In addition, today's corporations conduct their core business with great efficiency and alignment of sustainability with these core operations ensures a rapid and effective trajectory for achieving outcomes in comparison to treating sustainability work as a form of philanthropy.

Systematic integration of sustainability into strategic initiatives is key to achievement of meaningful sustainability related goals, since the long-term changes required are likely to be drastic departures from today's status quo and the path to achieving them is often as-yet unclear and quickly evolving. A strategic focus on sustainability allows appropriate actions to be taken at the right moment, as the context of the sustainability discussion plays out over the long term. Life cycle thinking is a key to achieving this strategic alignment by allowing companies to understand their position within the broad context of sustainability. Today's leaders in the sustainability space are continually finding creative ways to adapt life cycle thinking to the whole organization as well as its products and services, thus leading to a better understanding of consumer preferences, stakeholder pressure, existing regulation and future trends.

Life cycle management (LCM) is "a flexible integrated management framework of concepts, techniques and procedures incorporating environmental, economic, and social aspects of products, processes and organizations" (UNEP 2006, UNEP/SETAC 2009) to achieve the integration of sustainable development into the company, along the whole value chain (O'Rourke 2014).

2 Value Creation and Life Cycle Management

Value creation and differentiation are keys for companies. Today, the strong correlation between good financial performance and sustainability is widely accepted and recognized (DB Climate Change Advisors 2012). It is clear that businesses strategy that includes sustainability programs contribute to strong corporate performances.

A growing number of examples show that sustainability initiatives are a great catalyst to creating profit as well as new business opportunities. How to link sustainability to business value is, however, still poorly understood by a large majority of companies (Accenture and United Nations Global Compact 2013).

Companies seeking to pursue the sustainability path need to ensure that they do so while creating value for the company itself and along its value chain. Sustainability for the sake of sustainability is well-meaning, but likely to be ineffective. Without expressing the value creation of sustainability, companies will remain in "pilot projects" or small-scale sustainability projects, only able to engage and motivate those internal sponsors and team members that "buy-in" to the sustainability mission as a matter of personal conviction and unable to find the right strategies to leverage to core power of the business. Identifying the business value created by sustainability and the way to get there is thus essential. The value has to be perceived along the whole value chain, i.e. managers and collaborators at different levels of the company as well as external stakeholders whose expectations and potential influence must be identified and accounted for. Finally, this value creation must be measurable and measured in order to be acknowledged and communicated.

A key step each company should consider before embarking on the path of sustainability is answering the question "What is the value this strategy will generate for our organization?" Recognizing that each company's path to capturing value from sustainability will be unique, Bonini and colleagues (2011, 2014) proposed a framework that can serve as a universal starting point to understand the relationship between sustainable initiatives and value creation. It captures value in three key areas:

- *Risk management* linked to sustainability – encompass *risks due to operational disruptions* such as, for example, resource scarcity, extreme events from climate change; *risk due to reduced reputation* from relationship issues with stakeholders along the value chain; and *regulatory risk* from current restrictions and regulations to come.
- *Return on capital* – by onsite *operational efficiency through improved resource management*, e.g. energy efficiency, water reuse and byproduct valorization; by *developing sustainable value chains* expanding improved resource management efficiency through the supply chain or downstream extending producer responsibility; by *increasing employee motivation* through internal involvement and identification to company and its values; and by *green sales and marketing* seeking increased revenue from sustainability attributes.

- *Growth* – by *innovation* and the development of new sustainability driven product/service development; by developing strategies opening the door on *new markets*; and by regularly revisiting the *composition of business portfolios* to determine trends and potential risks and improve appeal for investors, as well as by competing better with existing products/services, as customers and consumers place increasing emphasis on sustainability in their purchasing decisions.

It is very important to remember that sustainability in itself is not necessarily generating value for an organization, unless it becomes aligned with the company's core business strategy.

2.1 Defining the Strategy of Sustainable Value Creation

The first step when defining a sustainable strategy is to define the expected outcomes. The first and most important question to be answered is about the expected business value from implementation of the sustainability action plan. In order to answer this question, one should:

(a) *Align with the company strategy*: Is the priority of the company to grow and to develop? To increase its margin? How is this going to be achieved? Sustainability should be used to support the company's strategic priorities.
(b) *Assess stakeholders expectations*: What stakeholders have an interest in sustainability and are to be involved? What is their influence on the company and its value-chain? On what type of value can they act? For example, clients will influence sales, key opinion leaders will influence brand reputation and suppliers of strategic materials can influence the level of risk.
(c) *Perform a risk assessment*: A sustainability strategy can be key to support the risk management plan of a company.

A meaningful strategy to unlock sustainability within an organization remains the key to link each initiative to one of the key areas of value creation. We can think at the life cycle management toolbox being at the opposite end of the value creation path within a company. Often an appropriate LCM tool (or a combination of them) is first identified, and then implemented by a business unit that is expected to generate tangible or intangible business benefits. Proceeding in such a sequence, however, is risky. Organizations might realize that generated business benefits are not aligned with company's value drivers like two ends of a tunnel being drilled from both sides that fails to meet in the middle due to poor planning.

There are potentially many good reasons to launch an ecodesign project:

- Starting from consumers expectations and perspective, developing a new packaging that is differentiated by its lower impact on the environment. But the first questions to tackle are: What does sustainability mean for the consumers? How do they perceive it? How will they react to a new packaging?

- Generating new ideas thanks to the motivation of the team to work on sustainability. It is however important to understand the expectations of the team: How do they perceive sustainability? What would they like to learn during the process? How is their work valued?
- Limiting risks by choosing appropriate materials. Which material price will be less influenced by the future price of energy? Which materials have an availability that is ensured for the coming 20 years?

The following example shows that one first needs to define the value to be created with sustainability, before launching a sustainability action.

2.1.1 The Liberté Case Study

Liberté, a dairy company based in Québec implemented in 2008 an ecodesign strategy, working on various products, but in particular on a drinkable organic yogurt-based beverage that was to be sold in a cardboard packaging. Liberté discovered through a life cycle assessment that this new packaging had a significantly lower impact on the environment than the plastic bottle traditionally found on the market.

The product, launched in 2008, was positioned as an ecodesigned product (organic milk and packaging optimized with a life cycle assessment). Unfortunately, the launch of the product was not the commercial success that the company's management had anticipated. Several reasons can be identified:

- In some cases, retailers and consumers were negatively surprised by the packaging design that was different than the industry norms for such a product. This is even though the product was sold among the organic products, where buyers are relatively well informed on sustainability issues.
- The *Producteurs de lait du Québec* (Milk Producers of Liberté) were subsidizing single portion dairy products with an additional premium for drinkable products packaged in a resealable plastic bottles favouring "on-the-go" consumption.
- Consumers (especially young children) were not used to the new packaging and had some difficulty opening them.

The learnings for Liberté were the following. They were able to determine hotspots linked to the environmental impacts of their product and were capable to mobilize the necessary resources and collaborate with suppliers to address them. However, although meeting the objective of a more sustainable packaging, the company failed to understand consumer's perceptions and needs. A better link between R&D and marketing needs to be established to test market perceptions in respect to these new products. This is especially true where such products counteract existing market trends, and in such cases, it is necessary to understand the likely consumer acceptance and whether consumer education or other actions need to be focussed on.

2.2 Sustainability Value Creation Framework

In order to minimize such implementation "failures", the authors propose a *sustainability value creation framework* that builds on life cycle management, but ensures at the same time connectivity between the selected LCM tools and the value drivers of the company (Fig. 3.1).

The framework describes the links between life cycle management tools and value creation through different steps: toolbox, business implementation, business benefits, value drivers and business value. One proceeds from left to right by asking the question "for what purpose", or "what do we want to achieve". From right to left, one can ask the question "how?" The value creation framework has to be defined for each company, and in many cases, the path toward value creation might be longer and more sophisticated than what is presented in Fig. 3.1.

Although one can easily start from anywhere within the framework, addressing the two questions and clarifying the chain is the starting point to ensure the linkage between sustainability actions and business value creation for the company. In Sect. 3, several of these paths to sustainable value creation are illustrated and discussed.

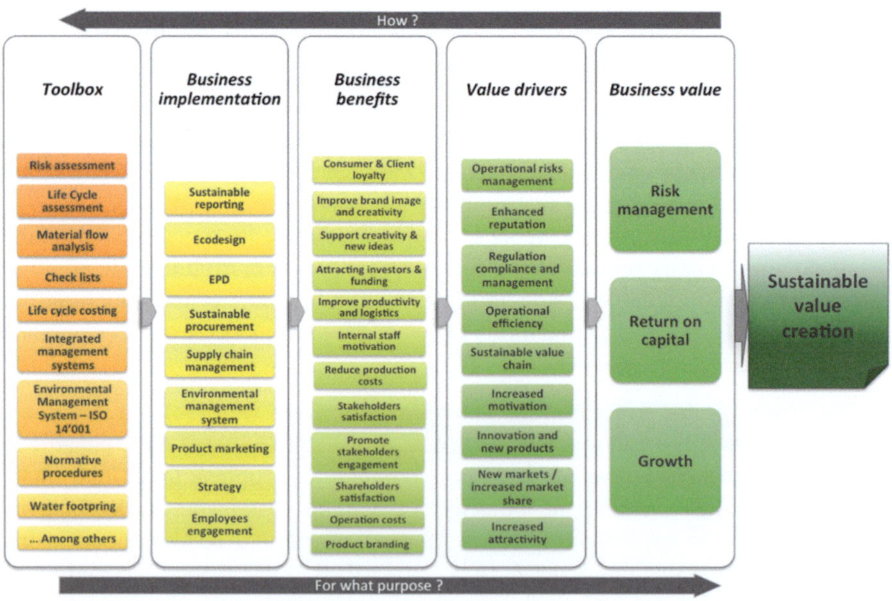

Fig. 3.1 Sustainability value creation framework linking life cycle management tools with business value creation of an organization

3 Paths to Operationalize Sustainability Through Life Cycle Management in a Company: Illustrated Applications

3.1 Sustainability as a Motivation Generator

3.1.1 What Leading Companies Are Saying About Sustainability Value Creation Through Employees Engagement?

Employees are increasingly considered as a key stakeholder group in driving companies to take action on sustainability issues (Accenture and United Nations Global Compact 2013). For example. **Beiersdorf's** Sustainability report clearly illustrates this pathway of value creation for companies: *Nowadays, employees express a strong interest in sustainability issues. They want to work for a company that shares their values and actively engages them in its sustainability commitment. This is becoming an ever more important requirement to attract, retain and motivate employees. We firmly believe that we can only increase our sustainability commitment with the support of our employees. In order for our sustainability strategy to be effective and achieve its targets, our employees need to understand what sustainability means to them in private and in business terms and how they can actively contribute. This in turn leads to increased motivation and enhanced performance, which benefits our business.* (Beiersdorf 2013).

Starbucks is also a leading example illustrating the integration of employees, called "partners" by Starbucks, as key stakeholders for the company and its value: *The management of Starbucks considers that one of the major reasons for Starbucks success is the "partners" and the relationship they form each day with the customers, and not the coffee itself. Starbucks thus invests a lot in its employees through equal treatment, a strong internal communication and feedbacks politic as well as social measures. This creates an interactive structure with "partners" in the center that makes the employees commit and identify themselves with the company. Starbucks sells the concept of the brand to their employees first, making them part of the story and ensuring their personal investment in the company's development, as they believe in the company and contribute to its success out of self-interest. (Starbucks, Global responsibility report* 2013*). "We built the Starbucks brand first with our people, not with consumers. Because we believed the best way to meet and exceed the expectations of our customers was to hire and train great people, we invested in employees."* (Starbucks 2013).

3.2 Sustainable Supply Chain to Reduce Risks

3.2.1 What Leading Companies Are Saying About Sustainability Value Creation Through Risk Avoidance?

Nestlé Example

> Better management of risks that arise from sustainability issues begins with detecting key risks of operational disruptions from climate change, resource scarcity, or community issues (such as boycotts or delays in getting permits for manufacturing) (Carbon Disclosure Project 2013). Faced with potential supply constraints, Nestlé, for example, launched a plan in 2009 that coordinates activities to promote sustainable cocoa: producing 12 million stronger and more productive plants over the next ten years, teaching local farmers efficient and sustainable methods, purchasing beans from farms that use sustainable practices, and working with organizations to help tackle issues like child labor and poor access to health care and education. (Bonini and Schwartz 2014)

Michelin Example

> Around 40% of a truck tire by weight is natural rubber and tire manufacturing uses 70% of the world's output of natural rubber, whose exceptional physical properties make it irreplaceable for truck, aircraft, agricultural and earthmover tires. (Michelin 2010)

Demand is constantly rising and therefore developing and maintaining rubber tree farms is a major priority for Michelin, which uses nearly 10 % of the world's natural rubber output.

Michelin recognized the risk represented by the raw materials, energy and non-renewable resources required for its products and production, as these resources, like oil or natural rubber, are becoming scarcer and more expensive. *"In 2010, raw material costs represented 27 % of Michelin's net sales. Optimizing their use is essential if these resources are to be conserved over the long term and if tires are to remain affordably priced."* (Michelin 2010).

Michelin is therefore working on engineering lighter tires that require less raw material and improve energy efficiency by optimizing the rolling resistance. Besides this strategy, the tire maker is also investigating alternative solutions to limit the pressure on primary resource supply through regrooving and retreading. Michelin is applying the life cycle assessment approach to ensure that such alternatives will not create value for Michelin while shifting environmental burdens elsewhere.

3.3 Sustainability to Increase Brand Perception

3.3.1 What Leading Companies Are Saying About Sustainability Value Creation Through Brand Perception and Positioning

Natura Example

> Natura Cosmeticos has long been considered a leader in sustainability, known for materials and marketing innovations that aim to reflect its tagline of "well being/being well." (Natura 2013). The Brazilian cosmetics maker has been recognized by organizations such as Corporate Knights, the U.N. Environmental Program, SustainAbility and the Boston Consulting Group. (Greenbiz 2014)

A corporate brand is not about product, but about how stakeholders view the organization and particularly its culture and value. Life cycle management helps managing expectation of stakeholders along the value chain. Which stakeholders need to be targeted in priority is a strategic decision of the company. For example, a positive brand perception helps increase customer loyalty or employee retention. Many prospective employees evaluate environmental policies as a measure of corporate value, but also sustainability programs within companies can have a significant impact on retention (The Guardian 2013).

4 Conclusions and Outlook

Through this chapter the authors made clear that sustainability, to be unlocked within an organization, needs to be embedded in company's strategy and have a clear link to business value creation. One need to answer first the question "what do we want to achieve" before thinking at "how" to implement a given action. From right to left, one can ask the question "how?" Life cycle management offers an holistic think-thank to identify opportunities for value creation along the entire value chain, and it provides flexible toolbox that can serve implementing sustainable actions within different departments involving different hierarchical levels. In this context, sustainability managers play an important role. They are a key actor for connecting different stakeholders vertically and transversely within the company, but also outside and linking sustainability actions into business drivers for value creation. Our experience has shown that a few key recommendations can increase the chances of success of a sustainability action in organizations. We summarize below a few key characteristics that have been proven to play a key role in operationalizing sustainability through life cycle management.

Identify a clear value creation goal: never implement a sustainability action without first identifying the value creation goals the given initiative will contribute to achieve. The answer might be growth, enhanced reputation, improved risk management, or anything else, that is connected to value drivers generating business value within the organization. So, before thinking about implementing any LCM-based tools, first answer the questions "for what purpose?", or "what do we want to achieve?" and map a pathway that carries these actions all the way through to value creation for the organization.

Think how you can best achieve the identified business value by reverse-engineering the path to value creation. This should include links to value drivers, business benefits, implementation strategies, and ultimately to identifying the appropriate life cycle management tools to support a given sustainability initiative. One has also to keep in mind that an organization is not homogeneous and needs are different depending on the department and place in the organizational hierarchy.

The right team to reach the objective: sustainability actions are transversal to different departments, which is one of the challenges, and building the right team is important in order to ensure success. The team should include, at least a sponsor or a pilot from the department that is expecting final value creation (HR for employee engagement, marketing for product positioning, etc.) and a representative from each department involved in the value creation path. If necessary, also involve external stakeholders concerned by the value creation goal. Involve them from the beginning, i.e. from the project scoping. In large organizations, where sustainability actions are developed company-wide and are too large to be supported by a single team, they need to be subdivided into smaller actions and subtasks. Though each one of these sub-projects might not have a clear understanding of the final goal of the company, they all support the global goal and are to be monitored by the overall project sponsor, through KPI and follow-up means.

Measurable indicators: define appropriate KPIs for all the different steps of the cause-effect chain, up to the value creation. The definition of indicators and target will first align all participants on results that can be expected from the action and will then help to follow the results.

References

Accenture and United Nations Global Compact (2013) The UN Global Compact–Accenture CEO study on sustainability 2013: architects of a better world

Beiersdorf (2013) Our commitment to people. Accessed March 2015. http://www.beiersdorf.com/sustainability/people/why-people

Bonini S, Görner S (2011) The business of sustainability: putting it into practice. McKinsey&Company. Accessed Oct 2011. http://www.mckinsey.com/insights/energy_resources_materials/the_business_of_sustainability_mckinsey_global_survey_results

Bonini S, Schwartz S (2014) Profits with purpose: how organizing for sustainability can benefit the bottom line. McKinsey&Company

Carbon Disclosure Project (2013) Sector insights: what is driving climate change action in the world's largest companies – global 500 climate change report 2013

DB Climate Change Advisors (2012) Sustainable investing: establishing long-term value and performance. Deutsche Bank

Greenbiz (2014) Natura. Accessed March 2015. http://www.greenbiz.com/blog/2014/02/04/natura-cosmetics-sustainability-amazon

Michelin (2010) Annual and sustainable development report. From: http://www.michelin.com/eng/media-room/press-and-news/michelin-reports/(category)/149/(year)/2010

Natura (2013) Natura report 2013. From: http://natura.infoinvest.com.br/enu/s-15-enu.html

O'Rourke D (2014) The science of sustainable supply chain. Science 344(6188):1124–1127

Starbucks (2013) Global responsibility report: goals and progress 2013. From: http://www.starbucks.com/responsibility/global-report

The Guardian (2013) Natura commits to sourcing sustainably from Amazon. From: http://www.theguardian.com/sustainable-business/natura-sourcing-sustainably-from-amazon

UNEP (2006) Background report for a UNEP guide to life cycle management – a bridge to sustainable products. United Nations Environment Programme, Paris

UNEP/SETAC (2009) Life cycle management: how business uses it to decrease footprint, create opportunities and make value chains more sustainable

Chapter 4
How to Implement Life Cycle Management in Business?

Eskinder Demisse Gemechu, Guido Sonnemann, Arne Remmen, Jeppe Frydendal, and Allan Astrup Jensen

Abstract This chapter discusses how business can implement life cycle sustainability assessment into their management strategies. Life cycle management is a management approach that provides business a systematic way of managing their sustainability issues. The PDCA (Plan, Do, Check and Act) cycle is one of the quality management tools that can be used by companies to implement life cycle management initiatives in order to improve their sustainability performance. The relevance of the PDCA cycle is discussed to ensure a continuous performance improvement by setting and implementing a well-defined plan, checking whether the ambition goals are achieved or any adjustment actions are needed to continue the evaluation process.

Keywords Life cycle assessment • Life cycle management • Life cycle sustainability assessment • PDCA cycle (plan, do, check, act)

E.D. Gemechu (✉) • G. Sonnemann
University of Bordeaux, ISM, UMR 5255, 33400 Talence, France

CNRS, ISM, UMR 5255, 33400 Talence, France
e-mail: eskinder.gemechu@u-bordeaux.fr

A. Remmen
Aalborg University, Aalborg, Denmark

J. Frydendal
Ecolabelling Denmark

Danish Standards Foundation

A.A. Jensen
Nordic Institute of Product Sustainability, Environmental Chemistry and Toxicology (NIPSECT), Copenhagen/Frederiksberg, Denmark

G. Sonnemann, M. Margni (eds.), *Life Cycle Management*,
LCA Compendium – The Complete World of Life Cycle Assessment,
DOI 10.1007/978-94-017-7221-1_4

1 Introduction

Global companies recently have shown an increasing interest in being engaged in sustainability initiatives and integrating it to their business management strategies through broadening their accountability beyond economic performance to include social and environmental aspects (Labuschagne et al. 2005). There are a number of driving forces behind their engagement: government regulation and intervention, stakeholder pressures, economic profit, globalization, business concern for society and the environment, technological advancement, social activism and so on (Estein and Buhovac 2010). The direct intervention of governments at regional, national or international level is one of the drivers. Governments are highly encouraging companies to improve their environmental and social performance while maintaining their economic benefits (Simpson et al. 2004). A number of legislation and regulatory initiatives have been established to promote technological advancement (Carraro and Galeotti 1997).

The development of methodological tools, databases, guidelines and procedures are being supported by governments so as to promote sustainability practices in businesses that can contribute to the transition towards a more sustainable economy. Some examples from the European Union (EU) are the EC directive on disclosure of non-financial and diversity information by large companies, which requires companies with more than 500 employees to include information about their environmental and social performance in their annual reports (EC 2014), the EC's strategy on corporate social responsibility (CSR), which encourages companies to have in place a system that integrates consumer concerns, environmental, ethical human rights and other social aspects into their business operations and core strategy with close collaboration with their partners (EC 2011). A number of national governments have also established policy initiatives to promote sustainability practice by businesses.

Besides governments' interventions, there are also other driving factors for companies' commitments to sustainability initiatives. One is the change in consumers' behavior towards sustainable consumption patterns. Consumers have become more concerned about the environmental pressure associated with products for their consumptions. They are showing commitments to buy products with relatively less impacts and they would like to be linked with companies that are environmentally and socially responsible (Perrini et al. 2010; Cherian and Jacob 2012). Companies that place sustainability initiatives into their business strategies are attracting more consumers and at the end making more profits. This phenomenon could stimulate and may lead companies that are not active in sustainability practice to be engaged so as to improve their competitiveness (Ginsburg and Bloom 2004; Lacy et al. 2010).

Corporate sustainability reporting (CSR) initiatives are also serving as an internal and external driver for companies' sustainability initiatives. They encourage employees and stakeholders to be engaged in sustainable business practices at the same time they could also increase competition and threats within and across industries (Porter and Kramer 2006). Economic globalization, which is characterized by its global, liberal and open economy (Dinda 2004), technological advancement that reduces material intensity and pollutions, social activism that creates awareness about the environmental and social pressure and force governments to set a regulation (Ginsburg and Bloom 2004) are also among the drivers that brought sustainability innovation into business context.

Sustainability is becoming an agenda for most business nowadays, but how to implement and integrate it with other strategies remains challenging. Life cycle management (LCM) is an approach that can be used by business with the aim to operationalize their sustainability initiatives: to have better environmental, economic and social performance simultaneously. This chapter, therefore, discusses the way how business can implement LCM into their management strategies.

2 Implementation of Life Cycle Sustainability Management

LCM is a management approach that provides business a systematic way of managing their sustainability issues. The PDCA (Plan, Do, Check and Act) cycle is one of the quality management tools that can be used by companies to implement LCM initiatives in order to improve their sustainability performance. This section introduces the PDCA cycle and discusses in detail its relevance to successfully implement LCM in any business.

The PDCA cycle, for Plan, Do, Check and Act, which is also known as the Deming cycle, was first proposed by Walter A. Shewhart (Shewhart 2011) in the 1950s and further conceptualized by W. Edward Deming (Deming 1952) to analyze and measure business processes and identify the main causes that affect products quality from customer requirements. The PDCA cycle has its root in scientific methods that have evolved for more than 400 years (Moen and Norman 2006). The cycle is a systematic series of steps and continuous feedback loop in which managers are able to identify and take the required measure to change the parts of the process that need improvement (Sokovic et al. 2010).

The PDCA cycle is an iterative process of a four-step management method that can be applied by any business with an ambition of continuous processes and products performance improvement Fig. 4.1, Table 4.1. The "Plan" step comprises different activities and identifies the main goal and purpose of the assessment and

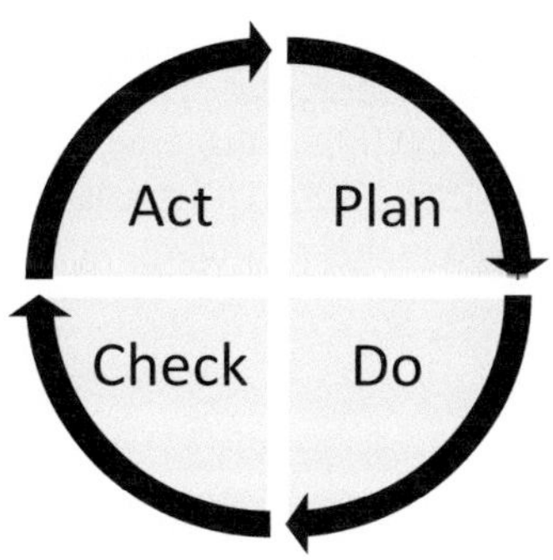

Fig. 4.1 Phases of the step-by-step approach

Table 4.1 LCM relevance of each stages of the PDCA cycle (Remmen et al. 2007)

PDCA cycle phases	LCM relevance
Plan	Set policies – set goals and determine the ambition level
	Organize – get engagement and participation
	Survey – overview of where the organization is and wants to be
	Set goals – select areas where the efforts will be directed, determine goals and make an action plan
Do	Make environmental and social improvements – put the plan into action
	Report – document the efforts and their results
Check	Evaluate and revise – evaluate the experience and revise policies and organizational structures as needed
Act	Take it to the next level – set up new goals and actions, more detailed studies, etc.

establishes the processes that are required to deliver the outcome. The "Do" step is a process of implementing the plan through performing a process to make the target output product. It also involves activities such as collecting and preparing data inputs to the next steps. The third stage of the PDCA cycle is the "Check" step. In this stage, the main results from the "DO" step are analyzed whether or not they meet the expected outcome set in the "Plan" step. The main factors that affect the performance of the product are identified and possible improvement solutions are suggested. The comprehensiveness of the plan to allow a proper execution of the product is checked in this step. The final stage is the "Act" step. The action is based on the output results from the "Check" stage. If the results from the "Check" reveal that the established new plan allows for an improved performance, then the action will be how the company sustains the improved performance as a new standard. On the other hand, if the plan fails to ensure an improvement, then the existing standard will remain in place. In both cases, the "Act" step closes the first cycle and continues the process by incorporating the knowledge accumulated during the entire process to be used to set a new goal, adjust the method used and so on.

LCM is a dynamic management process. Hence it considers the PDCA cycle approach among other management tools that enable a continuous business improvement over time and satisfy employees, customers and other stakeholders (Remmen et al. 2007). The PDCA cycle provides an organization with a systematic approach to management along the product life cycle. The recommendations of the PDCA cycle to the LCM are backed by the ISO management system standards for environment (ISO 2004) and quality (ISO 2008). The relevance of each PDCA element to the context of LCM is discussed in this section.

2.1 *Plan*

The PLAN step identifies the current existing level of sustainability ambition of the companies and ensures whether they have the required resources. This step generally covers the following areas: police setting (determining the level of business's ambitions), organize (get engagement and participation), survey and research (identify key environmental and social impacts and opportunities and decide what the business wants to be and where it would like to go) and finally goal setting (select an area/s where the efforts will be directed, determine goals and make an action plan).

2.1.1 Policy Setting

To consider LCM among the key success factors through all levels of any organization, it needs to be a part of the organization's policy. Therefore, it should possess the basic characteristics of an organization's policies. These are, for example: the policy has to be well known and also well understood by all stakeholders that are directly or indirectly affected from its implementation, it should be stable and not subjected to change frequently, it should be consistent with the company's organizational structure at all levels, it has to be formulated within the context of company's objectives, and more importantly, it has to be goal-based and visionary with a long-range focus, while also being realistic and concrete, grounded by its ambition level. Setting goals according to the levels of ambition ensures conformity between policy and actions. There are different levels of sustainability ambitions, just to cite a few (Table 4.2):

- Internal readiness and commitment to continuous sustainability improvements. This level signifies awareness that environmental and social improvements can be made using management frameworks such as ISO 9001 ((ISO 2008), ISO 14001(ISO 2004) and/or corporate social responsibility (CSR) (McWilliams and Siegel 2001), and that a commitment to improvements of product performance is the first step towards LCM.
- Life cycle sustainability performance improvement of products. This includes a safe and resource efficient process, product or system that provide better service to customers while reducing environmental and social impacts. This can be addressed through broadening companies' system boundaries beyond their manufacturing level, by including the full supply chain from raw materials selection and acquisition, use of product, distribution and end-of-life – based on a screening of the product's performance.
- Reduce climate change impact and be energy efficient by using low-carbon and renewable energy sources.
- Maximize the social benefits by engaging a wide range of stakeholders and also respect the rights of the social community whom the company works with.
- Create competitive advantages and maximize economic benefits from promoting products with better environmental and social profile.

Establishing such ambition levels can help an organization to understand where to begin the sustainability initiative. Depending on its past sustainability practices and current challenges, an organization can begin with a certain level of ambitions and goes to relatively high level once it acquires knowledge and familiarity on how to make its plan into action successfully.

Table 4.2 Sustainability ambitions – examples from global leading companies

Company	Sustainability ambitions
SONY (2015)	Curbing climate change – zero emissions of greenhouse gases by 2050
	Conserving resources – maximize the use of recycled plastic and other materials
	Controlling chemical substances – strict control over chemicals in raw materials and parts worldwide
	Promoting biodiversity – protect biodiversity, both through its business activities and through conservation
BASF (2015)	Improve the energy efficiency in production process by 35 %
	Reduce greenhouse gas emissions per metric ton sales product by 40 %
	Reduce the emissions of organic substances and nitrogen to water by 80 % and heavy metals by 60 %
	Reduce the withdrawal of drinking water for production by 50 %
	100 % introduction of sustainable water management at production sites in water stress areas
	Reduce emissions of air pollutant by 80 %
	Safety, security and healthy
	Reduce transportation accidents during shipments
	Minimize workers lost-time injuries
	Assess the risk of products sold by BASF worldwide
3M (2014)	Manage their environmental footprint
	Provide solutions that address both environmental and social challenges for the consumers and society
	Assure the safety of their product for the intended use through assessing their entire life cycle
	Appropriate management of any 3M health and safety issues that may impact customers, neighbors and the public
	Maintain a safe and healthy workplace
	Satisfy customers through providing superior quality and value
	Support the local community where 3M employees live and work
	Provide an attractive return for our investors

In whatsoever ambition level is set in the sustainability policy, it should provide a clearly defined and well-grounded strategies and objectives that can ensure environmental, social and economic improvements in all relevant stages of a product profile. The product profile should include impact assessment throughout the product's life cycle, from cradle to grave, taking into account interests from different stakeholders. The product profile should also answer questions such as: Where is the position of the organization compared to its important competitors? Whether the organization has relevant supply chain information on input parts and raw materials? Whether the product provides the demanded environmental quality and social concerns by the consumers and other interested parties? Whether the organization has the necessary resources to achieve its goals?

When the policy has been set, the next steps are to organize the effort and set up specific targets.

2.1.2 Establish a Team for Life Cycle Management Initiatives

LCM has to be part of an organization management processes in which it involves the participation of different parts of the organization and also different actors in the product chain when relevant. LCM initiative can be organized in a coordination group or team. The responsibility within a coordination group should lie with a team leader, who is responsible for ensuring that the group functions, meetings are arranged, minutes are taken, etc. The members of the coordination group should be selected so that all relevant departments or functions are represented, including top management, product development, production, product distribution, sales, marketing, and purchasing.

One possible option could be establishing a cross-functional team. It is a team made up of individuals from different departments within an organization. They should be not only environmental expertise, but from different backgrounds with different expertise. Cross-functional team for the implementation of LCM could be composed of members from relevant departments such as product development, production and distribution, sustainability and environment, finance and procurement, etc. Some team members may be engaged in full-time bases in order to secure continuous improvement of the environmental performance until the targets are achieved and they may return to their routine roles once the time frame is over. Other members could act as an ad hoc team in order to implement specific tasks, which could be, for example, a supplier evaluation scheme with the involvement of people from purchasing, product development and marketing. For small- or medium-sized (SMEs) organizations a single but permanent cross-functional team might be the best solution, but it depends on the specific structure and culture of the organization. Whether the cross-functional team last long or short, its members should maintain a strong link to their major responsibilities in their main department as well as maintain good relations with their managers. Although it is important to establish a cross-functional team in order to successfully implement LCM into business practice, it faces some challenges such as priority due to several responsibilities, lack of motivation due to additional tasks and others.

The establishment of a cross-functional team is a key to place LCM into action. A number of factors are crucial to successful cross-functional teams (McDonough III 2000). Setting clear and well-defined goals offer several benefits to the team members. They provide a common frame of reference, which at the end results in facilitating the team cooperation and organize the tasks. Empowerment in cross-functional team is another success factor. Allowing individual team members a decision-making responsibility help them to be more committed to the project and strive to meet the target goals. It also increases their satisfaction of being a part of the team members. Supports from senior management, such as demonstrating team commitments, providing help to overcome challenges and encouraging team members can also have a direct effect on the performance of the cross-functional team. Another key success factor is creating cross-functional team of interested people (McDonough III 2000).

Experience with LCM and other management systems shows the value of having interested, enthusiastic people at all levels of the organization.

In general, successful implementation of LCM framework in an organization requires a proper allocation of human and other relevant resources, assignment of responsibilities and accountabilities based on each team member's role for the different tasks, building expertise based on practical experience as well as procedures and instruction to ensure that activities are running properly (Remmen et al. 2007).

2.1.3 Review the Sustainability Status and Set Objectives and Targets

LCM tries to implement life cycle sustainability assessment (LCSA) (UNEP 2011) into a real world decision-making process by applying a life cycle management concept with the aim of maximizing both the social, environmental and economic aspects of individual organization along its products lines (Finkbeiner 2011). Therefore, reviewing the organization's products is an important step to identify when and how to start the LCM process. The review helps to have an overview on the sustainability status of the organization and also where it wants to go for. The review process mainly covers gathering information associated with a product life cycle, market situation and external stakeholders. Information about important suppliers, business associations, authorities, retailers, research institutions, etc. needs to be included so that important aspects are covered.

Many companies already possess information on the impacts of their production processes and operations, but are uncertain of how to expand their understanding to the product life cycle and to think in terms of products rather than processes in order to identify material flows through the entire product life cycle. This would include impacts associated with suppliers, purchased materials or components, storage and distribution, use of the product and waste streams.

Important aspects that need to be reviewed are summarized in Table 4.3. They are categorized into aspects for environmental and social impacts, for market or commercial conditions and for stakeholders. Aspects for environmental and social impacts include understanding of impacts associated with the company's processes or products. The aspects should not be limited to the organization's territory, rather they have to go beyond the facility boundary to include the whole supply chain, both internal and external stakeholders. ISO 14031 – environmental management – which is an environmental performance evaluation standard and guideline (ISO 2013) can be applied to asses companies in evaluating their performance against their policy, objectives, targets and other criteria associated with their products. It can also be adapted to be applied to address other management aspects, e.g. quality or health and safety, and from a sustainability perspective economic and social aspects as well. For companies engaged in sustainability actions, the information may be already available in their corporate and environmental reports. In general, this information is presented in terms of the production process, rather than the single product.

Table 4.3 Aspects of a survey (Remmen et al. 2007)

Areas of concern	Aspects to be reviewed
Environmental and social impacts	Life cycle stages – where are the most important environmental and social impacts?
	Technology – is there a new technology available or being developed that can reduce the impacts?
	Do the competitors have the same impacts and how do they address them?
Market/commercial conditions	Supply – what are the product profile's characteristics?
	Demand – how important is the social and environmental awareness of consumers and customers?
	Value – what advantages are achieved by adding positive environmental and social characteristics as an extra product quality?
Stakeholders	Product chain actors – are suppliers, retailers or others interested in collaboration on environmental and social initiatives?
	Authorities – what are the demands of authorities?
	Within sector – what are competitors doing? Codes of conduct within the trade?
	Which is the main pressure group, and what are their main concerns and priorities?
	What are the main concerns and demands of consumers in export markets?

In addition to information on environmental and social impacts of their products or processes, a better understanding is necessary of both internal and external stakeholders, the current market conditions and the future opportunities linked with sustainability achievements. Discussion with all relevant stakeholders helps to get information on the legal and other requirements including environmental and social concerns. In addition, such discussion is also important when setting up objectives and targets. Goals to improve products' environmental and social performance will also have an effect on customers' expectations and responses.

Once the status overview is performed, then the next step will be to set objectives and targets. The criteria to select possible areas of product improvement are based on the responses from the questions listed in the survey. In some cases, the social and environmental problems identified by an organization may be solved by another's efforts, for example, the invention of a new technology or a supplier phasing out harmful substances due to pressure from other stakeholders. Therefore, the decision on prioritizing areas of concern for LCM initiatives is based on the finding from research on the current situation and knowledge. The following aspects are usually considered in the decision process (Remmen et al. 2007):

- Areas of considerable environmental and social impacts along the products' life cycle (Relevance). Impacts occurred within the organization facility may be easily identified. But a holistic approach may be required to identify potential areas of environmental and social problems along the supply chain.
- Areas for possible environmental and social improvements (Potential). An organization can have several opportunities for improvement, and therefore

chooses different initiatives for improvement and also to involve all the relevant departments.
- Areas where the organization can make a tangible differences (Influences). High relevance and high influence could yield high potential for real improvements.

In the objectives and goal setting, different initiatives should be established and for each initiative a corresponding action plan that explicitly states the goals, main responsibilities and the time frame needs to be defined so that plans and initiatives are clear both for the management and employees.

2.2 *Do*

2.2.1 Put the Plan into Practice

In the first stage of the PDCA cycle, the layout of an implementation of LCM initiatives has been established. The LCM is in place as a part of the strategic policy of an organization, teams that are responsible to carry out the initiatives are formed, the environmental and social impacts associated with the product's life cycle are specified, areas for possible improvements are identified, the goals and objectives are set. These are the preparatory processes under the "Plan" step. Planning is an important stage, but implementation is vital, as practical results create credibility, enthusiasm and active support for a product initiative. By considering the entire life cycle of a product, a number of possible improvements can be identified, for example, reducing the total environmental burden by increasing the use of paper in the office with high recycle content, increasing the use of renewable energy sources, reducing water consumption and wastewater discharge in the production process or dematerializing a product recycle content (UNEP/SETAC 2013). For all possible improvements, corresponding goals are important. Then what comes next is to place them in practice, which is the "Do" step.

In addition to performing the possible improvements in the products life cycle, the step also includes new challenges to tackle, which might include addressing new or potential issues. Environmental regulations are becoming tightened and companies need to be ready for any regulation changes, so they can adapt their goals and action plans. The company could implement life cycle thinking to engage a redesign of a product to meet the requirements of the legislation, including easier recycling, simple dis-assembly, new materials selection, etc.

For the sake of securing a continuous sustainability improvements of the product, written procedures or instructions should be established. The procedures or instructions explain methods of operation to be followed, which guarantee the continuity of the performance improvements especially when people change jobs or new staffs are hired.

Existing practices and the view of key persons have to be taken into account, in order to make the procedures work and to be part of the daily practice. The procedures should be established to support activities that are agreed to be 'standard' in the organization.

2.2.2 Communicating the Life Cycle Management Initiatives Efforts and Results

The results of the initiatives, which are the outcome of an organization's commitment, need to be documented and also communicated to stakeholders and other interested parties. An official report that communicates relevant information with regard to the organization's sustainability ambitions, its managements practices in making their goals and strategies to happen and their overall contribution towards economic, environmental and social performance improvement is very crucial. Reports on initiatives can be used as a marketing tool as it demonstrates to customers, suppliers and other stakeholders how the organization is engaged in sustainability management, which in turn provides credibility, confidence and trust. The report can also serve as an internal communication tool for setting new targets, performance evaluation and re-defining strategies.

There is no mandatory format for reporting on sustainability initiatives. However, an organization can adopt different sustainability reporting frameworks to its context depending on its ambition level, for example, the Sustainability Reporting Framework established by the Global Reporting Initiative (GRI) (GRI 2013), Corporate Sustainability Reporting (McWilliams and Siegel 2001), etc. It is advantageous for an organization to have some insight into the type of environmental and social impacts that the stakeholders prioritize, so that communication can be tailored to meet these demands.

The report can cover several issues that are addressed in the LCM practice, it may include, but is not limited to (Remmen et al. 2007):

- Trends of resources consumption such as energy, water, fossil fuels and other resources
- Resources consumption reduction achievement after the introduction of LCM initiatives
- Achievement on increasing the recycle content of input raw materials in the production process
- Any effort made to reduce environmental impacts associated with distribution and transportation
- Information about the amount of solid waste production, wastewater discharges and emissions generation
- Resources supply risk and also opportunities due to geopolitical, social, relational, and other environmental factors such as climate change
- Any initiative towards increasing the social benefits of the local community, employees and other stakeholders.

Results may also be made visible by calculating key figures, for example, energy consumption during production of the product. The key figures and the practical results of the initiatives can be made public via, for instance, a leaflet, an eco-label or an environmental product declaration. Internal as well as external communications are necessary precondition for achieving the benefits of LCM initiatives.

2.3 Check

Once an organization evaluates its sustainability position and sets ambition goals to move forward to an improved performance, once it initiates a plan to achieve its targeted goals and once it performs a LCM practice and finally reports the outcomes from the implementation, then the next step is to evaluate and revise its achievements and learn from its mistakes. The latter is the "Check" step of the PDCA cycle. In this stage, the effectiveness of the proposed solutions is measured. Depending on the success of the LCM initiative, once the areas of improvement are identified and the general scope of the main project are evaluated, it may be possible to incorporate the improvements by repeating the "Do" step.

The types of questions one might ask at this phase would include (Remmen et al. 2007):

- What went well? And what did not?
- Which risks were identified?
- What preventive actions should be taken?
- Were the goals met?
- How can the effort be improved?
- Should more employees be involved in the initiative?
- Should the efforts be focused in a different direction?
- Were the appropriate means and methods applied?
- What was the influence on sales and customer demands?
- Should more external stakeholders be involved?
- Should the level of ambition be raised?

This evaluation may be conducted once a year and it can provide a platform for an adjustment of an organization's policy. Based on the evaluation results, the organization may decide to continue with the same sustainability objectives and goals or even set a higher ambition level that comprises more extensive environmental and social commitments.

A step-by-step approach of the LCM ensures interaction between knowledge acquired about the social and environmental impacts of a product throughout its life cycle and the possible implementation of product-oriented sustainability performance improvements. The "Check" step of the PDCA cycle evaluates this interaction and also involves the measurement, monitoring and evaluation of products and services as well as of the management system itself. Important elements of the "Check" step are (Remmen et al. 2007):

- Monitoring the performance of the processes and products in view of the defined objectives and targets with the support of indicators
- Feedback and criticism from customers and other parties are an important information source for organizations to improve products as well as the product development process
- Establishing preventive and corrective actions for potential and actual nonconformities with requirements

- Conducting internal audits to determine and provide information to management on whether LCM conforms to plans has been properly implemented as well as to identify improvement options.

Once the "Check" stage of the cycle is performed, the next step will be followed.

2.4 Act

Answers for the questions raised in the "Check" stages determine the possible actions in the "Act" stages of the PDCA cycle. If the target is met, then a new target for further sustainability improvement will be established or if the objectives are over ambitious and do not consider the potential of the company's resources, then a new target with less ambition level could be set. Therefore, the "Act" stage in any case ends the previous cycle and begins a new cycle with a new target. Whether the LCM initiatives were successful or not, the annual cycle needs to be accompanied with a top management review and setting directions (ISO 2004; Remmen et al. 2007). The top management review basically addresses if there is any need to change or modify the policies, objectives and other elements of the system in the light of audits, evaluations, changing circumstances provides management with the opportunity to continuously improve the organization's and its products' performance (Fig. 4.2).

Through the experience from the first round of improvements, an enterprise has likely identified areas for further investigations or initiatives.

If an organization's customers request documentation on the impacts of a product then a simple environmental assessment is appropriate. If it becomes evident

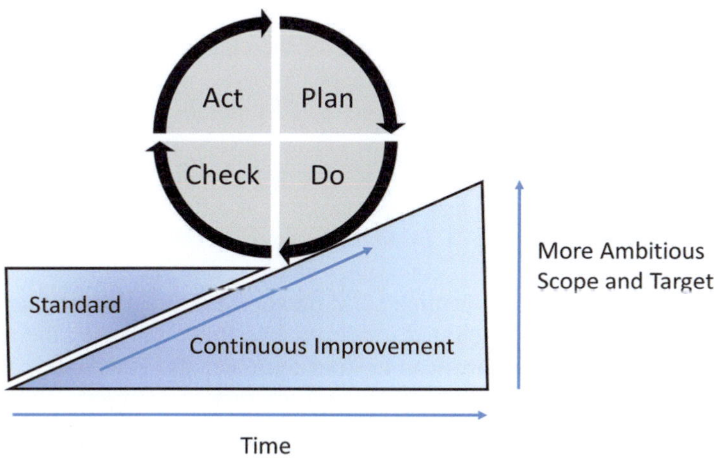

Fig. 4.2 PDCA cycle – a continuous learning and improvement process

that there are significant environmental impacts in the use stage of a product, then an investigation of consumers' desires and demands would be an obvious departure point of further consideration. If an enterprise uses chemicals or materials, which are on the list of undesirable substances, it would be sensible to begin phasing them out.

On the basis of experience, the initial area and goal(s) are redefined and a new round of efforts begins with plans, improvements, etc. Focus should remain on achieving specific environmental and social improvements to the product profile, while realizing results achieved throughout the improvement process.

During this and subsequent stages, the organization can begin (or continue) to broaden its relationship in the product chain. It is much easier to develop a base of knowledge if there is cooperation and an atmosphere of trust among producers, suppliers, retail store owners, disposal facilities and other stakeholders in the product chain.

3 Concluding Remarks

With a number of drivers both from governments, consumers, social activists, suppliers of technological advancement, internal CSR strategies and so on, companies are becoming more engaged in sustainability practices. LCM is an approach to help companies set up Life Cycle Management initiatives, to achieve environmental, economic and social benefits at the same time through implementing a step-by-step quality management tool. The relevance of the PDCA cycle is discussed to ensure a continuous performance improvement by setting and implementing a well-defined plan, checking whether the ambition goals and targets are achieved or any adjustment actions are needed as part of the evaluation process.

LCM has been identified as the way to operationalize sustainability challenges into business practices; however, its implementation faces significant challenges. Setting clear and measureable goals is one of the challenges. The focus of LCM initiatives is different from the usual business strategies, which are mainly focusing on maximizing the profit as the ultimate goal. LCM initiatives have a wider scope in order to have both social and environmental benefits along with maintaining the economic advantages. The divergent priorities between the financial and sustainability focuses are challenging tasks for managers at different organizational level. A successful implementation of LCM then needs a full integration across the organization.

Another challenge to implementing LCM into business practices is when there is a cost associated with the suggested improvements. In some cases, the economic benefits of being sustainable are seen in short-term actions. This is the case, when initiatives significantly reduce the energy and resource consumption, which internally provide financial benefits to the company. Such initiatives are easy to implement as they provide the company with environmental, social and economic benefits simultaneously. But this is not always true. In some cases, sustainability actions come with costs and fail to provide short-term financial incentives. Companies need to perceive the long-term positive consequence of their sustainability endearment, which can improve their attractiveness to consumer and increase their competitiveness in the long run.

References

3M (2014) Innovation improving lives – sustainability report

BASF (2015) Sustainability. https://www.basf.com/en/company/sustainability/management-and-instruments/our-goals.html. Accessed 10 Mar 2015

Carraro C, Galeotti M (1997) Economic growth, international competitiveness and environmental protection: R&D nd innovation strategies with WARM model. Energy Econ 19:2–28

Cherian J, Jacob J (2012) Green marketing: a study of consumers' attitude towards environment friendly products. Asian Soc Sci. doi:10.5539/ass.v8n12p117

Deming WE (1952) Elementary principles of the statistical control of quality: a series of lectures, 2nd edn. Nippon Kagaku Gijutsu Remmei, Tokyo, 103

Dinda S (2004) Environmental Kuznets curve hypothesis: a survey. Ecol Econ 49:431–455. doi:10.1016/j.ecolecon.2004.02.011

EC (2011) A renewed EU strategy 2011–14 for corporate social responsibility. European Commission, Brussels

EC (2014) Directive 2014/95/EU of the European Parliament and the Council – amending Directive 2013/34/EU as regards disclosure of non-financial and diversity information by certain large undertakings and groups

Estein M, Buhovac A (2010) Solving the sustainability implementation challenge. Organ Dyn 39:306–315

Finkbeiner M (2011) Towards life cycle sustainability management. Springer, Dordrecht/Heidelberg/London/New York

Ginsburg JM, Bloom PN (2004) Choosing the right green marketing strategy. MIT Sloan Manag Rev 46:79–84

GRI (2013) G4 sustainability reporting guidelines: reporting principles and standard disclosures

ISO (2004) ISO 14001:2004 Environmental management systems – requirements with guidance for use

ISO (2008) ISO 9001:2008 Quality management systems

ISO (2013) ISO 14031:2013 -Environmental management – environmental performance evaluation – guidelines

Labuschagne C, Brent AC, van Erck RPG (2005) Assessing the sustainability performances of industries. J Clean Prod 13:373–385. doi:10.1016/j.jclepro.2003.10.007

Lacy P, Cooper T, Hayward R, Neuberger L (2010) A new era of sustainability. UN global compact-accenture CEO study 2010. Routledge, London

McDonough EF III (2000) Investigation of factors contributing to the success of cross-functional teams. J Prod Innov Manage 17:221–235. doi:10.1016/S0737-6782(00)00041-2

McWilliams A, Siegel D (2001) Corporate social responsibility: a theory of the firm perspective. Acad Manage Rev 26:117–127. doi:10.2307/259398

Moen R, Norman C (2006) Evolution of the PDCA cycle. Accessed 1 Feb 2015. http://pkpinc.com/files/NA01MoenNormanFullpaper.pdf

Perrini F, Castaldo S, Misani N, Tencati A (2010) The impact of corporate social responsibility associations on trust in organic products marketed by mainstream retailers: a study of Italian consumers. Bus Stratege Environ 19:512–526. doi:10.1002/bse.660

Porter ME, Kramer MR (2006) Strategy and society: the link between competitive advantage and corporate social responsibility. Harvard Bus Rev 84:78–92

Remmen A, Jensen AA, Frydendal J (2007) Life cycle management. A business guide to sustainability. UNEP/SETAC Life Cycle Initiative

Shewhart WA (2011) Statistical method from the viewpoint of quality control. Dover Publications

Simpson M, Taylor N, Barker K (2004) Environmental responsibility in SMEs: does it deliver competitive advantage? Bus Strategе Environ 13:156–171

Sokovic M, Pavletic D, Kern Pipan K (2010) Quality improvement methodologies – PDCA cycle, RADAR matrix, DMAIC and DFSS. J Achiev Mater Manufacturing Eng 43/1, Nov. 2010

SONY (2015) Towards zero carbon emissions. http://www.sony.co.uk/hub/eco/sustainable-business-processes/zero-emissions

UNEP (2011) Towards a life cycle sustainability assessment: making informed choices on products. UNEP/SETAC, Paris

UNEP/SETAC (2013) LCM training material. http://www.lifecycleinitiative.org/resources/training/. Accessed 7 Feb 2015

Chapter 5
Life Cycle Sustainability Assessment: A Tool for Exercising Due Diligence in Life Cycle Management

Bernard Mazijn and Jean-Pierre Revéret

Abstract Starting from the output 'The Future We Want' of the Rio+20 conference 2012, the main focus of this chapter is on social responsibility (SR) in the value chain. The historical context of SR is discussed, related to the international standards as are the Guidance on Social Responsibility and the Global Reporting Initiative, linked with the management of organizations and enterprises. It is emphasized that due diligence along the value chain is seen as a requirement for claiming 'social responsibility'. Life cycle sustainability assessment (LCSA) contributes to the assessment and life cycle management (LCM) to the follow-up of exercising due diligence, all within the context of sustainable development. The over-arching LCSA is a combination of three different life cycle assessment techniques allowing to assess the impacts along the value chain: environmental LCA, social LCA and life cycle costing.

Keywords Life cycle assessment • Life cycle management • Life cycle sustainability assessment • Life cycle thinking • Rio+20 conference • Social responsibility

1 Introduction

The Rio+20 conference (2012) addressed the themes of 'a green economy in the context of sustainable development and poverty eradication', and of 'the institutional framework for sustainable development' as referred to in the output of the conference titled *The Future We Want* (UN 2012a, b).

B. Mazijn (✉)
Ghent University and IDO vzw, p/a Michel Van Hammestraat 76, 8310 Bruges, Belgium
e-mail: bernard.mazijn@ugent.be

J.-P. Revéret
CIRAIG – Université du Québec à Montréal,
Case postale 8888, succ. Centre-Ville, Montréal H3C 3P8, QC, Canada

G. Sonnemann, M. Margni (eds.), *Life Cycle Management*,
LCA Compendium – The Complete World of Life Cycle Assessment,
DOI 10.1007/978-94-017-7221-1_5

The adoption of the 10-year framework of programs on sustainable consumption and production patterns was confirmed. In the additional document (UN 2012c), several functions are listed to be included, amongst others "Promoting the engagement of the private sector in efforts to achieve a shift towards sustainable consumption and production, particularly sectors with a high environmental and social impact, including through corporate environmental and social responsibility."

Furthermore, the resolution itself is explicit on "renewing political commitment" and calls for the engagement of major groups and other stakeholders, inter alia the private sector. Therefore, the General Assembly of the United Nations states, e.g.: "We support national regulatory and policy frameworks that enable business and industry to advance sustainable development initiatives, taking into account the importance of corporate social responsibility. We call upon the private sector to engage in responsible business practices…"

2 Taking Up Social Responsibility in the Value Chain

2.1 About Social Responsibility of Organizations

2.1.1 Historical Context

The modern concept of corporate social responsibility (CSR), born in the USA, has developed over more than the last century. It can be traced back in a series of business practices in the late nineteenth century such as the philanthropy of some particularly rich capitalists and its development into a doctrine during the twentieth century that began to be theorized in the 1950s (Pasquero 2013). Many authors agree that the book by H. Bowen in 1953 'Social Responsibilities of the Businessman' is the seminal contribution to this field that has shaped its development for several decades. He offers an initial definition that reads: "Social responsibility refers to the obligations of the businessman to pursue those policies, to make those decisions, or to follow those lines of action which are desirable in terms of the objectives and values of our society." (Bowen 1953). During the sixties and seventies, there was a proliferation of definitions which built on Bowen's work. Joseph W. McGuire (1963) brought some more precise elements about the extension of the responsibility by writing "The idea of social responsibility supposes that the corporation has not only economic and legal obligations but also certain responsibilities to society which extend beyond these obligations." (McGuire 1963). However, these new visions of the role of businessmen were strongly challenged by liberal thinkers such as Milton Friedman, see his paper in the New York Times Magazine (1970) "A Friedman doctrine – The social responsibility of business is to increase its profits" (Friedmann 1970). This illustrates rather clearly the opposition between the fundamental vision of neoclassical economics, where business corporations contribute to the general interest, and the new managerial vision developed through CSR, where business has also a social and/or societal mission. The stakeholder theory, initially developed by Freeman (1984), had the ambition to provide managers with a conceptualization of the civil society and a theoretical model of the company and social actors, even if it did not really work to explain the social dynamics in place (Gendron 2013).

The development of CSR was strongly reinforced with the emergence of the concept of "sustainable development" and the international summits that followed the publication of the Brundtland Report in 1987, starting with the Rio summit in 1992. It is now a well-established field of research, of teaching in universities and of actions in companies. Nowadays as Capron (2013) suggests, implementing CSR and practices that contribute to sustainable development is usually presented as an approach derived from a strategic decision by the company to answer societal expectations but also, and especially in the European Union, more and more as a way to insert companies into national strategies defined by public authorities.

2.1.2 International Standards

Over the years CSR has been defined in different ways. Dalhsrud (2008) concluded from his analysis that "the existing definitions are to a large degree congruent", and "the confusion is not so much about how CSR is defined, as about how CSR is socially constructed in a specific context".

"A specific context" is determined to a large extent by the stakeholders which are an important dimension when taking up corporate social responsibility. Therefore, it could be expected that definitions of CSR at the international level are co-designed in a process involving stakeholders.

The overview of Dahlsrud could not yet take into account the final version of ISO 26000 – Guidance on social responsibility (2010) –, but the process behind did fulfil the "requirement" of co-design. It can be observed that the international standard has "social responsibility" in its title without reference to corporations or enterprises. The reason is "The view that social responsibility is applicable to all organizations emerged as different types of organizations, not just those in the business world, recognized that they too had responsibilities for contributing to sustainable development." Under the "Terms and definitions", "organization" is defined as "entity or group of people and facilities with an arrangement of responsibilities, authorities and relationships and identifiable objectives".

The widely accepted definition of ISO 26000 on social responsibility goes as follows "responsibility of an organization for the impacts of its decisions and activities on society and the environment, through transparent and ethical behaviour that

- contributes to sustainable development, including health and the welfare of society;
- takes into account the expectations of stakeholders;
- is in compliance with applicable law and consistent with international norms of behaviour; and
- is integrated throughout the organization and practised in its relationships."

It is noted in ISO 26000 that "activities include products, services and processes" and "relationships refer to an organization's activities within its sphere of influence".

Furthermore, to define the scope of social responsibility, seven core subjects are identified: organizational governance, human rights, labour practices, the environment, fair operating practices, consumer issues and community involvement and development. Each core subject includes a range of relevant issues.

There are three other leading international initiatives that are contributing to the uptake of social responsibility in business:

- The UN Global Compact is an initiative that started in 1999 and it can be seen as "a strategic policy initiative for businesses that are committed to aligning their operations and strategies with ten universally accepted principles in the areas of human rights, labour, environment and anti-corruption".[1] Networks of businesses that have committed to respect the ten principles and work extensively on capacity building among the enterprises have been set up at the international and national level. The last revision took place in 2010.
- The Global Reporting Initiative (GRI) is a multi-stakeholder initiative that was launched in 1997 by the Coalition for Environmentally Responsible Economies (CERES) and the United Nations Environment Programme (UNEP).[2] The GRI Guidelines are developed as a sustainability reporting framework that standardizes enterprises' reports on environmental, social and economic dimensions. For each dimension, performance indicators are formulated covering similar concerns as the issues of ISO 26000 (GRI 2011). The fourth revision of the guidelines has been launched in May 2013.
- The OECD Guidelines for Multinational Enterprises, one element of the OECD Declaration on International Investment and Multinational Enterprises, are "recommendations on responsible business conduct addressed by governments to multinational enterprises operating in or from adhering countries".[3] The latest revision of the guidelines was conducted in 2011.

These international initiatives, including their revisions over the years, illustrate the importance given to the uptake of social responsibility. However, the question arises if by these guiding initiatives enterprises are stimulated to develop a systematic and coherent due diligence approach in their corporate strategy regarding the value chain.

2.2 *Linking with Management*

With a long history behind, a stronger legitimacy and a set of structuring frameworks, CSR has now a clear role in the sphere of management. This appears in practice more rapidly and efficiently in sectors where it "fits" naturally with a set of

---

[1] See http://www.unglobalcompact.org/AboutTheGC/ (last accessed February 2015).

[2] See https://www.globalreporting.org/ (last accessed February 2015).

[3] See http://www.oecd.org/daf/internationalinvestment/guidelinesformultinationalenterprises/ (last accessed February 2015).

values that characterizes the domain. This is the case of "social economy" and cooperatives where the notion of social utility is already present. They are a fertile ground for placing CSR and related tools at a central place in the company's management strategy. The same holds true for the socially responsible investment sector. It is also a fact that large corporations are more prone to adopt CSR and adjust their management accordingly than small enterprises.

From a managerial perspective, CSR is the company's response to societal interpellations by producing different strategies, management tools, methods of control, evaluation and reporting. This implies that the company deals with societal issues such as public health, security, environment, which usually belong to the public sphere and therefore call for a political democratic debate (Capron and Quairel-Lanoizelée 2012). It therefore creates a kind of positive competition between companies and the State to produce public values (Bozeman 2007). The company will have to operationalize the concept of stakeholders and know their stakeholders through a mapping exercise. They will have to consider the conditions of production not only on their sites but also with their suppliers. This is why supply chain management often starts through the adoption of a responsible procurement strategy, one of the fast developing new management tools.

As there is a clear obligation of transparency, societal reporting has become a public objective that constraints companies to develop measuring tools for the social and environmental impacts of their activities. The numerous new standards and labels analyzed above are becoming new management tools to certify certain characteristics of products or processes that allow to act on different dimensions of production, in different parts of the world, through economics.

2.3 About "Due Diligence"

"Due diligence" is defined as a "comprehensive, proactive process to identify the actual and potential negative social, environmental and economic impacts of an organization's decisions and activities over the entire life cycle of a project or organizational activity, with the aim of avoiding and mitigating negative impacts" (ISO 2010). Another important consideration relates to "rule of law" versus "international norms of behaviour", mentioned as one of the points of attention in the ISO 26000-definition: "An organization should respect international norms of behaviour, while adhering to the principle of respect for the rule of law" (ISO 2010). In the international standard, this principle is linked with the notion of "complicity", indicating that this has both legal and non-legal meanings: "In this context, an organization may be considered complicit when it assists in the commission of wrongful acts of others that are inconsistent with, or disrespectful of, international norms of behaviour that the organization, through exercising due diligence, knew or should have known would lead to substantial negative impacts on society, the economy or the environment. An organization may also be considered complicit where it stays silent about or benefits from such wrongful acts." Furthermore, note that three forms of complicity can be described (ISO 2010):

- Direct complicity, when organizations (incl. enterprises) knowingly assist in the commission of wrongful acts
- Beneficial complicity, when organizations (incl. enterprises) benefit directly from these wrongful acts committed by someone else
- Silent complicity, when this involves the failure by an organization to raise with the appropriate authorities the question of systematic or continuous wrongful acts

It is not surprising that in the ISO 26000 these three forms are related and exemplified with the avoidance of human rights violation.

From these considerations, it is clear that "due diligence" is indeed "a comprehensive, proactive process" for an organizations' governance in addressing the issues of social responsibility.

In the Implementation Manual of the GRI Sustainability Reporting Guidelines, the notion of "due diligence" is used the first time as part of the "governance" requirements: "Report the highest governance body's role in the identification and management of economic, environmental and social impacts, risks, and opportunities. Include the highest governance body's role in the implementation of due diligence processes."

For the purposes of the OECD Guidelines for Multinational Enterprises, "due diligence is understood as the process through which enterprises can identify, prevent, mitigate and account for how they address their actual and potential adverse impacts as an integral part of business decision-making and risk management systems." In the paragraphs related to the General Policies, it is stressed that "enterprises should" "avoid causing or contributing to adverse impacts on matters covered by the Guidelines, through their own activities, and address such impacts when they occur". The Commentary on the General Policies is explicit on stressing that "own activities includes their activities in the supply chain". Furthermore, a Deming wheel approach is strongly recommended: "In the context of its supply chain, if the enterprise identifies a risk of causing an adverse impact, then it should take the necessary steps to cease or prevent that impact." However "The Guidelines recognise that there are practical limitations on the ability of enterprises to effect change in the behaviour of their suppliers." Note that this is very much related to the sphere of influence mentioned above.

For UN Global Compact "due diligence" is in the first place related to the human rights principles where "identifying and managing human rights risk will help business respect human rights and avoid complicity in human rights abuse".[4] The concept of "sphere of influence" has been introduced as well and is seen to "help map the scope of a company's opportunities to support human rights and make the greatest positive impact". But it is recognized: "While these opportunities may be

[4] See http://www.unglobalcompact.org/Issues/human_rights/The_UN_SRSG_and_the_UN_Global_Compact.html (last accessed February 2015).

greatest with respect to a company's own operations and workers, the ability to act gradually declines as consideration moves outward to the supply chain, to local communities, and beyond." Therefore, UN GC is developing further guidance on how to take a more proactive approach to integrate the Global Compact principles into supply chain management practices. Supply chain sustainability is an important work stream of UN GC.

2.4 Due Diligence: Assessment Through LCSA?

Referring to the expectations of stakeholders when taking up social responsibility, exercising due diligence means an identification of "the actual and potential negative social, environmental and economic impacts of an organization's decisions and activities over the entire life cycle of a project or organizational activity"[5] (ISO 26000). The following questions arise: (1) what does it mean actual and potential negative social, environmental and economic impacts"? and (2) how can "the entire life cycle of a project or organizational activity" be described?

2.4.1 The Social, Environmental and Economic Impacts

The reason for societal concerns about the social, environmental and economic impacts is because of the externalities produced by "activities that affect the well-being of people or damage the environment, where those impacts are not reflected in market prices. The costs (or benefits) associated with externalities do not enter standard cost accounting schemes" (Valdivia et al. 2011). Figure 5.1 shows a matrix illustrating the distinction between private costs and externalities and reflecting what is at stake when assessing the value chain producing goods or services within the context of sustainable development. "An externality occurs when a decision within the value chain imposes costs or benefits on others which are not reflected in the prices charged for the goods and services being provided by the value chain. Externalities are sometimes referred to as spill overs. An externality may also result in private costs, even though it might not be accounted for in the decision-making" (Benoit and Mazijn 2009).

The solid black line in Fig. 5.1 delimits the private costs and benefits reflected in the market price. Sometimes external relevant costs and benefits anticipated to be privatized, such as increasing prices of CO_2 emissions, are taking into account in monetary terms: see dashed line. However, it is illusory to think one can reflect all

[5]This is a quote coming from ISO 26000. Note that 'life cycle sustainability assessment' tries to provide a more comprehensive picture of the positive and negative impacts along the product life cycle. However, this is as such not contradictory because in terms management ISO 26000 recommends: "*An organization can exercise its influence with others either to enhance positive impacts on sustainable development, or to minimize negative impacts, or both*".

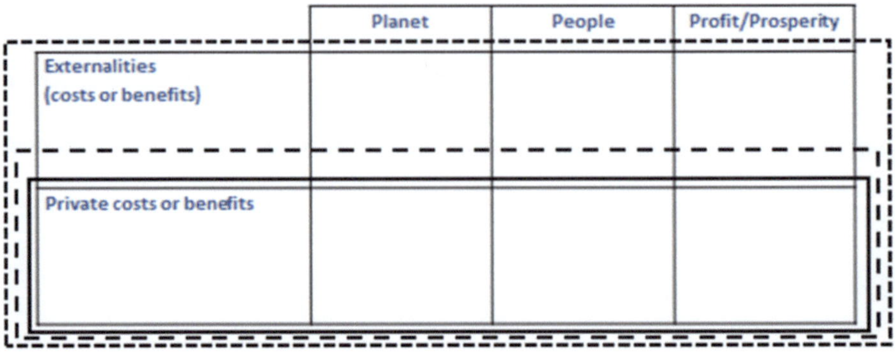

Fig. 5.1 Detailing the assessment of the value chain producing goods and services within the context of sustainable development[6]

externalities, within the limits of the dotted line, in the costs of goods and services produced by the value chain.[7] Therefore other indicators to take social, environmental and economic impacts into account – "in consistency with international norms of behaviour" (cf. CSR definition of ISO 26000) – are needed. In fact, the purpose is to detail the already mentioned seven core subjects of social responsibility.

2.4.2 Life Cycle Sustainability Assessment

The reference to "the entire life cycle" in ISO 26000 is linked to "a project or organizational activity" and as has been noted "activities include products, services and processes". In fact, this is closely connected with what has been phrased in Agenda 21 – Chapter 4 to "develop criteria and methodologies for the assessment of environmental impacts and resource requirements throughout the full life cycle of products and processes".[8] Later on, this "life cycle thinking" was explained as follows by UNEP: "Life Cycle Thinking is about understanding environmental, social and economic impacts into people's hands at the time they are making decisions. It offers a way of incorporating sustainability in decision making processes and can be used by decision makers in both the public and private sector for the development

[6] Note that making a distinction between the three dimensions of sustainable development (environment, economy and society) is often referred to as the "triple bottom line" concept (TBL) as coined by John Elkington in his 1997 book *Cannibals with Forks: The Triple Bottom Line of Twenty-First Century Business*, a concept which can be seen as similar to the 3P approach: people, planet and profit. However, since people and planet imply a collective interest, profit can be interpreted as private interest. Therefore, it is not surprising that the World Summit on Sustainable Development, Johannesburg 2002, referred instead to "people, planet and prosperity".

[7] The reason for stating this is related to the problems of having a scientific method of calculating the price at each stage of the value chain, being accepted by all stakeholders.

[8] See http://sustainabledevelopment.un.org/content/documents/Agenda21.pdf (last accessed February 2015).

of policies and products, as well as for procurement and the provision of services."[9] It has been the start at the time of the United Nations Conference on Environment and Development, Rio de Janeiro 1992, for a comprehensive effort to present 20 years later a tool for life cycle sustainability assessment (LCSA).

The precursor of "life cycle assessment (LCA)" goes back to the late 1960s. The development of the technique throughout the 1970s and 1980s was stimulated by eager enterprises and policy makers who wanted to have a better understanding of the environmental impact of packaging and energy content of products. Later on, LCA was applied to an increasing variety of product types, and methods for life cycle environmental impact assessment began to be developed. It resulted initially in the publication of the "Code of Practice" (Consoli et al. 1993), followed by the development of four ISO standards (ISO 14040–14043) published in 1997–2000, all of which were replaced in 2006 by two standards, ISO 14040 and ISO 14044 (ISO ISO 2006a, b). These standards describe the requirements and formulate recommendations for elaborating an LCA. At first it was meant to address the environmental aspects of a product and their potential impacts throughout that product's life cycle. However, the picture is not complete within a context of sustainable development unless social and socio-economic impacts on all actors along the life cycle, including workers, local communities, consumers and society, are analyzed. Discussions on how to deal with social and socio-economic dimensions of products throughout a life cycle started at the end of the 1980s. In 2004, the UNEP/SETAC Life Cycle Initiative[10] established an international Task Force to "to convert the current environmental tool LCA into a triple-bottom-line sustainable development tool": by 2009 the "Guidelines for social life cycle assessment of products" were published with a set of (sub-)categories of impacts (Benoit and Mazijn 2009). The subtitle of the publication is relevant within this context: "A social and socio-economic LCA code of practice complementing environmental LCA and Life Cycle Costing, contributing to the full assessment of goods and services within the context of sustainable development." Indeed, LCC or life cycle costing is regarded as the third LCA technique aiming at "the assessment of all costs associated with the life cycle of a product that are directly covered by 1 or more actors in the product life cycle (supplier, manufacturer, user or consumer, and/or End of Life actor), with the inclusion of externalities that are anticipated to be internalized in the decision-relevant future" (Hunkeler et al. 2008).

These different life cycle assessment techniques can be combined as part of an over-arching LCSA and allow to assess the impacts of the value chain. Recently, the methodology has been presented in two publications (Valdivia et al. 2011, 2012) where it is emphasised that LCSA "helps to organise complex environmental, economic and social data in a structured form; clarify the trade-offs between the three sustainability dimensions, life cycle stages and impacts; provide guiding principles

[9] See for more information: http://www.lifecycleinitiative.org/starting-life-cycle-thinking/ (last accessed February 2015).

[10] See www.lifecycleinitiative.org (last accessed February 2015).

to achieve sustainable production while stimulating innovation (by identifying weaknesses and enabling further improvements over the product life cycle); help to raise credibility by communicating useful quantitative and qualitative information about their products and process performances (which can also be used to inform labelling initiatives); and show how to become more responsible by taking into account the full spectrum of impacts associated with their products and services. LCSA can support decision-makers in prioritising resources and investments, and in choosing sustainable technologies and products. Finally, LCSA could support consumers in determining which products are cost-efficient; have a low environmental impact and are socially responsible; and, in general, promote awareness in value chain actors on sustainability issues." Note that it is not the aim of assessing in all details the life cycle, but to focus on the so-called "hotspots", i.e. the important impacts. It can be compared with the "materiality" exercise in ISO 26000 in which relevance, significance and priority is looked at in a systematic and coherent manner.

Finally, it is interesting to note that (methodological) developments around "life cycle thinking" were not limited to present tools for analysis or assessment. In fact, LCSA can be regarded as part of the PDCA-cycle[11] – where the "assessment" correspond clearly with the "check" and should be followed by action (before a renewed planning is set up). This is all about management, i.e. "life cycle management" which is another important area of work over the last 20 years (see e.g. Remmen et al. 2007).

2.5 *Implementation and Follow Up of Due Diligence Through LCM?*

As life cycle management (LCM) is still a domain in development, it is relevant to begin by providing some definitions based on the recent scientific literature.

The SETAC Europe Working Group on LCM defined it as "an integrated framework of concepts, techniques and procedures to address environmental, economic, technological and social aspects of products and organizations to achieve continuous environmental improvement from a life cycle perspective" (Hunkeler et al. 2004). UNEP brings the collaboration and stakeholder perspective, they see LCM "as a product management system aiming to minimize environmental and socioeconomic burdens… during the entire life-cycle…relying on collaboration and communication with all the stakeholders in the value-chain" (Balkau and Sonnemann 2010).

[11] PDCA stands for 'Plan, Do, Check, Act' in which 'Check' is sometimes replaced by 'Study' and 'Act' by 'Adjust' (see this volume, Chap. 3).

Obviously definitions vary depending on stakeholders who use or promote it. For this paper, the authors refer to the definition by Balkau and Sonnemann (2010) who see LCM "as an umbrella framework for combining and applying other management instruments in a more holistic life chain perspective".

They offer a classification of LCM approaches into three broad categories:

(1) Organization of a holistic form of sustainability management within individual companies using, for example, supply-chain management and product design (Five Winds International 2009)
(2) Government life-cycle policies and regulations to address system dysfunctions or to deal with certain product issues such as chemical contamination.
(3) Multi-stakeholder voluntary codes to manage sustainability issues for selected commodity materials and products.

Nilsson-Linden et al. (2014) provide an interesting view into the theoretical aspects of LCM and LCM in practice. For them "the review of the LCM literature indicates that it provides many normative prescriptions of what LCM is, including what tools, methods, and approaches to use". But they also tell us that in fact this literature indicates in fact what *ought to be* considered, but without providing compelling descriptions and analysis of the difficulties involved in organising LCM in practice. This is a common critique about the state of development of LCM.

However, the authors consider LCM as the most appropriate framework to integrate and organize adequately the large tool box that was developed piece by piece without a pre-existing integrating framework to contribute to CSR and sustainable development in a coherent and consistent manner.

Taking up social responsibility is much more than caring for the environment. At least six other core subjects such as human rights, labour practices, etc. are important as well. Therefore, it is fair to state that CSR should be framed within the broader context of sustainable development. Furthermore, the social responsibility is not limited to the management of the facility or plant of an enterprise; "due diligence" should be exercised all along the value chain. In other words, if it is not part of the corporate strategy, an enterprise may be considered complicit because it does not cope with the broad societal expectations of behaviour. But due diligence is a process and it needs frameworks and tools.

Our pieces come together here. CSR through ISO 26000 and through its links with sustainable development calls the organization to consider upstream and downstream of its activities and to look for aspects that were not even monitored recently by traditional organisations. One needs for that a perspective that is multidimensional and along the life cycle of a product or service: this is called life cycle thinking.

Life cycle sustainability assessment, taking into account the three dimensions of sustainable development, and using a toolbox of different techniques, will provide the enterprise with the analysis and the evaluation of the impacts along the value chain. It forms in its turn an input for life cycle management.

3 Outlook

To paraphrase Nilsson-Linden et al. (2014): "many normative prescriptions of what LCM is, indicate in fact what ought to be considered, including what tools, methods, and approaches to use". Indeed further research is needed at several levels.

The authors are confident that organizations should not wait to take up social responsibility by using the approach as outlined here above. However, they consider that there are needs for research and practice, inter alia:

- Collaboration between the world of (C)SR and LC(S)A: researchers and practitioners should learn to interact with each other, in particular on the issues "materiality", "due diligence", "value chain", "life cycle"
- Streamlining of the approach of "due diligence" in the different international standards (ISO 26000, GRI, OECD MNE Guidelines, UN GC)
- Incorporation of stakeholder involvement practice in (C)SR into LCSA and LCM
- (Better) integration of three LCA-techniques under the over-arching LCSA; details regarding methodological issues have already been identified: see e.g. Benoit and Mazijn (2009) and Valdivia et al. (2011)
- Design LCSA for due diligence within the context of social responsibility of (different) organizations (incl. enterprises)
- Implementation of LCM in various realities considering this adoption as a strategic change in management

It should be emphasized that this list is not exhaustive and that progress can be based on research as well as best practices. The Social LC Alliance (www.social-LCA.org), in which the authors take part, aims to contribute in meeting these challenges.

References

Balkau F, Sonnemann G (2010) Managing sustainability performance through the value chain, Corporate Governance. Int J Bus Soc 10(1):46–58

Benoît C, Mazijn B (eds) (2009) Guidelines for social life cycle assessment of products – a social and socio-economic LCA code of practice complementing environmental LCA and life cycle costing, contributing to the full assessment of goods and services within the context of sustainable development. UNEP-DTIE, Paris

Bowen HR (1953) Social responsibilities of the businessman. Harper and Brothers, New York

Bozeman B (2007) La publicitude normative: comment conciliar valeurs publiques et valeurs du marché. Politiques et Management Public 24(4):179–196

Capron M (2013) La RSE: Un danger ou un atout pour la démocratie. In: Gendron C, Girard B (eds) Repenser la responsabilité sociale de l'entreprise. L'école de Montréal, Armand Colin Recherches, Paris, pp 123–134

Capron M, Quairel-Lanoizelée F (2012) La responsabilité sociale d'entreprise, Coll. 'Repères. La découverte, Paris

Consoli F, Allen D, Boustead I, Fava J, Franklin W, Jensen AA, de Oude N, Parrish R, Perriman R, Postlethwaite D, Quay B, Seguin J, Vigon B (1993) Guidelines for life-cycle assessment: a 'Code of Practice'. SETAC, Brussels/Pensacola

Dahlsrud A (2008) How corporate social responsibility is defined: an analysis of 37 definitions. Corp Soc Responsib Environ Manage 15(1):1–13

Friedman M (1970) A Friedman doctrine – The social responsibility of business is to increase its profits. New York Times Magazine, 13 Sept 1970

Freeman RE (1984) Strategic management: a stakeholder approach. Pitman, Boston

Gendron C (2013) La firme: portrait d'un objet ambigu. In: Gendron C, Girard B (eds) Repenser la responsabilité sociale de l'entreprise, Armand Colin Recherches. L'école de Montréal, Paris, pp 135–148

GRI (2011) GRI and ISO 26000: how to use the GRI Guidelines in conjunction with ISO 26000. Global Reporting Initiative, Amsterdam

Hunkeler D, Saur K, Rebitzer G, Finkbeiner M, Schmidth W-P, Jensen A, Stranddorf H, Christiansen K (2004) Life cycle management. SETAC, Pensacola FL

Hunkeler D, Lichtenvort K, Rebitzer G (eds) (2008) Environmental life cycle costing. SETAC Press/Taylor & Francis Group, New York

ISO (2006a) ISO 14040 – environmental management – life cycle assessment –principles and framework. International Organization of Standardization, Geneva

ISO (2006b) ISO 14044 – environmental management – life cycle assessment – requirements and guidelines. International Organization of Standardization, Geneva

ISO (2010) ISO 26000 – Guidance on social responsibility. International Organisation for Standardisation Geneva

McGuire J (1963) Business and society. McGraw-Hill, New York

Nilsson-Lindén H, Baumann H, Rosén M, Diedrich A (2014) Organizing life cycle management in practice: challenges of a multinational manufacturing corporation. Int J Life Cycle Assess. doi:10.1007/s11367-014-0818-y

Pasquero J (2013) La responsabilité sociale de l'entreprise: trajectoire d'une idée. In: Gendron C, Girard B (eds) Repenser la responsabilité sociale de l'entreprise. L'école de Montréal, Armand Colin Recherches, Paris, pp 47–62

Remmen A, Jensen A, Frydendal J (2007) Life cycle management – a business guide to sustainability. UNEP-DTIE, Paris

UN (2012a) Report of the United Nations conference on sustainable development – Rio de Janeiro, Brazil 20–22 June 2012. Document nr. A/CONF.216/16, United Nations, New York

UN (2012b) Resolution adopted by the General Assembly – The future we want. Document nr. A/RES/66/288UN General Assembly, New York

UN (2012c) A 10-year framework of programmes on sustainable consumption and production patterns. Document nr. A/CONF.216/5, United Nations, Rio de Janeiro

Valdivia S, Ugaya C, Sonnemann G, Hildebrand J (eds) (2011) Towards a life cycle sustainability assessment – making informed choices on products. UNEP-DTIE, Paris

Valdivia S, Ugaya C, Hildenbrand J, Traverso M, Mazijn B, Sonnemann G (2012) A UNEP/SETAC approach towards a life cycle sustainability assessment – our contribution to Rio+20. Springer, Int J Life Cycle Assess, November 2012, online

Chapter 6
Life Cycle Management: Labelling, Declarations and Certifications at the Product Level – Different Approaches

Frieder Rubik

Abstract The focus of this chapter is on *external communication of product features intended to provide professional, commercial and private consumers with information on the characteristics of products and services*. Mandatory approaches are distinguished from voluntary ones; the chapter is focused on the latter. Based on ISO standardization work, this chapter differentiates between qualitative, quantitative and self-declared voluntary approaches. Section 2 presents an overview of different concepts and approaches as tools applicable within Life Cycle Management. Section 3 deepens relevant approaches by describing some characteristic elements. Section 4 elaborates on a hierarchy, whereas the final Sect. 5 summarizes the outcomes and draws some conclusions.

Keywords External communication of product features • Consumers • Life cycle assessment • Life cycle management • Product-related information • Sustainability

1 Introduction

Life Cycle Management (LCM) is an umbrella term denominating a business management concept for sustainable products. It can be applied in the industrial and service sectors with the aim of minimizing environmental, social and economic burdens linked to a company's product, product portfolio and brand during its entire life cycle to enhance their overall sustainability performance and value chain. Thus LCM facilitates continuous improvements of product/systems in terms of their economic, social and environmental sustainability. The focus of this chapter is on *external communication of product features intended to provide professional,*

F. Rubik (✉)
Institute for Ecological Economy Research (IÖW),
Bergstr. 7, D-69120 Heidelberg, Germany
e-mail: frieder.rubik@ioew.de

G. Sonnemann, M. Margni (eds.), *Life Cycle Management*,
LCA Compendium – The Complete World of Life Cycle Assessment,
DOI 10.1007/978-94-017-7221-1_6

commercial and private consumers with information on the characteristics of products and services. Such communication from manufacturers towards their clients is based on the insight that product information is – in most cases – asymmetrically allocated between buyers and sellers (Karl and Orwat 1999: 114). According to Nelson (1970) and Darby and Karni (1973), consumers are not able to judge all qualities of products. In order to cope with asymmetric information, consumers need support in their purchasing activities provided by different tools. There is a widespread arena of different approaches to transmit this information, there are *qualitative approaches* using symbols and logos, and there are *quantitative approaches* presenting quantitative and numeric information in different units. They intend to fill the information gap so-called credence goods leave behind, providing information transmission. They aim to establish a reliable and trustworthy information system regarding product features.

Section 2 presents an overview of different concepts and approaches as tools applicable within Life Cycle Management. Section 3 deepens relevant approaches by describing some characteristic elements. Section 4 elaborates on a hierarchy, whereas the final Sect. 5 summarizes the outcomes and draws some conclusions.

2 Overview on Different Approaches

The transmission of information between sellers – i.e. industry and business – and their clients is not only motivated by coping with asymmetric information, but by a series of driving forces (see UNEP 2006: 43) depending on the target audiences:

- *Private consumers* to get competitive advantage in emerging or new markets
- *Commercial business purchasers* to respond to requests of business in the supply chain or to compete on the business-to-business market
- *Public purchasers* to demonstrate compliance with Green Public Procurement (GPP) requirements
- *Societal stakeholders* to respond to requests and pressures from NGOs
- *Banking and finance* which are keen to judge technical and environmental risks of companies and their products
- *Policy makers and public administration* to deliver information and data to support them in policy decisions and to favour reasonable decisions

Communication between manufacturers and the mentioned target groups might take on different forms: oral, written, formalized, informal, standardized, etc. Some of them are mandatory, prescribed by national or international regulations, some are voluntary; Fig. 6.1 provides a classification of different approaches to transmitting information.

Mandatory approaches request that every producer or retailer introducing products in the market is obliged to fulfil prescriptions on the provision of product information. Compulsory product information refers often to the health and safety aspects of products, giving details of chemical substances contained within the

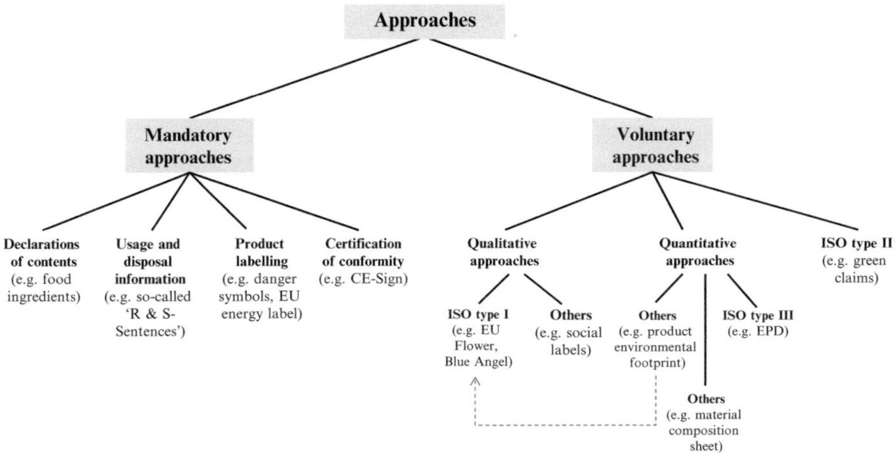

Fig. 6.1 Classification of different information transmission approaches (Source: Rubik and Frankl 2005: 34 (modified))

product or information on the proper usage and disposal of the product, other types are certificates of conformity of products with specific regulatory requirements. In contrast to compulsory approaches, *voluntary ones* leave to market actors the decision of whether to use it or not. There is a wide range of such approaches. Much effort has been made by the International Organisation of Standardisation (ISO) to structure environmental approaches which were subdivided into three types of voluntary labels:

- *ISO Type I labels (Eco-label):* "Voluntary, multiple criteria-based third party programs that awards a licence authorising the use of environmental labels on products. These labels provide qualitative environmental information" (ISO 2000: 1). They are covered by ISO 14024 published in April 1999, last reviewed and confirmed in 2009.
- *ISO Type II labels:* "Self-declared environmental claim made by manufacturers, importers, distributors, retailers, or anyone else likely to benefit from such a claim without independent third-party certification" (ISO 1999: 3). They are covered by ISO 14021 published in 1999.
- *ISO Type III labels:* "Quantified environmental data using predetermined parameters and, where relevant, additional environmental information. Note 1: The predetermined parameters are bases in the ISO 14040 series of standards (…). Note 2: The additional environmental information may be quantitative or qualitative" (ISO 2006: 2). They are covered by ISO 14025 published in 2006.

Comparing these types of labels (see Table 6.1) several significant differences could be recognized: In general, the different schemes claim to fill the information gap by "condensing" information. The number criteria – the metrics – depend on the type: Type I and III cover multiple areas, whereas Type II one single area. The same

Table. 6.1 Comparison of the three ISO labels

Criteria areas/metrics	**Life cycle consideration**
Type I: multiple Type II: single Type III: multiple	Type I: yes Type II: no Type III: yes
Selectivity	**Third party verification/certification**
Type I: yes Type II: no Type III: no	Type I: yes Type II: preferred Type III: yes

Source: GEN (2004: 12)

refers to the consideration of life-cycle: its examination is a core element of the types I and III and not requested by Type II. The symbol of the Type I indicates that requirements, which are (nearly) not visible on the label itself, were fulfilled. By doing so, it is selective: It "translates" quantitative and qualitative information and transmits them to the target groups. This means that a label allows them to distinguish between products with, and without, the label. Type II and III are not selective. The third party verification is another request of the types I and III, but not strongly requested by Type II.

3 Some Exemplary Information Transmission Approaches

In this section, we focus on three different approaches, namely qualitative approaches by labels, self-declared environmental claims and quantitative approaches.

3.1 Qualitative Approaches

3.1.1 Eco-Labels

Addressed Issues Eco-labels according to ISO type I should consider the entire life-cycle of a product based on scientific evidence, their application is voluntary and up to the decision of the applicants. They refer to environmental issues, like energy consumption, material composition, emissions, use of dangerous substances etc. They are intended to label products with considerable less environmental impacts than the market average along the life-cycle – i.e. the "best in the class'. These last few years, environmental topics have been supplemented by the integration of social criteria into some requirements of the eco-labels, e.g. working conditions, fair-trade issues.

Institutional Issues For each product group, criteria have to be developed and fixed. The criteria development is carried out in an open participatory process, e.g. by boards, committees, panels, expert groups representing different economic and social interests (e.g. trade, industry, consumer and environmental organisations). However, the final decision on requirements has to be taken by an institution independent from manufacturers and their interests. The fulfilment of the requirements has to be proven by a third-party verification procedure. Having passed the requirements, applicants receive the allowance to use the symbol of an eco-label scheme, which is restricted to a predefined period of some years. This restriction is intended to review the requirements and to update them, if needed, taking into account new technological developments, new information and other challenges.

Target Groups Mainly private consumers.

Status Eco-labels have been in place since 1978, when the German Blue Angel became the first voluntary eco-label scheme worldwide, followed just over a decade later (1989) by the Japanese Eco-Mark. Altogether, the labelling landscape has become more and more complex, and also confusing. According to Ecolabel Index,[1] 458 eco-labels in 197 countries covering 25 product groups exist: some are applied to a vast range of product groups whereas others are restricted to a single and specific product group. Globally, providers of eco-label schemes co-operate in the "Global Ecolabelling network" (see: http://www.globalecolabelling.net/).

Examples European eco-label "EU-Flower", German Blue Angel, Scandinavian "Nordic Swan", Australian "Good Environmental Choice", Japanese "Eco Mark Program", US "Green Seal" or the "Green label Thailand".

3.1.2 Social Labels and Standards

Addressed Issues The consideration of environmental challenges is only one challenge, but due to the increasing "length" and complexity of supply chains, to the globalisation of markets and supply or production chains, and due to pressures from stakeholders, the social features along the chain gain increasing importance. Beside company and workplace related standards like ISO 26000 and SA8000, some labels cover social issues such as ban of child labour, social rights, labour union laws, fair prices, working conditions. However, a common international standard like the ISO 14020-series does not exist.

Institutional Issues The institutional characteristics depend on the requirements label scheme, in general reliable labels are independent from business and request an independent certification of the fulfilment of their requirements.

Target Groups Mainly private consumers, but also business and public purchasers.

[1] http://www.ecolabelindex.com/ (accessed March 5, 2015).

Status The increasing importance of social issues could be observed by the increasing number of labels dealing with this topic. The webpages of the Sustainability Compass (http://www.sustainability-compass.com/) or of the Standards Map (http://www.standardsmap.org/) offer a broad overview on social (and sustainability) labels.

Examples "Rugmark" label, "Fairtrade" label.

3.1.3 Certificates of Conformity

Addressed Issues The issues addressed are diverse and refer to specific needs. They might document for example sustainable forestry, fishery, and agriculture. The certificates document fulfilment of specific environmental requirements, which are often based on upstream challenges during resource extraction. The right to use a certificate allows their holders to distinguish their certified products from those of competitors and might offer market opportunities by positive discrimination.

Institutional Issues The institutional characteristics depend from the requirements label scheme, in general reliable labels are independent from business and request an independent certification of the fulfilment of their requirements.

Target Groups Private consumers, but also business and public purchasers.

Status A lot of different certificates of conformity have been developed, an overview is hard to get, but there are several webpages providing some overviews, e.g. the already mentioned ones of the Sustainability Compass, of the Standards Map or of the Ecolabelindex. The Forestry Stewardship Council (FSC) provides certificates for companies which fulfil a number of forestry requirements; the requirements have been elaborated by the Forest Stewardship Council (FSC) which is an international organization with business, NGOs, trade unions and representatives of indigenous people. Applying companies need an independent verification of a certifier accredited at FSC.

Examples "FSC" (Forrest Stewardship Council) label, "MSC" (Marine Stewardship Council) label, "Rainforest Alliance" label.

3.2 Self-Declared Environmental Claims

Addressed Issues Self-declared environmental claims according to ISO 14021 depend from the interests of the business/industrial associations in charge of label. Beside environmental issues, also social issues might be highlighted.

Institutional Issues There are not specific institutional prescriptions; however the relevant national/international legislation (e.g. competition laws) has to be respected.

Target Groups Mainly private consumers.

Status Self-declared labels could be regarded as a business marketing approach to inform on the environmental qualities of their products by self-declaration. According to the ISO 14021 standard such labels do not require an independent third-party registration. The number of such labels has grown continuously. The perception and "reputation" of the labels depend on some strong parameters, like credibility of the creator, product group, market competition, etc. In general, especially NGOs suspect self-declared environmental labels and do not support them.

3.3 Quantitative Declarations

3.3.1 Environmental Product Declarations (EPD)

Addressed Issues In line with the ISO standard 14025, Environmental Product Declarations (EPD) provide quantified environmental data for a product with pre-set categories of parameters (product category rules, PCR). The data should be based on LCA tools and calculations and consider supply chains. They might be also highlight and restricted to some (or one single) environmental challenges – single-issue EPDs. EPD intend to compare a product of the information provider with other products of the specific product group.

Institutional Issues The product category rules have to be elaborated in a participatory consultation process involving stakeholders like business, NGOs etc. Companies presenting EPDs of their products have to verify the data according to the rules of the ISO 14040 series. The verification of data has to be carried out by independent verifiers.

Target Groups Mainly business (commercial procurers, public procurers, retailers).

Status In 1998, the Japan Environmental Management Association for Industry (JEMAI), with the support by the Japanese Ministry of Economy, Trade and Industry (MITI), started an experimental program for Type III environmental declarations which resulted in the EcoLeaf's official launch in 1999. A global network of EPD-organizations and practitioners pushes the development of EPD's (see http://gednet.org/). Recently, climate-related EPD's focusing on climate relevant data came up.

Examples Japanese "Eco-Leaf", "International EPD® System".

3.3.2 Product Footprint

Addressed Issues The addressed issues depend on the objective of the footprint concept. An encompassing footprint refers to different environmental challenges, whereas the water footprint, for example, is restricted to water-related challenges.

Institutional Issues The institutional issues are still under development.

Target Groups Private consumers, commercial procurers.

Status The origin of different footprints is the concept of the ecological footprint (Rees and Wackernagel 1996) which were followed by different other footprint approaches (see Fang et al. 2014), like product water footprint, land footprint or carbon footprint.

However, of increasing importance are the ongoing efforts of the European Commission to create a Product Environmental Footprint. They have their origin in a communication of the European Commission (2008), which called for the elaboration of a product carbon footprint. Later, the Commission decided to extend the work to other environmental aspects resulting in a product footprint. The "Communication Building the Single Market for Green Products" (European Commission 2013a) and methodological recommendations (European Commission 2013b) pushed the further development. The Commission's product footprint should be based on LCA, mentioned are the corresponding ISO standards of the 14040 series and some other concepts, it might consider 14 different impact categories. Like for EPD, for each product group so called "Product Environmental Footprint Category Rules" should be prepared and used for the proliferation of data.

In autumn 2013, a 3 year two stages-pilot phase started to elaborate and test the Commission's approach.[2] Actually pilots[3] run for 25 different product groups like wine, household detergents or thermal insulation materials. The product footprint is intended to be applied in different context, business internally, business to business and business to consumer. The final format of the product footprint is not decided, different examples of communication vehicles have been provided[4] and will be tested during the pilots.[5] If the target audience is the final consumer, product footprint might come close to symbols and might be interpreted as a qualitative approach – however this is still an open issue of consideration of the Commission.

Example European "Environmental product footprint" approach.

3.3.3 Material Composition

Addressed Issues Without reference to any ISO standard, business in the supply chain provide information on the composition of their products, especially with regard to the material they consist (UNEP 2014). The objective is to use an agreed data sheet which fulfils information needs of business clients and does not request case-by-case adoption.

---

[2] See the webpage http://ec.europa.eu/environment/eussd/smgp/product_footprint.htm (accessed March 6, 2015).

[3] See http://ec.europa.eu/environment/eussd/smgp/pef_pilots.ht (accessed 9 March 2015).

[4] See Mugdal et al. (2012) and a Background Paper (2013).

[5] See Finkbeiner (2014) for a strong critique of the Commission's efforts.

Institutional Issues There are no specific institutional prescriptions; however, cooperation between competitors (horizontal cooperation) and clients (vertical cooperation) supports the unification of the data sheets.

Target Groups Mainly business (commercial procurers, public procurers, retailers).

Status The push for the development of material composition sheets is based on requests from clients (commercial or public ones) asking for more information on the products they purchase, e.g. by questionnaires, which ask manufacturers to transmit information about the products/pre-products they sell. Requests and information needs are diverse, often very heterogeneous. Therefore, several branches took the initiative to elaborate unified composition sheets. This is the case in a couple of branches, e.g. electronics industry, car industry. The Consumer Electronics Association (CEA) representing branches from several continents elaborated such a document. Besides industry branches, also global players – focal companies – use their strategic position in the chain to ask their suppliers to deliver information according to unified sheets.

Examples "Material Composition Declaration for Electrotechnical Products" of CEA.

4 Selection of Approaches

In previous research (cf. Rubik and Frankl 2005), we looked for key influencing factors for a successful application of different approaches. Beside general factors like credibility of a scheme and its costs and fees, product group specific factors are the key, see Fig. 6.2. The latter ones could be separated into factors related to environmental challenges, to the market situation, to the relevance of different stakeholders and to the type of approach chosen.

If the product group is the key, which product groups might be distinguished? Rubik and Frankl (2005: 265f.) argues for six categories:

- *Non-recoverable consumable goods:* e.g. tissue papers detergents, soil improvers
- *Recoverable consumable goods:* e.g. copying and printing paper, packaging
- *Energy-consuming durable products with main impact during the use phase*: e.g. cars, IT-equipment, washing machines, refrigerators, dishwashers
- *Energy-passive durable products*: e.g. furniture, textiles, footwear
- *Simple services:* e.g. car washing, laundry services
- *Complex services*: e.g. tourist accommodation

The environmental challenges and impacts of products within these six categories are very different along their life cycle. This means that the consideration of different environmental are not homogeneous, but heterogeneous, also the type and approach preferred as well as the target groups differ. Therefore, Rubik and Frankl (2005: 266ff.) elaborated a guide for the selection of approaches depending on the product category and the target audience. Figure 6.3 shows the proposed allocation

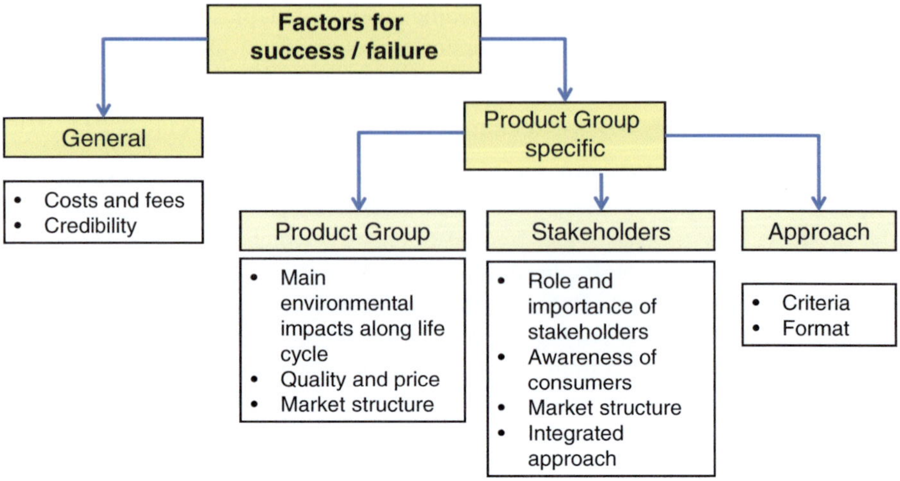

Fig. 6.2 Factors for success or failure of an information transmission approach (Rubik and Frankl 2005: 264; slightly modified)

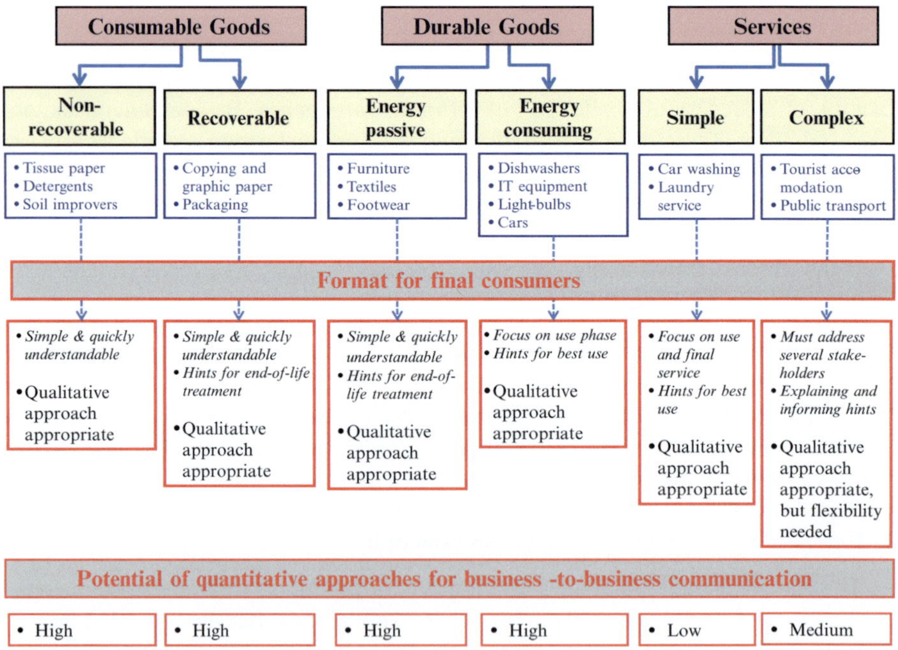

Fig. 6.3 Format of information transmission tools with regard to six product categories (Rubik and Frankl 2005: 273; slightly modified)

of product groups into the six product categories as well as the life cycle phase in which the main environmental impacts occur. It distinguishes among final consumers and business-to-business communication. The latter one is of special interest for LCM-reflections. In general, we rank the potential of qualitative tools in the case of services as restricted. The potential for consumables and durables are assessed as high in the business-to-business communication, although the criteria differ according to the concrete product group and the main environmental challenges along their life-cycles.

However, with regard to final consumers, we judge that qualitative labelling approaches are the main tool to be applied. But the requirements behind the label differs according to the product category, e.g. with regard to energy-passive or energy active durables. Beside the label, also hints for end-of-life treatment are needed to explain consumers an appropriate environmental (more) benign product removal.

5 Conclusions

The proliferation of information as part of LCM is a strong request to support actors downstream with appropriate information. We focused on actors external to the company generating the information (in this case other internal tools are needed).

The tools presented in this section intend to reduce the information seek costs for consumers. However, the target audiences are very different:

- *Private consumers* ask for easy to use and understand information tools, qualitative approaches like the ISO type I approaches (e.g. the EU Flower or the Nordic Swan) are the promising tools which differentiate products within the same product group. Their successful reception by consumers might increase the sale volume of the labelled products and result in reductions of environmental burdens.[6]
- *Business clients* commercial purchasers, public purchasers or retailers – have different information needs, some are requesting quantitative information whereas others need "condensated" information as provided by labels. These different needs require an appropriate strategy of sellers to transmit information towards their clients. The basics are quantitative information based on product category rules agreed and unified within the branch in consensus with the competitors. The information transmitted might support the clients to compare products within the same product group and to priorise them according to different criteria, among them environmental ones. Depending on the type of product – we proposed six categories – additional quick to understand and easy to recognize information might be needed and in this case qualitative labels play a prominent role.

[6] We hint to the discussion on rebound effects which might have some converse effects (see, e.g., Santarius 2012 or Maxwell et al. 2011).

But beside the sellers and buyers, business and consumers, policy makers are keen to push labels as a prominent tool of environmental and, consumer policy. The example of the European Commission highlights labelling issues in its SCP/SIP Action plan (European Commission 2008) and argues for a broad getting-the-information-right-strategy (European Commission 2013a). The outcome of this process – product environmental footprint – is still early in the pilot phase (PEF 2015). But we might expect an instructive – but controversial – outcome of this process and a ripening of these efforts resulting in an encompassing right to know and duty to inform policy strategy. Therefore, it is up to responsible life cycle management to proactively shape the future.

References

Darby MR, Karni B (1973) Free competition and the optimal amount of fraud. J Law Econ 16:67–88

European Commission (2008) Communication from the commission to the European Parliament, the council, the European Economic and Social Committee and the Committee of the Regions on the Sustainable Consumption and Production and Sustainable Industrial Policy Action Plan. http://eur-lex.europa.eu/legal-content/EN/ALL/?uri=CELEX:52008DC0397. Accessed 8 Mar 2015

European Commission (2013a) Communication from the Commission to the European Parliament and the council building the single market for green products – facilitating better information on the environmental performance of products and organisations. http://eur-lex.europa.eu/legal-content/EN/TXT/PDF/?uri=CELEX:52013DC0196&from=EN. Accessed 8 Mar 2015

European Commission (2013b) Commission recommendation of 9 April 2013 on the use of common methods to measure and communicate the life cycle environmental performance of products and organisations. http://eur-lex.europa.eu/legal-content/EN/TXT/PDF/?uri=CELEX:32013H0179&from=EN. Accessed 5 Mar 2015

Fang K, Heijungs R, de Snoo G (2014) Theoretical exploration for the combination of the ecological, energy, carbon, and water footprints: overview of a footprint family. Ecol Indic 36:508–518

Finkbeiner M (2014) Product environmental footprint – breakthrough or breakdown for policy implementation of life cycle assessment? Int J Life Cycle Assess 19:266–271

GEN [Global Ecolabelling Network] (2004) Introduction to ecolabelling. http://www.globalecolabelling.net/docs/documents/intro_to_ecolabelling.pdf. Accessed 9 Mar 2015

ISO (1999) ISO 14021. Environmental labels and declarations – self-declared environmental claims (Type II environmental labelling). International Organization for Standardization, Geneva

ISO (2000) ISO 14020. Environmental labels and declarations – general principles. International Organization for Standardization, Geneva

ISO (2006) ISO 14025. Environmental labels and declarations – type III environmental declarations – principles and procedures. International Organization for Standardization, Geneva

Karl H, Orwat C (1999) Economic aspects of environmental labelling. In: Folmer H, Tietenberg T (eds) The international yearbook of environmental and resource economics 1999/2000. Edward Elgar Publishing

Maxwell D et al (2011) Addressing the rebound effect. Final report for the European commission DG environment. http://ec.europa.eu/environment/archives/eussd/pdf/rebound_effect_report.pdf. Accessed 9 Mar 2015

Mudgal S, Muehmel K, Kong MA et al (2012) Study on different options for communicating environmental information for products. Final report http://www.ecologic.eu/sites/files/publication/2014/different-options-for-communication-environmental-information-for-products-2012.pdf. Accessed 10 Mar 2015

Nelson P (1970) Information and consumer behavior. J Polit Econ 78:311–329

PEF (2015) Background document for the testing of communication vehicles in the environmental footprint pilot phase 2013–2016. Brussels. http://ec.europa.eu/environment/eussd/smgp/pdf/Comm_bgdoc_v1.1.pdf. Accessed 10 Mar 2015

Rubik F, Frankl P (eds) (2005) The future of eco-labelling. Greenleaf, London

Santarius T (ed) (2012) Green growth unraveled. How rebound effects baffle sustainability targets when the economy keeps growing. Heinrich Boell Foundation, Berlin

UNEP (2006) Background report for a UNEP guide to life cycle management. A bridge to sustainable products. Paris. http://www.lifecycleinitiative.org/wp-content/uploads/2013/09/UNEP_Background_document_LCM_2006_Febr.pdf. Accessed 13 Mar 2015

UNEP (2014) Sustainability of supply chains and sustainable public procurement. Paris. http://www.unep.org/10yfp/Portals/50150/downloads/Final_report_Sustainability_of_supply_chains_SPP_140630_aug19.pdf. Accessed 10 Mar 2015

Wackernagel M, Rees W (1996) Our ecological footprint. Reducing impact on the earth. New Society Publishers, Philadelphia

Chapter 7
Mainstreaming the Use of Life Cycle Management in Small and Medium Sized Enterprises Using a Sector Based and Regional Approach

Naeem Adibi, Vanessa Pasquet, Aubin Roy, Alice Salamon, Jodie Bricout, Catherine Beutin, Quentin Renault, Marie Darul, François Xavier Callens, Marc Haquette, Patrick Orlans, Jeanne Meillier, Joanne Boudehenn, Sophie Reynaud, Sophie Cabaret, Christophe Bogaert, and Christelle Demaretz

Abstract Although Life Cycle Management (LCM) is becoming commonplace in larger corporations it is far from mainstream. To achieve sustainable production and consumption patterns, LCM needs to be taken up by whole supply chains that include small and medium sized enterprises (SMEs). From a business perspective, this represents a competitivity issue, as these SMEs are increasingly under pressure from clients and legislators to provide more information about the environmental impacts of their products, and to take responsibility for them both up and down the value chain. Therefore a sector based and regional approach is needed to foster the implementation of LCM in SMEs. This has been done in Northern France, where professional support organizations, including clusters, business federations and Chambers of Commerce, have come together under the auspices of the [avniR]

N. Adibi (✉) • V. Pasquet • A. Roy • A. Salamon • J. Bricout
[avniR] Platform – cd2e, Rue de Bourgogne, Base du 11/19, 62750 Loos en Gohelle, France
e-mail: n.adibi@cd2e.com

C. Beutin
Aquimer competitivity cluster, 16 rue du Commandant Charcot,
62200 Boulogne-sur-Mer, France

Q. Renault
Agroe, excellence cluster, Quartier des Trois Parallèles, La Citadelle Avenue du Mémorial des Fusillés, 62000 Arras, France

M. Darul • F.X. Callens • M. Haquette
cd2e excellence cluster, Rue de Bourgogne, Base du 11/19, 62750 Loos en Gohelle, France

P. Orlans
CETIM Technical Centre for the Mechanical Industry,
40 rue Eugène Jacquet, 59708 Marcq-en-Baroeul, France

G. Sonnemann, M. Margni (eds.), *Life Cycle Management*,
LCA Compendium – The Complete World of Life Cycle Assessment,
DOI 10.1007/978-94-017-7221-1_7

LCA Platform to explore ways to help businesses to adopt LCM. Nine pioneer sectors, textile, seafood, packaging, mechanical, food, wood, construction, recycling and renewable energies, have undertaken an ambitious project to integrate LCM into their business. The methodology for all nine sectors follows five major steps: benchmark, sector maturity assessment, needs identification, action plan and implementation.

Keywords Life cycle assessment • Life cycle management • Regional development • Sector-based approach • Small and medium sized enterprises • SMEs

1 Introduction

Life Cycle Management (LCM) has been defined by the SETAC working group as "an integrated framework of concepts, techniques and procedures to address environmental, economic, technological and social aspects of products and organizations to achieve continuous environmental improvement from a life cycle perspective" (Hunkeler et al. 2001). Supply chain actors have been pursuing to integrate LCM within their activities and collaborations for years.

Therefore, now, many major global companies have internal and external programs to assess and manage the sustainability performance of goods and services across the life cycle. LCM-related initiatives include sustainability parameters, like Key Performance Indicators (KPIs) and existing additional processes (Nilsson-Lindén et al. 2014). Several private companies also finance collaborative life cycle research through initiatives such as SCORELCA, CIRAIG, Sustainability Consortium and UNEP/SETAC Life Cycle Initiative.

J. Meillier
UP-tex competitivity cluster, 40, rue Eugène Jacquet, 59708 Marcq-en-Baroeul, France

J. Boudehenn
Nord Picardie Bois, 34 B Rue Emile Zola, 59650 Villeneuve-d'Ascq, France

S. Reynaud
MATIKEM competitivity cluster, Parc scientifique de la Haute Borne, 9 rue Héloïse, 59650 Villeneuve d'Ascq, France

S. Cabaret
CCI Grand Lille, Place du Théâtre, 59000 Lille, France

C. Bogaert
ADEME Nord-Pas de Calais, Centre tertiaire de l'Arsenal, 20, rue du Prieuré, 59500 Douai, France

C. Demaretz
Nord-Pas de Calais Regional Council, 151 Avenue du Président Hoover, 59000 Lille, France

LCM is also increasingly impacting public policy related to resources use and recycling, and more recently product-related environmental policies. For example, the European Commission launched "Building the Single Market for Green Products – Facilitating better information on the environmental performance of products and organizations" in April 2013. The Product/organizational Environmental Footprint (PEF/OEF) method (European Union 2013), published in April 2013, includes Commission recommendation on the use of common methods to measure and communicate the life cycle environmental performance of products and organizations. Within the related PEF/OEF pilot projects the Commission has engaged mainly big companies and industry associations.

Only few initiatives have been developed recently to help integrating LCM within SMEs in a sector. In 2013 the UNEP/SETAC Life Cycle Initiative supported eight pilot projects using Life Cycle Management Capability Maturity Model (LCM-CMM) in Cameroon, Uganda, South Africa, India, Brazil, Colombia, and Peru. As an example, results in Colombia showed that the companies could apply the LCM concepts to their organizations with limited technical support (Moreno et al. 2015).

Whilst the existing initiatives help to make significant progress towards sustainable production and consumption patterns, LCM needs to be taken up by whole supply chains that, by definition, include many small and medium sized enterprises (SMEs). From a business perspective, this represents a competitivity issue, as these SMEs are increasingly under pressure from clients and legislators to provide more information about the environmental impacts of their products, and to take responsibility for them both up and down the value chain. (Bricout et al. 2012)

2 Context for the Study

In this chapter we present a case study on how the use of Life Cycle Management can be mainstreamed in SMEs using a sector based and regional approach. A methodology for this purpose has been developed and applied in Northern France. It is a highly industrial region facing important challenges for sustainable development. The strong industrial past has affected the region through a complete reconversion from coal and heavy industry to other industrial and service sectors. Social challenges include the high population density (324 people per km^2 vs. 113 in France) with 95 % living in urban communities. The environmental challenges this region is facing are diverse. The greenhouse gas emissions per capita are 30 % higher compared to the French average. The share of renewable energy consumption is four times less than the national level. Only 1.9 % renewable electricity (17.7 % France) is produced in the region and 16.4 % of the surfaces area is artificial (8.8 % France). Despite the loss of industrial activity over the past 50 years, Northern France remains the third largest industrial region in France with the fourth largest economic turnover. It is also a significant transport and logistics hub, with the densest road

Fig. 7.1 The northern France region

network in Europe and significant rail and canal infrastructure (Nord-Pas de calais Regional Council 2013) (Fig. 7.1).

The Northern France region has been actively pursuing a transformation towards more sustainable economic models for many years. The non-for profit organization cd2e was established in 2002 to support this "eco-transition". As early as 2007, cd2e and their partners identified Life Cycle approaches as a needed decision-making tool to help local industry to implement Life Cycle Management. Cd2e created the [avniR] platform in 2009 to bring together multiple stakeholders around this challenging issue. This collective approach was needed to foster the supply and demand for LCA at the same time and has evolved from a focus on assessment to the support of companies in their efforts to improve their sustainability performance using LCM. On the LCM demand side, [avniR] works with industry clusters and public authorities to explain life cycle approaches and develop collaborative projects ([avnir] 2015). On the supply side, [avniR] provides training and a "hub" for academics and consultants to improve their capacity in Life Cycle Assessment, ecodesign and LCA based communication (Bjørn et al. 2013). Over 300 people have been trained in LCA/LCM through the platform, and more than 90 individual projects have been directly and indirectly supported.

Key milestones for the [avniR] platform include:

- 2011: first annual international [avniR] conference organized in Lille
- 2012: first ecodesign prize organized with different regional actors, in partnership with the Institut de Développement de Produits and Novae in Québec
- 2012: sectorial LCM studies launched in the textile, seafood, packaging and mechanical sectors

- 2014: launch of a new governance based on different working groups to strengthen collaboration between the different stakeholders within the platform

The sectorial approaches launched in 2012 have the ambitious objective of moving beyond applying LCA studies in an isolated fashion, to mainstreaming life cycle management across entire sectors at the regional level, including the numerous SMEs present. The first "wave" of four sectors in 2012 (textile, seafood, packaging and mechanical) enabled the development of a common methodology to establish strategic action plans to mainstream LCM into business, education and research organizations. A second wave of sectors (food, wood, building and recycling/end-of-life) undertook the process in 2013 and the renewable energies sector is developing their strategy in 2015.

This chapter presents the overall approach and the first results obtained in the mainstreaming of business Life Cycle Management practice and public Life Cycle Thinking awareness in a region through this, to our knowledge unique sectorial approach, whilst developing the necessary competence to transform awareness into action by stakeholders at different levels of influence.

3 Approach

Most SMEs lack the financial capacity or human resources to implement LCM on their own. Therefore, they need to work with support organizations at the regional level, and in general they are ready to work with other companies of their sector.

In Northern France, professional support organizations, including clusters, business federations and Chambers of Commerce, have come together under the auspices of the [avniR] LCA Platform to explore ways to help their businesses adopt LCM. The strategy for achieving this is based on the following key concepts (Adibi and Bricout 2012):

1. LCM capacity should be built into existing support organizations, so that businesses receive advice and tools from organizations that they already know and trust
2. Tools and actions to support LCM integration need to be adapted to different sectors to make them as relevant as possible to SMEs
3. Training and research capacity, also within universities and public research centers, needs to be developed to respond to needs of businesses

Before starting the process, there is an identification of the most strategic sectors to help integration of LCM within the region. The economic, environmental and social characteristics of the sector are considered jointly with the difficulty of mobilizing the sector (level of motivation/awareness and the existence of well identified actors for dissemination).

Fig. 7.2 Sector strategic action plan development methodology

For the sectors, existing support organizations are identified in a second step. We selected competitivity and excellence clusters, as they include innovation in their scope to help emerging new and innovative products and markets based on the implementation of LCM approaches. Once these organizations are identified, one reference person is designated for each sector. The different reference persons from the sectors, called "Life Cycle Champions", are creating an active network.

Champions are first trained on LCA and then on LCM approaches. A specific training of two days has been developed and delivered by the [avniR] platform. The training covers an advanced introduction to LCA, a detailed presentation of sectorial approaches and different steps and critical points to better integrate LCM within sectors based on the first sectorial experiences.

Based on [avniR] experience, it is strongly recommended to assure the presence of the same persons during the entire process through the implementation phase. When a change of person occurs, the step-by-step capitalization of the results becomes essential and the hierarchy within the structure (e.g., the president of the cluster) has to be extremely motivated to assure the continuation of the process.

All sectors follow the same general process for developing their LCM Action Plan, as shown in Fig. 7.2. The approach as presented in this chapter is elaborated within mid-long term vision and is not applicable within a short-term prospective.

Five major steps are followed within the process:

1. Benchmark
2. Sector maturity assessment
3. Needs identification
4. Action plan
5. Implementation

3.1 *Benchmark*

The benchmark process starts with the identification of the sector specific characteristics and a detailed mapping of the sector actors. For some of the sectors with very complex structure, the mapping phase is essential (e.g., building, recycling and packaging). The mapping helps to identify the major players including, businesses, education and research organizations as well as the interactions with institutions and authorities.

The sectors are very different from one another regarding the characteristics and the specificities that need to be identified during the benchmark phase. The major differences to be considered are detailed below.

3.1.1 Structure of the Sector

Some sectors include mostly very small and medium companies, whilst others may be dominated by big or medium enterprises or a mix of both. The type and size of organizations influence the way they will apply LCM. Also attributed resources are often related to the size of the structures and the turnover.

3.1.2 Organization of the Sector

Big companies that may not physically be present within the region sometimes drive sectors, for example, the mechanic sector has a very significant presence of purchasing/supplier relations. For some other sectors there might be a very logical and complementary chain of actors, varying from one country to another or from one region to another, as, for instance, for the building sector. In some sectors, companies (small, medium or big) may actually control the overall supply chain of their product.

3.1.3 Product, Organization and Process Oriented Vision

Another very import factor that may change completely the way sectors are responding to LCM is the way they look at the product. In some sectors, products are in the center of the business efforts, such as for the case of textile, fish and food. In others, processes are dominant, such as for the recycling sector. In most cases, for those sectors with a dominant process and organization vision, the role of actors and the relevance of products need to be clarified to help them uptake LCM.

Once a detailed mapping has been done, the next step within the benchmarking is to identify LCM initiatives, tools and case studies relevant to each sector, focusing on, but not limited to West Europe. This phase also includes a hotspot analysis based on existing LCAs within the sector. Figure 7.3 shows a list of major points considered within the benchmarking phase.

3.2 Sector Maturity Assessment

The maturity assessment aims at understanding the maturity of the actors of a sector in relation to LCM practices. In this way each sector and their respective actors can be trained corresponding to their needs in order to build LCM capabilities as quickly and efficiently as possible.

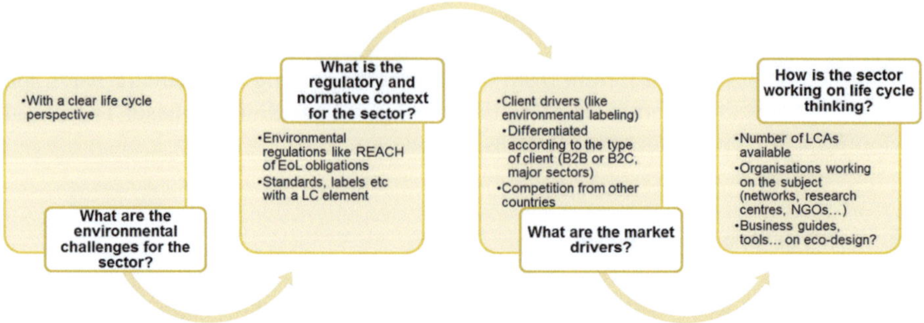

Fig. 7.3 Benchmarking

The first step for the maturity assessment is the sampling phase. Considering the mapping of actors (throughout the supply chains) and the structure of the sector (small, medium and big companies) a sample of the actors is selected. The sample covers different product/material groups within a given sector (e.g., food, meat, fruits, vegetables, drinks, etc. or the recycling of plastics, metals, wood etc.) to the extent possible within the region. In most cases, a sample covering 20–40 % of the actors is sufficiently representative to formulate conclusions. Table 7.1 provides the key figures of sampling done for the textile, packaging, seafood and mechanical sectors.

Table 7.1 Key figures of sampling

	Textile	Packaging	Seafood	Mechanical
Companies concerned by the study	200	90	130	750
Companies mobilized	53 (26 %)	32 (36 %)	37 (28 %)	201 (27 %)
Surveyed	17	–	17	-
Interviewed	28	21	20	201
Active participation (workshops)	8	11	–	–
Researchers, universities, training organizations, etc.	7	4	16	8
Institution partners, federations, etc.	10	14	8	15
Total	70	50	61	224

The maturity assessment of businesses, education bodies and research centers in the region in relation to LCM practices is undertaken via interviews with key stakeholders. To assess the maturity of the sectors, a baseline method has to be adapted for each sector and applied. We developed an adaptable baseline method called "[avniR] sectorial framework" (Adibi et al. 2012) based on the first wave of sectorial approaches and other methods developed for a similar purpose such as the Life Cycle Management Capability Maturity Model used by The UNEP/SETAC Life Cycle Initiative to promote LCM capability in different parts of the world.

The four methods used in the first wave of sectorial studies (textile, packaging, seafood and mechanical sectors) were developed independently and without applying any common framework. Based on the results of these studies and the feedback of responding companies and organizations, more sub-criteria were added. In the same time, the study of additional methods (CMM UNEP, etc.) provided a more exhaustive understanding of existing assessment methods. The [avniR] sectorial framework was finalized considering both experiences from the first wave pilot and other maturity assessment methods. The framework of the assessment method takes into account five main criteria:

1. Company global strategy for LCM
2. Challenges associated to LCA based approaches
3. Eco-design/LCA maturity approach
4. Resources dedicated to LCM
5. Continual improvement and communication efforts

The framework was applied within the second wave and was very successful to give a clear overview of the sectors' maturity. The method also helped through an indirect harmonization of the approach within different sectors.

The interest to develop the method was also to provide a common baseline that would allow later to monitor the advance of the sectors, once the implementation phase starts. In addition, the maturity assessment helped different sectors to better understand the action plans of other sectors, since not all of them started at the same maturity level, nor had the same need for training and capacity building support.

3.3 Needs Identification

The need identification considers two major aspects:

1. Benchmarking (national and international on the existing LCM capacities: LCM initiatives, tools and case studies relevant to each sector at global level and results of the hotspots analysis)
2. Sector maturity assessment results based on Sect. 3.2

For the needs identification, the LCM champion involves relevant stakeholders to build up a SWOT matrix to identify and categorize significant internal factors (i.e. strengths and weaknesses) and external factors (i.e. opportunities and threats) to integrate LCM in their sector. In order to better prepare and facilitate this phase, some preparatory documents are sent in advance to different stakeholders to facilitate their participation during the SWOT preparation.

Stakeholder engagement is a key aspect of the needs identification and a solid baseline for the action plan development that fits the needs of the sector. It is especially important for the implementation phase as the actions are "owned" by the participants. The step is also very important to educate the new stakeholders joining the process.

By the end of this phase the results of the first three phases are presented in an open meeting with all stakeholders. The aim is to make more and more stakeholders familiar with the process and findings, and motivate them to contribute to a successful implementation in the later stages.

3.4 Action Plan

Each sector develops very detailed action plans based on the SWOT matrix and the identified needs and expectations. The experience with the first wave helped to develop major common indicators related to each action.

A non-exhaustive list of major common indicators is provided here: primary target, title description, goals, costs, time frame, potential action leader, starting time and duration, frequency, priority and management indicators.

In addition and to help cross cutting actions to be developed, actions are grouped in three major categories:

1. Sector specific actions
2. Replicable actions (to one or more sectors)
3. Cross cutting actions (joint actions of two or more sectors)

3.5 Implementation

The LCM action plans are developed in parallel, enabling the sectors to identify cross cutting actions. This process is managed by the regional LCA Platform [avniR], through the network of "Life Cycle Champions". Champions have been trained in the nine sectors; they meet regularly to exchange experiences in implementing life cycle approaches, identify cross cutting projects and to monitor sustainability performance advances within different sectors. The study was done through different waves. The latest sectors incorporated learnings from the first wave of sectors are shown in Fig. 7.4. Several actions are ongoing and some are finalized within different sectors. The results show significant improvements in SMEs within most sectors. The efficiency and effectiveness of the actions are monitored within a scoreboard developed together with LCM champions. Unfortunately, due to confidentiality restrictions, at the moment, no detailed results can be shown to highlight the improvements of the sustainability performance of products, processes and organizations achieved within the case study of Northern France.

4 Conclusions and Outlook

The chapter describes an innovative approach for mainstreaming LCM that is able to leverage sectorial and regional networks in order to help overcome barriers to the implementation of Life Cycle Management. From a business perspective, integration with existing professional organizations confirmed that SMEs access advice

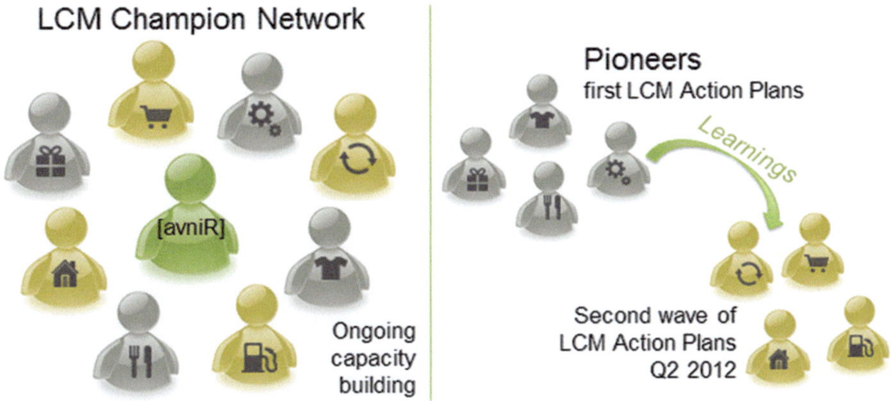

Fig. 7.4 The LCM champion network

and tools through organizations that they already know and trust. Working with several sectors in parallel through the Life Cycle Champion network encourages a multidisciplinary approach, essential to improving decision making across entire supply chains.

The results that can be reused elsewhere include new developments, such as a methodology to evaluate the maturity of sectors adaptable for different sectors and other regions in order to foster the implementation of Life Cycle Management worldwide.

In addition a very detailed action plan is developed for large scale roll-out of Life Cycle approaches in business and higher education. Each action plan aims to plan for the generalized roll-out of LCM in businesses across all industry sectors in the region. The engagement of stakeholders ensures that the benefits and effects will be perpetuated beyond the implementation phase.

In order to widely test and improve the methodology, a large scale project integrating regions from other European countries (Belgium, Portugal and Spain) is ongoing (Life Cycle in Practice – LIFE+ Funds). It aims to apply this approach in different regions, to improve the method and to validate the conclusions.

It is planned to publish the improvements achieved with regard to the sustainability performance of products, processes and organizations of at least one sector within the case study of Northern France, once confidentially agreements have been finalized.

Acknowledgements Authors would like to express sincere thanks to Christian Traisnel and Eric Kniaz for all their valuable feedbacks. The approach described is supported by the Nord-Pas de Calais Regional Council and the French environment agency, ADEME.

References

[avniR] (2015) Etudes filières. http://www.avnir.org/FR/Etudes-filieres-227.html. Accessed 12 Mar 2015

Adibi N, Bricout J (2012) Mainstreaming life cycle approaches in northern France through the collaborative platform [avniR]. SETAC 18th LCA case study symposium, Copenhagen, p 164

Adibi N et al (2012) Introducing a sectorial framework to better evaluate life cycle thinking maturity. In: Proceedings of the 2nd international [avniR] annual conference, Lille, 2012

Bjørn A et al (2013) Mapping and characterization of LCA networks. Int J Life Cycle Assess 18(4):812–827

Bricout J et al (2012) Mainstreaming life cycle management: using a sector based and regional approach in northern France in the textile, seafood, packaging and mechanical sectors. In: Proceedings of the SETAC Europe annual meeting, Berlin

European Union (2013) PEF OEF methods. Off J Eur Union 56

Hunkeler D et al (2001) Life cycle management: bridging the gap between science and application. Int J Life Cycle Assess 6(6):384–390

Moreno OAV (2015) Implementation of life cycle management practices in a cluster of companies in Bogota, Colombia. Int J Life Cycle Assess. doi:10.1007/s11367-015-0875-x

Nilsson-Lindén H (2014) Organizing life cycle management in practice: challenges of a multinational manufacturing corporation. Int J Life Cycle Assess. doi:10.1007/s11367-014-0818-y

Nord-Pas de calais Regional Council (2013) La strategie regionale CLIMAT volet CLIMAT du SRADDT

Part II
Advancing the Implementation of Life Cycle Management in Business Practice

This part focuses on the LCM progress made with regard to implementation processes of life cycle approaches and its linkage to business value creation.

Chapter 8
From Projects to Processes to Implement Life Cycle Management in Business

Martin Baitz

Abstract In general, companies start using a life cycle approach to manage the sustainability challenges and opportunities of their products through projects using life cycle assessment or other tools of the life cycle management toolbox like greenhouse gases accounting. As companies gain experience the way they manage the life cycle of their products matures, it becomes less about implementing projects and more about putting in place organization-wide procedures. The latter allow companies to address systematically the identified business challenges and opportunities of their whole portfolio or at least their key products with much less effort than carrying out multiple individual projects.

Keywords Life cycle assessment (LCA) • Life cycle management (LCM) • Sustainability • Process integration

1 A Brief History of Life Cycle Approaches

In the beginning of a new thinking some aspects dominated the discussion. Energy – in the way it is technically used so far – is limited and environmental impacts can be a threat, if ignored. As the supply chain has to cooperate to solve (economic and technical) tasks and harvest its opportunity, it was quite evident environmental aspects needed a supply chain approach as well.

Life cycle thinking (LCT) was borne in the 1980s. In the 1990s, scientific bodies like SETAC (Society of Environmental Toxicology and Chemistry) recognized a need to draft rules to harmonize the new way of thinking, calculating and analyzing products systems in a way that international cooperation and exchange is fostered. This "Code of Practice" (Consoli et al. 1993) may be understood as an important yardstick to turn thinking into doing.

M. Baitz (✉)
thinkstep AG, Hauptstraße 111-113, 70771 Leinfelden-Echterdingen, Germany
e-mail: martin.baitz@thinkstep.com

G. Sonnemann, M. Margni (eds.), *Life Cycle Management*,
LCA Compendium – The Complete World of Life Cycle Assessment,
DOI 10.1007/978-94-017-7221-1_8

Life cycle assessment (LCA) was about to be borne (Fava et al. 1990) and became an international standard under ISO in 1997 (ISO 14040: 1997). Building on this new standard industry, research and academia was able to use this method in a consistent way. Due to differences in goal and scope between stakeholders the interpretation of the systems and the results can be different; however keeping to ISO ensures that the differences remain understandable and interpretable. This empowers consumers or users of information to check against their own (technical, political or private) motivations and boundaries. The method remains consistent and transparent. The possibilities within this new assessment method reached beyond engineers and scientists. The call to consistently measure, control and target against life cycle results combined with related communication measures lead to a further evolution.

Life cycle management (LCM) developed from a professional interpretation and decision with the help of life cycle assessment. For most professional users of life cycle based sustainability approaches in industry, these three evolutionary steps belong together. LCT is the required mindset, LCA the method and LCM the process.

1.1 From Pioneers to Industry Standards

There are many individuals mentioned if talking about scientific or societal thought leaders in life cycle aspects and the roots of sustainability. However, thoughts stay basically fictional until they are applied and measurable.

Pioneer companies like, e.g., Volkswagen, Daimler, Renault, BASF, DOW, Wacker, ThyssenKrupp, and Amcor recognized quite early on the necessity and opportunity to apply the topic in a professional manner using a standardized method, related software tools, a blend of own in-house data, specific supplier data, realistic upstream data and justified background data.

During the 1990s, pioneer companies and organizations pace their way towards international standards. Without these pioneers, LCM probably would not have (or at least significantly later) matured into applied LCA about two decades ago (see also Hunkeler et al. 2004).

1.2 From Trial to Maturity

After the international standards were in force, the number of studies with questionable claims decreased, whereas the number of studies with reasonable results increased. Most likely due to better identification of unjustified claims. Reducing the arbitrary application of the method, combined with clear rules, described life cycle assessment being implemented in a number of companies and organizations in the mid-1990s to early 2000s to drive product innovation.

Even having now the standards and related reviewers available, some stakeholders or believers in specific schools of thought still pretend today LCA is arbitrary: However, in most cases the preferred method of these critics are completely non-standardized and sometimes even their private invention. Even more astonishing, the most arbitrary results originated rather not from industry, maybe because of self-protection.

However, refocusing on the important aspects, it may be summarized that LCA took its chance to mature from scientific into professional applications in industrial organizations, while some non-standardized approaches, some experimental databases and trial software disappeared.

2 From Project to Process

LCA is recognized as the best available methodology to investigate environmental sustainability performance in a reliable and transparent way and for communication, along the value chain and throughout one's own organization; to support and help to cross-check development and strategic decisions. LCA in practice must be time, cost and resource efficient. LCA results and the underlying data are only the basis for communication and decision making and are to be converted into a technical conclusion, the nature of which is determined by the recipient of the result (e.g., product engineers, executive management, marketing, suppliers, or consumers). LCA is applied for quantitative environmental management and should reflect the industrial reality adequately. This sounds trivial, but often enough data and modelling approaches are used to produce results, which have a goal and scope that is not suitable for decision support. Benchmarking one's own processes and products against the competition – commonly on a cost basis – is a common practice in industry. Evaluation of the results within the competitive landscape is needed to take suitable decision for your own optimization approach. Aside from the internal use of LCA as an internal planning tool, another potential lies in connecting partners along the value chain. By collaborating on an LCA, suppliers and customers strengthen their relationship, gain valuable insights in markets and their success factors, and enhance an overall exchange of experiences. This fosters innovation. LCA is a business imperative today. Therefore successful LCA application is the main aim in industry (Baitz et al. 2012a).

As described earlier (Baitz et. al. 2012b), the topic of LCA needs proper implementation in the companies: It must be manageable. Management of tasks basically calls for two main aspects: measurable information and properly installed processes. What gets measured gets managed.

Proper management of life cycle aspects needs a shift in the mindset: From project thinking towards process thinking. Alignment of methods, data and communication throughout the company is an important step. But why restrict the degree of freedom in doing LCA in companies? Simply to gain quality of results, harvest the full potential of aligned application, communication of LCA results and to be able to measure company targets against consistent actuals.

Companies realize that an extended use of LCA can foster value creation, if implemented as part of a professional management process.

In (Baitz et. al. 2014) it is discussed that successful companies basically create value through sustainability in four discrete ways:

- In sales over increasing market share or new market entry by quantifying of benefits B2B or B2C or by promoting innovation and new products based on solid facts.
- Through cost reduction due to increase of value chain and operational efficiency as well as employee productivity.
- Risk mitigation like operational risk management or regulatory management which supports business continuity.
- The brand value is increased due to reputation as well as employee attraction and retention, which lowers new employee hiring costs.

To generate value on sustainability in a company, a platform combining IT technology (software and data exchange systems), content (adequate databases) and a positive user experience is key. The positive user experience is most efficiently implemented by an adequate and balanced share of technology and service support.

In sequence, companies want to determine more quickly what the social, environmental and economic tradeoffs of their products are (screening)? How they can provide more value by combining multiple data sources (scoping)? And how the companies may integrate sustainability management with existing management systems, such as ERP, PLM, CAD or supply chain management (scale)?

Along this approach of "Screen, Scope, Scale" many companies identify their specific pathways, then grow their project phase and finally begin to prepare an evolutionary step towards process implementation.

2.1 Screening to Identify Pathway

Under the umbrella "innovation with sustainability" different topics exist that companies want or already need to cover; ideally with an integrated approach. The importance of the topics are related to the specific situation of the company. No matter what a company already does, it is valuable to build on existing activities and to get in-house people to support the "pathfinder mission".

Under the headline "innovation and cleantech" some companies do sustainability research and development, external research funding, product stewardship and sustainability solution marketing. In strategic energy management often e.g., office energy management, IT/telecoms energy management, industrial energy management, on-site renewable energy and energy management systems get clustered. Sustainable supply chains need to be managed e.g., retail or supply chain energy management or aspects of supply chain transparency.

Environmental performance has long been an important aspect in the companies' sustainability approaches including environmental remediation, occupational health and safety, pollution control and prevention, biodiversity and land stewardship, and waste water and water management.

Most of the above mentioned topics are done in pure project style. Some companies started different projects with direct links to individual tasks and needs that needed to be addressed. This might be tasks concerning suppliers, customers, cooperating industries or from governmental bodies or their own research and development topics.

Some companies selected so-called lighthouse projects on especially important topics to gain insights on intercompany cooperation while ensuring that the results are valuable for the company to guarantee a wide range of company stakeholders from project engineer to executive management are interested in the results. The selection of lighthouse project topics can be done monetarily (important revenue stream) or by image (specific product with specific message) or by any other company-relevant aspects.

However, some companies prefer to start with non-critical trial projects to test the approach through a shadow-project in a protected internal environment without any information exchange with suppliers, customers or the public on data gathering and results communication.

Typically in the screening phase the four aspects – materiality, practicability, quality and transparency – are addressed in a company specific way:

- Under materiality questions are answered like: Which data do I have? Which data is relevant for me?
- Practicability addresses questions like: Where do I get my company data and my supplier data from? Where do I get reliable background data?
- Concerning quality companies answer questions like: What does quality mean in my business? What are relevant quality indicators for my goal?
- Transparency in professional applications means: Are processes documented and change monitored? Are routines qualified and auditable?

In the screening phase companies start with initial workshops to clearly scope the work and define the goal. In most cases the first is step is to benchmark against the average in the industry. In a materiality assessment the inventory situation of in house data is checked and assessed. Life cycle based product and corporate environmental draft footprints are often a first screening result.

An example of doing a successful screening is a US based company which design and manufacture floors and ceilings for residential and commercial products. Their challenge was losing market share in an industry, where sustainability is a defining strategy. The solution was life cycle assessment showing the quantifiable and relevant savings of water, virgin material and electricity, supported by senior leadership to reach corporate awareness. The business value was cost reduction, brand enhancement, improved business environment as well as top line revenue growth.

Another example is a city development in the United Arab Emirates designed to rely on solar energy and other renewable energy sources. The goals are to monitor embodied carbon in supply chain and construction to achieve a "carbon neutral" goal of 475 g/m^2 and to set up a database of sustainable building processes and materials. Features are to produce environmental product declarations (EPDs) and carbon footprints, linked to a city portal and consistently setting targets, benchmarking and monitoring the performance. Main outcomes here is decision-making support e.g. in the progress of building completion vs. total carbon emissions to date.

A third example of a value adding screening phase activity is a national dairy service acting as a body for dairy farmers and the industry to help farmers adapt to a changing operating environment, and achieve a profitable, sustainable dairy industry. The challenge was to establish the industry's carbon footprint, from farm to a representative national dairy product. Further to generate a reliable basis for product carbon footprinting and environmental labelling. The solution was a web-based data collection and integrated analysis approach. The benefit are verifiable greenhouse gas footprints at industry level and an auditable and expandable platform reporting solution leading to customized greenhouse gas footprints for individual farmers.

2.2 *Scoping to Grow Project Phase*

After successful screening of the company specific pathway, the companies typically aim to merge many environmental and social topics under one common umbrella of life cycle based sustainability approaches.

Related ISO standards are in most cases the solid basis of environmental related work in companies. The companies' data foundation – consisting of in-house and own site data, specific supplier data, representative generic upstream and downstream data as well as background data – typically evolved and grew from the starter projects.

In the scoping phase the companies evaluate the magnitude and impact that can be reached within their given goal and scope.

The pathway successful companies follow is in principle comparable or often even identical. However, the speed and the needed measures differ and are specific to each company. Competency and persistence of the person in charge of the topic are decisive. However the most important differentiator is if a dedicated in-house team – dealing more or less exclusively with the topic – is available in the respective company or not. Without a dedicated in-house team the chances to succeed are equal, just the measures to reach it are different.

In-house teams are a huge knowledge pool and can work on sensitive aspects without any external interference, which is an asset for any company. However, depending on the company structure in-house teams can be quite expensive. Working with no dedicated in-house team might be cheaper and easier to manage, as the content and deadline responsibility is managed by a contractor. However, the quality of the contactors work is decisive for the quality of the companies end results.

It might be fair to say that in most of the successful companies in this topic established dedicated in-house teams along their journey in the topic and can work in many or most cases fairly independent from external consultant companies or experts.

The beauty of the topic is that businesses can define their own journey and mile stones towards sustainable success. No entry hurdle, no rush, no "point of no return" decisions in the process are necessary.

A successful sustainability approach simply means, to build on existing measures, to use as much as possible existing in-house information and to choose a professional software and data solution to create multiple (business) benefits from the approach.

Summarizing the scoping phase is characterized by (company and external) stakeholder engagement often with strategy workshops to align on strategy development and implementation, identification of suitable corporate reporting software solutions of, e.g., GRI (Global Reporting Initiative), Carbon, EH&S (Environment, Health, Safety), or Building Portfolios. Further LCAs and organisational environmental footprints are undertaken and Environmental Product Declarations (EPD) done. The growing data and resulting demand calls for professional product sustainability software solutions.

An example doing successful scoping is a German financial institution. They decided to solve the topic with and ISO 14001 compliant environmental management system. The idea was central sustainability performance management. Their aim was to quantify the sustainability performance. They reached reduced electricity consumption of € 111,000/year alongside with 29 % reduction in paper consumption, which equals 207 t. The gained business value was a total resource cost reduction of € 300,000 in 6 months and total cost reduction of another € 500,000 in the second year. Additional business value was reputation and brand enhancement.

Another example is an American multinational consumer goods company. Their defined goal was to integrate a product sustainability software tool that anyone in the organization could use. Boundary conditions were that no new hires were needed and that the staff did not need training on the product sustainability software tool. The solution was an LCA solution allowing product designers to understand the environmental impacts and deliver results of a proposed modification in less than 15 min. Further benefits were achieved by integrating the solution fully into the product design community.

A third example of successful scoping is from a German specialty chemical company. Very early the leading heads of the initiative recognized the goal of the endeavour must be to turn LCA from a project based business into something embedded into the core processes of the company to avoid implementation risks of new products or processes. The solution was to focus on dedicated and relevant products and projects with the support of an in-house team. This led to instant knowledge transfer to the companies' sustainability team. This was achieved with professional software and database solution and ad hoc on-demand consulting supplemented from the external partner. The benefits are extensively and reliably used and communicated LCA information throughout the company. The LCA group also experienced extraordinary internal and external visibility and success.

2.3 Scaling to Prepare Process Phase

After successful scoping, the company aims to scale the approach according to their specific product or business strategy. Appropriate communication of facts based on one core life cycle information system is essential. C-level, engineering department heads as well as research engineers need tailored communication packages based on the same facts and data.

Therefore software and database management and maintenance ideally moves into the core of the activity. Foreground data management and in-house data collection needs to be consistent with supplier data collection and integration as well as background data management. Data updates from the background data supplier and distribution of their own company data to subsidiaries is organized including quality assurance routines. Leading companies start to organize their work by client–server based team work approaches. To harvest information synergies similar activities are combined under one common umbrella of a life cycle based sustainability approach, incorporating corporate sustainability activities as well as compliance and material information topics.

Integration and automation is a core topic if companies aim to scale their life cycle based sustainability activities. Automated LCA generation using existing information like e.g. "bill-of-materials (BOM)" or recipe lists is another area where leading companies are scaling the process towards higher quality in less time. Linking to ERP systems and available in-house data systems is also a promising step in scaling. The scaling phase increasingly enables integrated monetary assessments, with approaches like life cycle costing (LCC) and environmental costing approaches (e.g. Trucost) along with options to quantify or evaluate social aspects.

Summarizing, it can be said that in the scaling phase enterprise sustainability performance software (GRI, Energy & Carbon, EH&S, supply chain, building portfolios), management systems (EMAS, ISO 14001/50001), product sustainability performance software (LCA, EPD, PEF, Eco-Design), EPD and LCA automation as well as PLM-integrated materials compliance management is most effective when operated over one integrated platform to gain maximum from synergy and consistency and to reduce risks from data gaps and inappropriate data.

An example of successful scaling is with an American multinational confectionery, food and beverage enterprise that had a challenge to understand the environmental impacts for products and packaging. The aim was to improve process efficiently and examine new promising technology. The solution was to conduct LCAs for core food products and to establish a professional LCA in-house team. Alongside these activities an extensive database of in-house process data as well as external upstream and background data was partly set up and partly bought on demand. The measures incorporated sustainability into many of its existing and new products. The business value was the company is able to select the best alternative with regards to economic and environmental aspects before finalizing a new product design. This includes the quantification of incorporated costs in analysis to realize savings.

Another example again with an American multinational, this time an information technology corporation. The challenge here was increased customer demand for environmentally preferable and professionally registered products. Further, and inevitably, the need for LCA-backed information to guide product development teams. A further challenge was the complex supply chains. The solution was a LCA platform with pre-configured LCA templates to allow to scale and run over 70 LCAs on their portfolio using an LCA solution. The business value was that the LCA platform realized savings of over USD$1 million and hundreds of hours of time. A further business value is the increased efficiency and the ease of registering products within the assessment standard. This means in the end a faster market access to maintain a competitive advantage.

Another example, this time a German automobile manufacturer needed its existing and new products to comply with legislations such as the EU Directive 2005/64/EC on the recyclability of motor vehicles. The company applied professional software and extensive data sets to analyze vehicle components in terms of their recyclability. Scenario calculations allowed the consideration and comparison of different material options, and to improve favourable options. For example, different front module concepts were compared and material options like steel, aluminium-steel, plastic-steel assessed. The analysis revealed different strengths of the concepts concerning primary energy demand, weight and recycling. Scenario calculation showed improvement potentials which have been applied accordingly. The business value is combined compliance and improvement.

Another German automobile manufacturer also achieved success with international production sites and brands. The challenge was to perform an LCA of every new vehicle already in the design phase, to be able to grasp and execute on design options and optimization potentials. The goal was to combine design for environment with marketing and communication. The solution was a software implementation and corresponding BOM import for efficient LCA work. The benefit is the automobile manufacturer now optimizes cost and environmental impact of cars already during product development.

3 From Process Integration to Sustainable Supply Chain and Product Management

Companies that successfully embedded life cycle approaches in business processes have increased the business value of sustainability. According to our approach, companies run through a maturity curve (Fig. 8.1), with more or less external support. To enhance stepwise success and business value with sustainability the maturity curve can be entered at any point. Importantly it is not where a company starts, but to know the pathway and the next step. Solutions are needed that are able to bring a company efficiently further without expensive rework or duplicating effort.

IT solutions are key to drive the value of sustainability approaches. Merging sustainability aspects, compliance aspects and supply chain aspects under one platform approach is most promising.

Data is the foundation. Actuals as well as targets are important. The business value must be very clear: top line, bottom line and risks involved. Sustainability must be executed; a tangible example is the approach presented: screen, scope, scale. It is important to converge organization and product performance. Therefore it is important to combine materials compliance, risk and sustainability management. Collaboration is important; not only internal, but also to stakeholders, the supply chain and customers. Design for sustainability must be done early and must be solved upstream, not downstream.

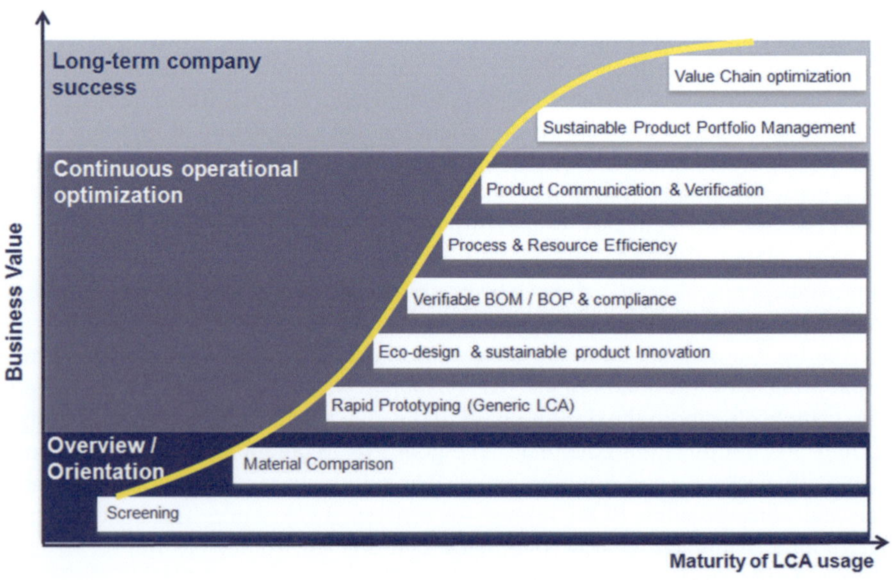

Fig. 8.1 The maturity curve of increasing business value

Fig. 8.2 Business value of sustainability

Life cycle related sustainability solutions are able to increase sales, improve the brand and reduce costs and risks, if these are integrated in business processes to do sustainable supply chain and product management of long-term *and* short term aspects of innovation and mitigation (Fig. 8.2).

Companies can rather freely decide where to start their journey and if they solve the topic in-house or with external support. However a suitable and smart IT and database solution is in any case indispensable.

References

Baitz M, Osumi M, Yukiyoshi H, Florin H (2012a) Symbiosis of science and application – LCA as business case. Ecobalance conference, Yokohama

Baitz M, Albrecht S, Brauner E, Broadben C, Castellan G, Conrath P, Fava J, Finkbeiner M, Fischer M, Fullana i Palmer P, Krinke S, Leroy C, Loebel C, McKeown P, Mersiowsky I, Möginger B, Pfaadt M, Rebitzer G, Rother E, Ruhland K, Schanssema A, Tikana L (2012b) LCA's theory and practice: like ebony and ivory living in perfect harmony? Int J Life Cycle Assess. doi:10.1007/s11367-012-0476-x

Baitz M, Osumi M, Yukiyoshi H, Florin H (2014) Life cycle based sustainability solutions: how successful companies turn sustainability solutions into company benefit and business value. Ecobalance conference, Tsukuba

Consoli F, Allen D, Boustead I, Fava J, Franklin W, Jensen AA, de Oude N, Parrish R, Perriman R, Postlethwaite D, Quay B, Séguin J, Vigon B (eds) (1993) Guidelines for life-cycle assessment: a 'code of practice', 1st edn. From the SETAC workshop held at Sesimbra, SETAC, Pensacola, 31 Mar–3 Apr 1993

Fava JA, Denison R, Jones B, Curran MA, Vigon B, Selke S, Barnum J (eds) (1991) SETAC workshop report: a technical framework for life-cycle assessment. Smugglers Notch, Vermont, SETAC, Washington, DC, 18–23 Aug 1990

Hunkeler D, Saur K, Rebitzer G, Finkbeiner M, Schmidt WP, Jensen AA, Stranddorf H, Christiansen K (2004) Life-cycle management. SETAC, Pensacola. ISBN 1-880611-77-5

ISO 14040 (1997) Environmental management – life cycle assessment – principles and framework. Geneva

Chapter 9
How to Make the Life Cycle Assessment Team a Business Partner

Mark Goedkoop, Eric Mieras, Anne Gaasbeek, and Soledad Contreras

Abstract In this chapter we explore the need and opportunities to make the life cycle assessment (LCA) team more relevant for the business. Sustainability trends and alternatives for LCA are analyzed to identify what makes them relevant for and appealing to business managers, the difficulties LCA practitioners face to get their message across have been identified, and a five-step approach to make the LCA team a business partner will be described. The goal is to empower LCA teams and practitioners to create sustainable value for the business they work in.

Keywords Champion for LCA • Circular economy • LCA community • LCA team • Life cycle assessment • Life cycle management • Product sustainability • Sustainability

1 Introduction: 20 Years of Life Cycle Assessment, Have We Understood the User Needs?

One may wonder why it is still a relatively small community that performs LCA and why LCA has not become more mainstream in business processes, while LCA methodology is used by thousands of companies and has proven to be a relatively robust tool for understanding the impacts of products. Most major multinational companies nowadays have something like an LCA department, which is usually a team of a handful of specialists that perform LCAs mostly to support internal decisions.

To explore this issue we first take a broad perspective of the developments in companies regarding product sustainability; these developments are too often missed by LCA practitioners. Next we will describe our findings from research we

---

M. Goedkoop (✉) • E. Mieras • A. Gaasbeek • S. Contreras
PRé Consultants, Amersfoort, The Netherlands
e-mail: goedkoop@pre-sustainability.com

G. Sonnemann, M. Margni (eds.), *Life Cycle Management*,
LCA Compendium – The Complete World of Life Cycle Assessment,
DOI 10.1007/978-94-017-7221-1_9

did among corporate LCA practitioners, and finally we will describe how we can link the LCA practitioners to corporate developments to make them more relevant in the business.

2 Understanding Major Product Sustainability Trends

About a decade ago, the LCA community was somewhat taken by surprise by the increasing popularity of the cradle to cradle approach. Multi-national companies followed suit; they did in most cases not abandon LCA, but became very active in cradle to cradle. Cradle to cradle now seems to be surpassed in popularity by Circular Economy, partially due to the lack of transparency in the cradle to cradle approach.

Interestingly enough, Circular Economy is also not a very concrete methodology; there is no ISO standard and there are no precise rules. In fact, it misses all aspects we in the LCA world find so relevant. Yet in spite of this, it has gotten a huge uptake with major companies that are even turning around the way their business works. Circular economy is of course more than an assessment method; it is a vision towards a desired future. It tries to transform linear models (produce, distribute, use and dispose) into a circular loop of products. For instance, instead of burning coal and selling electricity, major energy companies are now transforming themselves to be ready for the new reality: electricity is traded between small producers and individuals with solar cells on their roof. While this is a clear example with major sustainability benefits, many ideas in circular economy discussions are not really assessed, and often it is not clear at all what the sustainability merits are.

So what makes these methodologies popular if it is not primarily their ability to measure? What really works is that they present a vision; they tell companies, follow us, follow our guidance and we will not only help you to make products less bad, but we simply turn them into good products, as the cradle to cradle community often said, without ever substantiating what "good" is. And is vision not something which we as the LCA community are often too hesitant about?

2.1 Understanding "Alternatives" to Life Cycle Assessment

While LCA aims to be a neutral basis to measure sustainability impacts without having a vision for a desired future, we also see attempts to make LCA more meaningful (Kiron et al. 2015). In December 2013 a large conference around natural capital was organized in Edinburgh and over 500 people joined; many industry leaders and leaders of international organizations all shared their enthusiasm for this new concept called natural capital (or as we say in LCA: monetization). We spotted only one other representative from the LCA community. In the meantime, we have seen overviews of more than 100 initiatives to develop natural capital methodologies, all outside the LCA community, all reinventing the wheel and forgetting that the very first serious impact assessment method based on monetization was developed by Bengt Steen in 1989. The big idea is to develop metrics that are understood by the CFO (chief financial officer) and CEO (chief executive officer). Developing metrics that assure results are compelling for the business is probably something the LCA community could learn from.

Another development which is largely missed by the LCA community is the methodology developed by the sustainability consortium. They started out developing an LCA based method to assess products on a large scale, but failed to do so. Therefore, they switched to a procedure that focuses on hotspots and improvement opportunities in the lifecycle. This shift seems to be working very well and we should learn from this.

Coming from another angle, but with the same core idea to make results more meaningful and also to give guidance to the management is the recent discussion to link metrics to planetary boundaries. One idea is to develop a "planetary boundary enabled LCA method". The Stockholm Resilience institute identified eight planetary boundaries, or levels of impacts we should not pass. Initiated by Unilever, a "planetary boundary enabled LCA method" is being developed by a group of experts led by the University of Surrey, with the involvement of Unilever.

2.2 *The Risk of Ignoring These Trends*

In our vision we cannot ignore these trends if we want to ensure a relevant role in policy and business. The assumption in the LCA community is: What gets measured will get managed. This works if managers understand the measurements and can set goals. This works when they talk about revenues, ROI and strategic targets, but what to make of LCA results? Should they set a reduction target of 20 %? Why 20? And why not 5 or 50 %? They do not have a reference, do not have a gut feeling and often not a clear vision about what LCA results can mean for them. This is what these new concepts do so well; they come with a vision that is understandable, that is actionable and often simply "feels good", or they come with a financial metric that managers (think) they understand, or a reference to something like planetary boundaries. In the case of TSC, the idea is that KPIs and improvement opportunities are based on a general consensus from science, NGO and industry. All these "alternatives" seem often more attractive than an accountancy-like calculation procedure that reports indicators in incomprehensible midpoints. However, there is hope. LCA is the only systematic way to measure, or at least it is much more consistent and transparent than any of these alternatives.

Table 9.1 Partners and participants in round tables

Country	Partner organization	Remarks
Sweden	Miljögiraff	Mixture of companies and research institutes
Denmark	2.-0 LCA	High attendance: 10 companies
UK	SimaPro UK Ltd	Also consultants
France	EVEA	Focused on eco design
Italy	2B	Focus on luxury food companies
Germany	GreenDelta	In German language, only in 2012
USA	PRé North America	Only in 2014, several industry associations

3 Understanding the LCA Community Itself

To develop our understanding of the LCA community, especially the corporate LCA practitioners, we organized a number of workshops in various countries in 2012 and again in 2014.

For this research, we worked together with a number of business partners of our organization in Europe. We asked them to invite their corporate clients to come together for an open discussion on how they see the future of LCA, what their challenges are, and how they cope with them. Of course we also asked them what kind of features and tools would help them most. The meetings were organized in early 2012 and throughout 2014 with seven partners; see Table 9.1.

The majority of the participants were LCA experts working in large multinationals but there were also some researchers and consultants present.

3.1 The Overarching Message

Many corporate LCA experts are telling us they have difficulty communicating their findings to the internal stakeholders. One of the problems is that the results are not well understood by their colleagues in marketing, research and product development. It is as if they operate in different worlds. LCA results do not really connect to what their internal clients want to know. This means many are seeking ways to drastically simplify results or translate them into guidelines.

Getting the results out and understood is a problem, getting data in is another big problem. Often other departments are not really motivated to supply data. Getting access to data held by the purchasing department, for example, can be very difficult as they do not see the benefits.

The general picture that emerges is that the LCA department finds it difficult to connect and to be relevant, while at the same time companies have committed themselves to improve the sustainability of products in a rational way. This seems to imply that there is a disconnect between the ambitions of the companies and what the LCA experts can offer – do they speak a different language? Or is there another reason that they are not able to connect? This is an intriguing question. This conclusion of a gap between the LCA department and the business was also supported by the results of a quantitative research amongst 274 LCA practitioners in 28 countries conducted by PRé in the second half of 2014.

4 What's Next: How to Tackle This Chasm?

As a message to the LCA community and the researchers: the bottleneck is not in the lack of sophisticated tools and methodologies. Companies need LCA departments and practitioners that are better connected with other departments and show genuine interest in alternative approaches. The solution is not to make better LCAs but get a better understanding of what marketing, design, research, purchasing and other departments need.

Based on the round tables, quantitative research and existing scientific research (Frankl 2002) we developed a five-step approach to make the LCA department more relevant:

1. Become a champion for LCA
2. Assure long-term management commitment to sustainability
3. Link LCA with business objectives
4. Find a shared language
5. Jointly explore new applications

These five steps will help LCA departments and practitioners to increase their impact in their company.

4.1 Become a Champion for Life Cycle Assessment

Most LCA experts have been trained in technical skills and try to capture reality in a model to see what can make a change. However, technical skills are not sufficient to make a change. We learned from successful LCA practitioners how they are working to become an internal champion, an intrapreneur for LCA, someone who connects with other people and departments, who pleads the cause of LCA and that explores how LCA can contribute to the company's goals.

Business managers want to have as much information as possible to reduce the risk, but know that you never know 100 % for sure. The same holds true for designers, they often work on an idea, a hunch without knowing whether it will work; for instance, they know they have a cost target, but in the early sketch phase they have no way of checking the cost with any precision, so unlike many LCA experts, they are happy with any information that helps them to stay on course and consider the environmental impact. A large part of training the designers is about managing (and living with) uncertainties.

Intrapreneurship has some aspects that are quite contradictory to LCA. Perhaps the most important one being that you are not completely sure about the outcomes before you do something. You take a calculated risk. That requires courage and stamina as you do not know what the outcome will be. It also requires the courage to acknowledge that we are imperfect. In businesses, people are used to making decisions with an uncertain outcome, at least to some extent.

To understand how people use the results from an LCA it is important to look at yourself from their perspective. That can change a lot. Just as turning the camera on planet earth – as the astronauts from the Apollo space mission did – had a tremendous impact on how we look at what we do to (the environment on) the earth. A very simple way of doing that is meeting with the users from your LCA study without a specific purpose, drink a cup of coffee and listen to what they are working on, what drives them, what successes they have achieved, and what challenges they face. That will give you great insight in what opportunities LCA can help with. And seizing opportunities is what makes a change.

Knowing what your (internal) clients need – or call it your audience – will also help to present the results in a way that is appealing to them and is relevant for their goals and needs. If you learn how to tell your story, you can convince, motivate and stimulate people to do something with the results from your LCA study. And of course, facts play an important role in that story.

4.2 Long-Term Management Commitment to Sustainability

Many big companies have a long term commitment to sustainability. Research (MIT, UNEP and BCG 2015) shows that 42 % of the boards are committed to sustainability and that 31 % of the companies have operational KPIs related to sustainability. Still, support from senior management really is a prerequisite for having an impact; some experienced corporate LCA practitioners have been able to connect to the long-term goals of the company. They learned that management likes facts; facts can help them to make the right, informed decisions about sustainability. And facts are something LCA can deliver, sometimes with some uncertainty, which is something management has learnt to deal with.

Thus, LCA and LCM can play an important role on the road to achieving sustainability goals. We all know the examples of companies that use LCA or LCM in their sustainability strategy. At the same time there are still many companies that do not build their sustainability strategy on sound and robust figures. They follow what others are saying and doing or one of the latest trends with less consistent and transparent methods to measure. So, there is an opportunity for all LCA practitioners because they know the facts and can show something else than what you would expect or what everyone else is doing.

For companies that already use LCA, top management is not only committed to sustainability, but also to LCA. The survey PRé conducted in 2014 shows that in 57 % of the companies that use LCA, management is aware of LCA. So, LCA is on the agenda in the majority of the companies. Finding an internal sponsor in the (top) management can accelerate the use of LCA in the business. That will definitely help you to deliver value for the business.

4.3 Shared Language

Business and LCA have two totally different languages, or as we would say in LCA – nomenclature. As LCA practitioners we know what happens when you do not have the nomenclature right; things get mixed up. So, make sure you understand and capture the language from the people that use your studies (Fig. 9.1).

As Gregory Unruh (2014) puts it in a series *"Each functional area has its own conversation built on terminology and jargon suited to their specific business concerns. ....tap into these functional conversations and help managers develop what can be called a sustainability dialect that translates corporate sustainability goals into the local functional discussions and thinking."*

Speaking the same language also helps to embed LCA in the core processes of these departments. Only if it is embedded the full potential can be achieved. Only if it is embedded these departments will see and experience the potential of LCA or even sustainability. If not, LCA will probably stay a staff driven "exercise".

Fig. 9.1 Different needs and language of departments in a company

This transition is not something that can be achieved overnight. It requires a good radar to sense what is important for people, not only functional but also personal. Sharing – or at least understanding – the same language is really essential to pick up the signals. It also requires a step-by-step approach to link the results from LCA to the objectives and language of these departments or even the corporate strategy.

Next we will describe what such an approach could look like.

4.4 Link Life Cycle Assessment with Business Objectives

In most companies, sustainability goals are set top-down. In the corporate sustainability strategy, goals are set for energy, water and climate change, for example. Business units and departments are then given the task to achieve these goals and start various projects to realize these goals in the timeframe given. The projects are topic, location or process specific and do not take trade-offs into account. Based on the goals, companies want to measure the results to report on them and improve their performance. Often it turns out to be quite hard to report and improve the goals. They were not really substantiated by insights on a business and product level which is where the improvements have to be achieved (left column of Fig. 9.2), and let alone the negative trade-offs these improvements could have.

LCA is much more a bottom-up approach as it starts at a product level. In the product or service life cycle it is identified where a product has its impact. LCA can deliver insights in what impacts occur for each impact category and life cycle stage. It enables us to identify hotspots and improvement opportunities, from material selection to pinpointing the most impactful supplier, from process innovations to contributions of each phase. However, at the same time LCA struggles to link these hotspots and improvement opportunities to the KPIs of the company (see column 2 of Fig. 9.2), especially as these are often formulated in a different language. TSC has done some great work to link hotspots to KPIs, but adding the top-down route is essential to make it company specific and meaningful for management. An integrated approach provides a unique opportunity to achieve maximum alignment of product and corporate strategies (Furfori et al. 2014).

To increase the impact of LCA, the LCA practitioner and the LCA department need to connect the dots. To take the lead in this you can create an overview of business and project KPIs and combine these with the hotspots you identified through LCA studies (Fig. 9.2). Based on those insights and KPIs, projects for improvement can be identified. This approach helps to link the product level with the business/corporate level. It brings together the sustainability insights on a corporate and product level. By doing this you can identify sweet spots, overlap in hotspots as well as blind spots. It also brings together different departments when it shows they need to collaborate to achieve their goals. The same applies for suppliers; based on the hotspots the most relevant suppliers for the sustainability goals – which is not the same as the most important suppliers in terms of costs – can be pinpointed. By doing this the sustainability strategy gets more substantiated and goals become

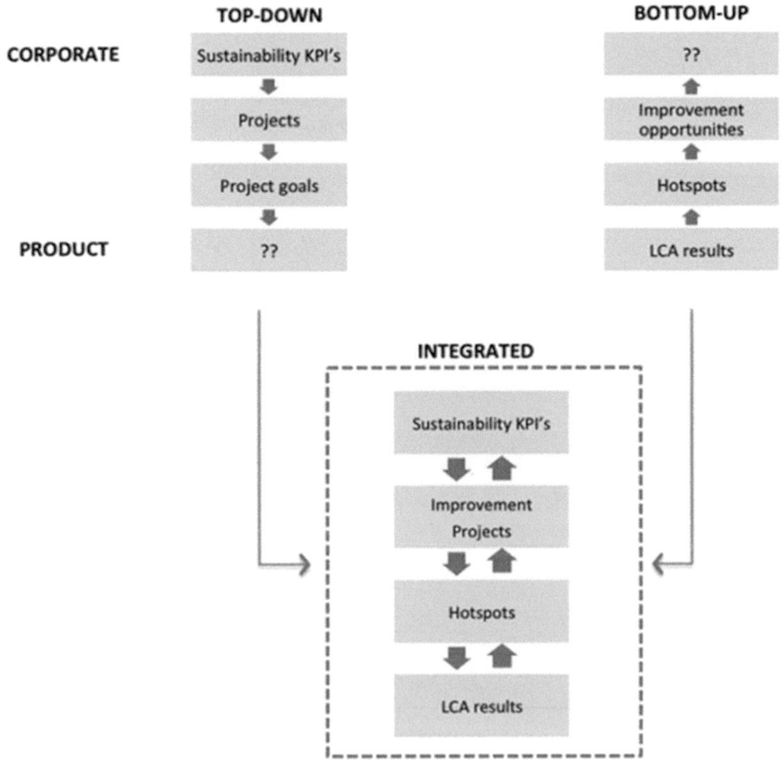

Fig. 9.2 A top-down and bottom-up approach to link LCA results and business objectives

more realistic. By forming new partnerships with suppliers or customers and stimulating cross-company collaboration, innovation can be spurred.

LCA can play an instrumental role in the further implementation of these improvement projects, but do not hesitate to use other tools if needed. Once it has been embedded in the processes, the next step is to set up the infrastructure to facilitate this and to explore new opportunities to create value. The latter can only be done if you are a trusted business partner within your company.

4.5 *Jointly Explore New Applications*

Over the past few years a number of studies have been published about the use of LCA in business (amongst others Chun and Lee 2013; Piekarski et al. 2013). These studies identified the several ways LCA can be used. Some of the studies also identified for which departments the studies could be used. What is missing in those studies is the combination of the two. So, what use is relevant for a specific department?

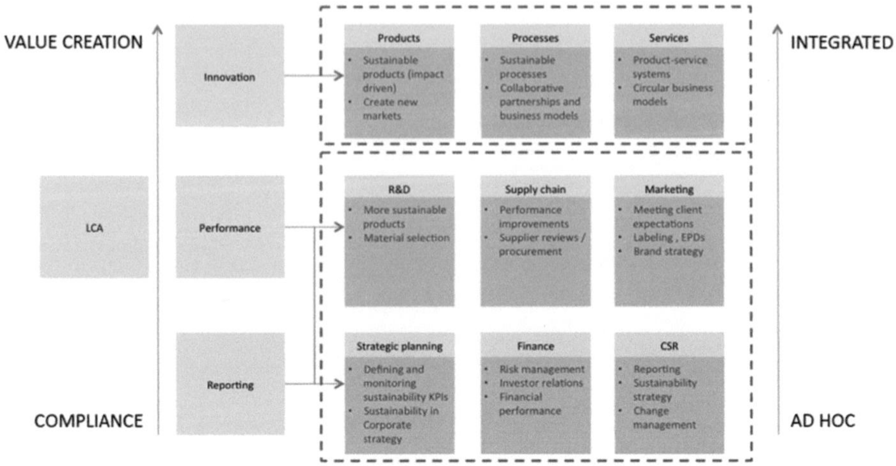

Fig. 9.3 Applications of LCA in (different departments of) a company, inspired by Moro Piekarski (2013)

We tried to represent that in Fig. 9.3 below, which was inspired by the study of Moro Piekarski. In your daily practice this is something you need to investigate.

Apart from the type of use, it is important to look at how it is used (ad hoc vs. integrated) and why it is used (reporting, performance improvement or value creation). The overall purpose – or the why question – is related to the corporate sustainability strategy: is it aimed at compliance and reducing risk or is the goal to create shared value. The more it is aimed at value creation, the more LCA should be integrated in the business. Is it more directed at compliance and reporting, LCA probably stays more ad hoc. Reporting and improving performance can be done on a department level, innovation requires a more integrated and holistic approach.

To develop these specific applications for LCA it is essential to link up with people from these departments to explore the needs and opportunities – remember the personas we presented earlier. In some companies they recognized this need and created a specific position to liaise between the LCA department and the internal client. In a transition phase this could be a good solution.

5 Conclusions

LCA has developed into a sound and robust methodology that is probably the best approach available to measure social and environmental impact. Within the LCA community there are a lot of talented and motivated people to make a change. We saw how LCA practitioners expand their own ecosystem and step out of their comfort zone based on facts and how this increased the impact they have.

We have given a five-step approach to become more relevant and step into a world that identifies opportunities to create sustainable value. However, our best advice is listening, and especially listening to those outside the LCA community, the people who are engaged, in what we have called "alternatives" and of course the departments that could benefit from LCA. What matters is whether you can provide a basis for rational decision-making by the business that will result in sustainable products and services as well as business value.

References

Chun Y-Y, Lee K-M (2013) Life cycle-based generic business strategies for sustainable business models. J Sustain Dev 6(8)

Frankl P (2002) Life cycle assessment as a management tool. In: Ayres RU, Ayres LW (eds) A handbook of industrial ecology. Edward Elgar, Cheltenham

Furfori S, Antonini V, Breedveld L (2014) The link between LCA and CSR with espresso coffee as an example. In: 9th international conference LCA of food, San Francisco. 8–10 Oct

Kiron D, Kruschwitz N, Haanaes K, Reeves M, Fuisz-Kehrbach SK, Kell G (2015) Joining forces, collaboration and leadership for sustainability. http://sloanreview.mit.edu/projects/joining-forces/

Piekarski CM, Mendes da Luz L, Zocche L, De Francisco AC (2013) Life cycle assessment as entrepreneurial tool for business management and green innovations. J Technol Manag Innov 8(1):44–53

Unruh G (2014) Sustainability dialects. http://sloanreview.mit.edu/article/sustainability-dialects/

Chapter 10
Sustainability Improvements and Life Cycle Approaches in Industry Partnerships

Peter Saling

Abstract Prioritizing sustainability as a key strategic focus and managing it similar to other parts of business such as marketing and sales, leading companies have been able to better identify and manage risks as well as enhance brand value and corporate reputation. With industry partnerships a holistic approach is possible. Common metrics, shared value chain data and joint sustainability management form the basis of a successful cooperation. Decision-making processes can be supported efficiently and influence the whole product system, thereby facilitating clear, measurable value creation throughout the supply chain. Industry partnerships with implementation of innovative business models are a key enabler for companies to realize more sustainable solutions.

Keywords Eco-efficiency analysis • Econsense network • Life cycle assessment • Life cycle management • PlasticsEurope • Sustainable solution steering • Sustainability

1 Introduction

Awareness of sustainability in business decisions has an increasing importance for different stakeholders. To develop more sustainable solutions which meet society's needs is a key challenge in industries along the whole value chain.

There is a growing awareness in the financial market that a company geared towards sustainable development in order to outperform peers over the long-term while minimizing risks.

P. Saling (✉)
BASF SE, Sustainability Strategy, Ludwigshafen, Germany
e-mail: peter.saling@basf.com

G. Sonnemann, M. Margni (eds.), *Life Cycle Management*,
LCA Compendium – The Complete World of Life Cycle Assessment,
DOI 10.1007/978-94-017-7221-1_10

2 Industry and Associations Initiatives

2.1 *Together for Sustainability (TfS)*

Several initiatives of industries and associations have been founded with the goal, to implement sustainability principles in supply chains and to improve product applications. There are well defined networks exchanging general information but as well industrial collaboration networks exchanging and offering LCI information to calculate complete LCA studies.

A good example for initiating a general information platform to improve sustainability is a new initiative of the chemical industry, the "Together for Sustainability" platform, where chemical companies work together for improving sustainability in their supply chain. As multinational leading companies, chemical companies strive for a sustainable development and support the principles of the United Nations Global Compact and Responsible Care. The companies take responsibility for their own operations and in the sphere of their influence for our supply chains to support adherence to existing regulations and to respond to the needs and expectations of consumers and society. To use resources more efficiently and reduce the bureaucratic burden for suppliers, the participating companies share supplier sustainability assessment and audit data.

Within the TfS assessment as well as audit, the supplier's sustainability performance is verified against a pre-defined set of audit criteria. These topics have been defined by TfS and are tailored to the requirements of the chemical industry.

Benefits for the companies but as well for other players in the supply chain are seen in avoiding double audits and assessments, improvement and assurance of quality of assessment and audit results. Sharing of assessment and audit results with multiple customers on one platform, high quality through selected and qualified partners are additional benefits. Engaging with customers on sustainability requirements and challenges for building up long-term business relationships will be a positive outcome as well as lowering risks in relation to sustainability requirements. Knowing sustainability performance allows to improve performance (TfS 2012, 2015). This can be achieved additionally by calculating facts about products and their precursors with an LCA approach to generate detailed information.

One example for a cross sector initiative is the econsense network. The goal of this initiative is an open dialogue, the members of econsense strive to further advance the implementation of economic, social and ecological objectives, with the awareness that business with strengths in innovation and investment also assumes certain responsibility for the success of sustainable development. At the same time, companies can only discern their corporate social responsibility when supportive and reliable political framework conditions offer them a sound environment. The objectives of econsense are:

- To pool corporate activities on sustainability topics, such as climate protection and demographic change, and to jointly further develop these projects
- To actively shape the political and social discourse
- To credibly communicate the solution competence of the economy

- To strengthen the open dialogue between political and social groups
- To highlight the possibilities and limitations of corporate responsibility
- To promote sustainability concepts and CSR in the business community and raise awareness of policymakers for framework conditions that promote innovation and competitiveness (econsense 2014)

2.2 Life Cycle Inventory Data Platforms of Associations

Associations provide more and more data for LCA practitioners. Single LCI information in a format that can easily be used and introduced to common LCA software systems and is very helpful to generate complete LCA studies based on average figures of the relevant industries.

PlasticsEurope (2011) promotes the use of life cycle thinking (LCT) to improve understanding about product benefits and to take more informed decisions. As a scientific method, life cycle assessment (LCA) is a technique to analyze the potential environmental impacts associated with a product, process or service. It involves:

- Compiling an inventory of energy and material inputs and environmental releases
- Assessing the potential environmental impacts associated with identified inputs and releases
- Calculating performance indicators to inform decisions

PlasticsEurope was the first industry organization to assemble and publish detailed environmental data on the processes operated by its member companies. The first Eco-profile reports were published in 1993. Since then, more reports have been added and continuously updated, so that there are now more than 70 Eco-profile reports freely available. In 2006, a complementary Environmental Product Declaration (EPD) programme was started. Eco-profiles and EPDs cover high volumes, bulk polymers, some of the more widely used engineering plastics and several common plastics conversion processes (Boustead 1993). Widely acknowledged among life cycle practitioners and other stakeholders worldwide as representative datasets, they have been included in various commercial life cycle databases as well as in the publicly available European Reference Life Cycle Data System (ELCD).

PlasticsEurope has clear objectives when compiling the Eco-profile reports, representing European production averages:

- The first is to place scientifically sound data in the public domain for use in product life-cycle studies, without compromising the confidentiality of detailed process data of the individual companies.
- The second is to encourage environmental improvements in production processes through benchmarking against a European industry average.
- The third key factor is that, given the large contribution of upstream effects to the Eco-profile of a polymer and in view of the distribution of input materials, such as ethylene or naphtha via the European pipeline network, industry averages are the most robust representation of polymer production systems.

Since the first Eco-profile reports were published, the LCA methodology, standardization and practice have undergone substantial changes. New concepts, such as EPD, Carbon Footprint or Product Environmental Footprint (PEF) have emerged. Downstream industries like the building and construction sector have their own standards and data needs. Hence, Eco-profiles need to change in response to best practices and stakeholder needs. To this end, PlasticsEurope periodically seeks stakeholder input on the Eco-profile methodology. Furthermore, in view of the need for globally harmonized practices and comparable results, PlasticsEurope welcomes and actively invites liaisons with other regional federations. As a contribution towards shared best practices, the Eco-profile methodology aligns with other material- or sector-specific standards.

2.3 Steering Product Portfolio to Foster Sustainable Solutions

The ambition to create sustainable products and services is being driven by a number of compelling business factors. New laws and standards regarding carbon emissions and other sustainability topics are being implemented all over the world. At the same time, there is growing market uncertainty about the cost of raw materials and the availability of natural resources. Finally, the end consumers are evolving their expectations about the goods and services they purchase. Increasingly, they are holding brand owners and companies to a higher account in terms of materials that go into consumer products and the way those products are made.

BASF started and will continuously analyze the complete portfolio from the viewpoint of the sustainability needs of their customers. Therefore BASF developed the evaluation process "Sustainable Solution Steering" that allows to gain enhanced internal transparency and consistency on the sustainable development performance in all areas of the business globally (BASF 2015). The process also acts as an early warning system that can identify where solutions are facing sustainability obstacles. It also helps to identify sustainability benefits for the company, customers, society and the environment. In addition, Sustainable Solution Steering provides the business units with the information they need to communicate opportunities for sustainable solutions. This insight can be used to differentiate in the specific markets, enter into dialog with customers about their sustainability needs and hence generate new business opportunities.

First a qualifier check is conducted, where each solution is evaluated based on companies and the value-chain-specific performance for economic, environmental and social criteria. After that, the solutions are ranked according to their sustainability performance in the respective application and clustered into one of the following four categories:

- Accelerator: A solution with a substantial sustainability contribution in the value chain

- Performer: A solution that meets the basic sustainability standards in the marketplace
- Transitioner: A solution for which a specific sustainability issue is actively addressed
- Challenged: A solution with a significant sustainability concern identified and for which an action plan is under development

LCA information or as well eco-efficiency analysis results can be used to support this evaluation process.

Ultimately, Sustainable Solution Steering will benefit customers by delivering new business opportunities through innovative solutions as well as providing support on their own sustainability needs. It is a life cycle management process that can be applied to other industries as well (Kicherer and Voeste 2014).

3 Examples of Assessments and Applications

3.1 Using Plastics Europe LCI Information

In order to produce plastic products, energy resources are consumed. Currently such energy resources are almost entirely obtained from non-renewable sources, and by using them, greenhouse gas (GHG) emissions are produced. Nevertheless, even more energy would be consumed and more GHG emissions emitted, if plastic products are to be substituted by alternative materials. This was established in a study by (Pilz et al. 2005).

The study generally follows an "80/20-approach", meaning that the authors aim to cover 80 % of influences with 20 % of effort that would be required for a more comprehensive study. As a result, a high degree of reliability was ensured for the general magnitude of the overall results, but not for every specific figure in the case studies investigated, where – based on the "80/20-approach" – many (reasonable) assumptions had to be made where data were not easily available.

Calculation of life cycle energy and GHG emission balances: Data for the production phase of plastic products were mostly taken from the "Ecoprofiles" as published by PlasticsEurope. Production data of alternative materials was taken from the database ecoinvent (2007) or comparable sources. In the use phase the calculation covers issues where plastic products have a different impact on energy and GHG emissions compared to alternative products. The effects considered are mainly fuel consumption for transportation, prevented food losses, differences in thermal insulation properties, and fuel savings due to the lower mass of plastic automotive parts.

For example, substituting plastics in the case studies throughout Europe (EU27+2) in 2007 would increase the life cycle energy consumption by around 2.140 million GJ per year and the GHG emissions would increase by 110 Mt CO_2 equivalents per year.

The energy savings that can be attributed to the use of plastics varies significantly according to the application area, with packaging being by far the most important. A conservative estimate of the impact of the total plastics market has been made by extrapolation using only half of the energy savings and GHG emission reductions of the quoted examples.

The results show that the total life cycle energy needed to produce, use and recover plastic products in Europe (EU27 + 2) is 4.300 million GJ/a and the total life cycle GHG emissions are 200 Mt/a.2 Furthermore it can be concluded that substitution of plastic products by other materials wherever possible would need around 57 % (1.500–3.300 million GJ/a) more energy than currently used in the total life cycle of all plastic products today. In the same way, substitution of plastic products up to the theoretical maximum would cause 78–170 Mt or about 61 % more GHG emissions than the total life cycle of all plastic products today.

In other words, the plastic products on the market today have enabled energy savings of 2.400 million GJ per year, equivalent to 53 million tonnes of crude oil carried by 205 very large crude oil tankers.

The GHG emissions saved (124 Mt per year) are equivalent to the total CO_2 emissions of Belgium in the year 2000 (UNFCCC 2009) and are also equivalent to 39 % of the EU15 Kyoto target regarding the reduction of GHG emissions (Pilz et al. 2010).

3.2 *Evaluating Product Sustainability, a Contribution from CEFIC*

Industry has already made significant achievements in sustainability, driving safe, environmentally sound operations through its Responsible Care® scheme and corporate social responsibility activities. These efforts were started well before the United Nations' Rio Declaration in 1992 and have become ever more important since.

Today, emerging regulatory and social trends around sustainability create both pressures and opportunities for chemical companies at global and EU levels. Legislative requirements, stakeholder expectations and companies' own business and Responsible Care strategies are driving the development of more sustainable chemical products and supply chains.

Clear trends are already surfacing:

- The introduction of REACH creates new pressures on specific substances.
- The emergence of eco-design, Green Public Procurement (GPP), Ecolabel criteria and waste prevention schemes is creating demand for more sustainable products.
- Rising consumer interest in sustainable goods is incentivising retailers to develop sustainability measures for their suppliers.

The move towards sustainable products will take many years to progress through legislative and business processes. During this time, retailers, consumers and non-governmental organizations (NGOs) will continue to call for transparency and clear statements about the constituents of the goods they purchase.

The Eco2chem project for eco-efficiency measurement set up at sector level in Belgium and actively supported by local authorities and several research organizations, aims to select the best-fit eco-efficiency measurement methods applicable to chemical processes and products. The main outcome is SUSCHEMCompass, a web-based tool to help companies, especially SMEs, select the eco-efficiency measurement method best suited to their specific needs.

The tool focuses on measurement methods for the economic and environmental aspects at company level and/or project level, but may be extended to social aspects in a follow-up project. A variety of methods have been identified, ranging from quick scans to thorough life cycle analysis and from freely available tools to proprietary tools. For each method, the web-based tool includes an information sheet summarizing the history and scope of the method, and what it can and cannot measure. The tool can be used by all interested parties free of charge (CEFIC 2012).

3.3 Sustainability Improvements Support with Eco-Efficiency Studies: Pavement Preservation Technology for Asphalt Roads

The society depends on roads as a vital component of their national economies. Ensuring that these roads are safe, long-lasting and cost effectively installed and maintained is thus essential to the sustainability of the transportation network. Pavement preservation is the systematic scheduling of nonstructural maintenance applications to protect engineered road pavements and extend their service life. This helps promote better road conditions, increases safe driving by minimizing surface deterioration and the potential for structural failure and is a more efficient use of tax payer money. The challenge facing many government agencies and key material specifiers is how can they decide which pavement preservation technologies and materials are the most eco-efficient? On what basis should they make their comparison and what metrics truly define the sustainability of road construction materials? BASF in collaboration with a key customer, Vance Brothers, utilized the eco-efficiency analysis to compare the relative eco-efficiencies of two of the more prevalent pavement preservation technologies for urban roads in the United States. The life cycle environmental and economic impacts of a polymer modified asphalt emulsion based micro-surfacing technology were compared to a two-inch polymer-modified hot mix overlay. The analysis was based on the environmental and economic impacts required to maintain a one-mile stretch of a 12 ft lane of urban road using best engineering practices for a 40 year lifetime. The question was whether it was more sustainable to install a more durable layer (hot mix overlay also known as Mill and Fill) that contained 10 % recycled materials but required overall more materials and extensive road work or to utilize a less cost and resource intensive maintenance technology like micro-surfacing more frequently (due to its shorter lifespan) in order to achieve the same desired road performance. The eco-efficiency study along the whole supply chain with a cradle to grave approach showed that micro-surfacing consumes about 40 % less primary energy and resources than hot mix overlays over the 40 year life cycle of the road. Hot mix overlays scored higher due to higher bitumen consumption, hotter production and application temperatures as well as increased fuel requirements for transporting larger amounts of materials to and from the job site.

Having detailed results depicting how the individual system components contribute to the overall impact category is essential for informed decision making. Of particular interest was the discovery of the significant environmental impact the road markings had over the life cycle for micro-surfacing. Thus to further improve the overall eco-efficiency of micro-surfacing it may be necessary to look at optimizing other aspects of the system components which make up the overall technology.

By using more sustainable products and solutions, it clearly was shown that the micro-surfacing technology has a significantly reduced environmental fingerprint. These benefits can be directly attributed to its more efficient use of resources, its lower energy consumption as well as lower overall emissions to the environment. By combining its preferred environmental profile with its reduced life cycle cost (25 % less than hot mix overlay), clearly places micro-surfacing as a more eco-efficient material for the base case analysis.

With any rigorous analysis that involves copious amounts of data, it is essential that the results are presented and communicated in a way that facilitates clear understanding as well as helps bring into context the significance of the findings. The eco-efficiency methodology by BASF through the use of the environmental fingerprint and portfolio is well suited to distilling vast amounts of data and presenting it in a concise, balanced format (Saling et al. 2002, 2005; Landsiedel and Saling 2002).

As many important stakeholders of LCA or eco-efficiency studies are not as well versed in many of the common units of measurements (e.g. grams SO_2 equivalents for assessing acidification potential or mega-joules for energy consumption) they are sometimes not able to adequately assess the relative significance of the measured impacts. Thus communicating the results in more commonly understood terms or equivalencies is an essential aspect to effectively communicating the results and ultimately facilitating strategic review and decision making.

Just considering the context of the micro-surfacing study which was for only a single mile stretch of urban road over 40 years, the advantages of micro-surfacing over hot mix overlay could additionally be expressed in more commonly understood equivalencies such as:

- Approximately 540,000 kg less material required
- 34 t less material sent to land fill
- Energy efficiency (EIA 2005):

  - Savings equivalent to the annual consumption of energy in 110 US homes
  - Over 42,000 L less oil consumed per every lane-mile

- Smaller carbon footprint (EPA 2015):

  - Reduction equivalent to taking over 20 cars off the road
  - Carbon sequestered annually by over 8 ha of pine forest

Through the use of eco-efficiency and the communication of results using common equivalencies, state agencies will be able to make more informed and strategic decisions related to promoting the sustainability of road constructions (Uhlman and Saling 2010).

4 Conclusions

The life cycle approach for partnerships of companies along supply chains enables different industries the identification of hotspots, improvements of their products and applications and finally the marketing of more sustainable solutions in the market.

Through collaboration toward common goals, business can address some of the critical environmental and social problems the world faces while strengthening their own resilience to global challenges. Different types of business solutions can be generated. Main conditions for them are to be impactful, measurable, scalable, replicable and beyond business-as-usual.

Scalable means that they can have a meaningful impact on the world.

Replicable enables them to be applied by many companies, in multiple sectors, regions and countries.

Measurable is important to know how they are making a difference. Beyond business as usual businesses and governments begin to work – and collaborate – differently.

In sum, the initiatives are good for business, so they have a commercial logic that contributes to the broader good and to the bottom line (WBCSD 2015, Vision 2020).

Action2020 is the WBCSD platform for sustainability in action. It is the roadmap for how business can positively influence environmental and social trends while strengthening their own resilience to issues like climate change, demographic dynamics and skills shortages. Based on the latest scientific consensus, action 2020 sets an agenda for business to take action on sustainable development to 2020 and beyond.

LCA tools, eco-efficiency analysis and other LCM tools and approaches can support the basic needs for the development towards more sustainable solutions in the market with scientific sound, detailed and meaningful information quite efficiently and will be used more intensively in the future. The collaboration between different stakeholders in the market will help to make significant improvements and foster more sustainable solutions. Analytical tools as LCA can support decision-making, visualization and marketing as well as the management of products in the value chains.

LCM tools enable companies to drive innovative product development focused on bringing more sustainable products to the market place. The tools and methods behind them clearly identify the factors whose optimization will directly translate into improvements in the sustainability profile, even during the early stages of product development. It facilitates clear understanding of trade-offs and helps in preventing inadvertently shifting environmental impacts from one area to another or between the economic and environmental pillars. By measuring the impacts on a system level and including a comprehensive approach to environmental impact assessment, it also safeguards against potentially reaching false conclusions that could result when only single metrics were considered.

Life cycle management tools applied in the collaboration and partnerships of companies and industries is also an effective market communication tool. Since the entire life cycle of a product is analyzed, the effects on customers along the supply chain can be quantified and evaluated and thus a more strategic value proposition can be developed. Communication can also go beyond direct customers with the results being used to support engagement and education amongst government agencies, regulatory bodies and NGOs.

References

BASF (2015) Sustainable investments. https://www.basf.com/en/company/investor-relations/sustainable-investments.html

Boustead I (1993) Eco-profiles of the European plastics industry. Report 3: Polyethylene and polypropylene. Report to The European Centre for Plastics in the Environment (PWMI) (later Association of Plastic Manufacturers in Europe (APMI)), Brussels May 1993

CEFIC (2012) Sustainability of products – what it's all about. http://www.cefic.org/Documents/IndustrySupport/RC%20tools%20for%20SMEs/Document%20Tool%20Box/Sutainability_of_products_What_its_all_about.pdf

econsense (2014) http://www.econsense.de/sites/all/files/Basisflyer_E_2014%20neu.pdf

EIA (2005) Energy Information Administration (EIA). Residential energy consumption survey. Energy consumption and expenditures. Tables: www.eia.doe.gov

EPA (2015) Greenhouse gas equivalencies calculator. United States Environmental Protection Agency. http://www.epa.gov/cleanenergy/energy-resources/calculator.html

Kicherer A, Voeste D (2014) Sustainable solution steering, our approach to support our customers. https://www.basf.com/documents/corp/en/sustainability/management-and-instruments/sustainable-solution-steering/BASF_Booklet_Sustainable_Solution_Steering.pdf

Landsiedel R, Saling P (2002) Assessment of toxicological risks for life cycle assessment and eco-efficiency analysis. Int J Life Cycle Assess 7(5):261–268

Pilz H, Schweighofer J, Kletzer E (2005) The contribution of plastic products to resource efficiency. PlasticsEurope – Association of Plastics Manufacturers, Brussels

Pilz H, Brandt B, Fehringer R (2010) The impact of plastics on life cycle energy consumption and greenhouse gas emissions in Europe, Summary report June

PlasticsEurope (2011) Eco-profiles and environmental declarations – LCI methodology and PCR for uncompounded polymer resins and reactive polymer precursor (version 2.0, April 2011). www.plasticseurope.org

Saling P, Kicherer A, Dittrich-Kraemer B, Wittlinger R, Zombik W, Schmidt I, Schrott W, Schmidt S (2002) Eco-efficiency analysis by BASF: the method. Int J Life Cycle Assess 7(4):203

Saling P, Maisch R, Silvani M, König N (2005) Assessing the environmental-hazard potential for life cycle assessment, eco-efficiency and SEEbalance. Int J Life Cycle Assess 10(5):364–371

TfS (2012) Brochure. http://www.tfs-initiative.com/download/TfS_Brochure_en.pdf

TfS (2015) Overview. http://www.tfs-initiative.com/benefits_and_criteria.html

Uhlman BW, Saling P (2010) Measuring and communicating sustainability through eco-efficiency analysis. Chem Eng Prog, CEPE 106(12):17–26

UNFCCC (2009) United Nations Framework Convention on Climate Change. In: UNFCCC conference, Copenhagen, Dec 2009

WBCSD (2015) Action 2020 overview. http://www.wbcsd.org/Pages/EDocument/EDocumentDetails.aspx?ID=16154&NoSearchContextKey=true

Chapter 11
Sustainable Value Creation with Life Cycle Management

B.M. Krishna Manda, Henk Bosch, and Ernst Worrell

Abstract Life cycle management has gained traction in the last decades. However, even today it is not yet implemented in all companies due to lack of the connection between sustainability and value creation. In fact, managers are pressed to deliver value, and their performance is measured on how well they deliver the value. In this chapter the authors contribute to bridging the gap between sustainability science and business management by application of life cycle assessment (LCA) in corporate sustainability and aligning it with business activities/functions and value creation. They illustrate the context of corporations, sustainable value creation opportunities and the role of different business functions in integrating sustainability in the core business. Two cases demonstrate how business functions can use LCA-based insights for business decisions and how they are directly connected with value creation opportunities.

Keywords Corporate sustainability • Life cycle assessment • Life cycle costing • Life cycle management • LCM toolbox • Product sustainability • Social life cycle assessment • Sustainability • Value creation

1 Introduction and Objective

The relevance of the various sustainability aspects differs from company to company depending on the context, the type of product systems, geographical scope, and related social and environmental problems/drivers. Thus, the integration of sustainability in business is difficult and inherently complex. It requires a

B.M.K. Manda (✉) • E. Worrell
Copernicus Institute of Sustainable Development, Utrecht University, Utrecht, The Netherlands
e-mail: b.m.k.manda@uu.nl

H. Bosch
DSM Nutritional Products, Kaiseraugst, Switzerland

G. Sonnemann, M. Margni (eds.), *Life Cycle Management*,
LCA Compendium – The Complete World of Life Cycle Assessment,
DOI 10.1007/978-94-017-7221-1_11

holistic understanding of the interdependence of industrial systems. To this end, sound tools are needed that can capture the complexity and provide metrics to embed sustainability in different business decisions. Systems thinking helps to understand the different parts within the system and their relation to other systems. Life cycle assessment (LCA) is a systems analysis tool that can assess and help improve the environmental performance (one of the three pillars of sustainability) of products and processes by providing powerful insights into the whole value chain (ISO 2006a, b; ILCD 2010). By doing this, LCA provides an understanding that allows avoiding shifting impacts from one process step/industry to another, from one impact category to another and from one place to another. LCA supports businesses in making various decisions such as the selection of processes, materials, and supply chains. By supporting these business decisions and actions, LCA can offer value creation opportunities to business and improves shareholder and stakeholder value simultaneously. Similarly, other tools such as Life Cycle Costing (LCC) and Social Life Cycle Assessment (Social-LCA) can be used to understand the economic costs (externalities can also be included) and social impacts and risks (both positive and negative) throughout product life cycles. The authors focus on environmental LCA and combine other tools wherever possible.

LCA has been applied in companies and in public policy making. When applied in companies, LCA has often been seen as a mere auxiliary technical tool and the insights were limited to the impact quantification, which is the major strength of the tool, without actively involving business functions. Many companies are not implementing LCA in their day-to-day business due to its resource-intensive nature, complexity and the difficulty of contextualizing the relevance of LCA for the circumstances of companies. There has hardly been any exploration of how LCA can offer advice to existing corporate structures through decision support of business functions. And most importantly, the insights of LCA have by far not been fully exploited for the potential value creation opportunities in companies. There is little research to understand the role of LCA in supporting business functions (Sandin et al. 2014) and consequently linking it with sustainable value creation opportunities (UNEP/SETAC 2009; Rebitzer and Buxmann 2005; Gloria et al. 2014).

The objective of this chapter is to bridge the gap between sustainability science and business management by contextualizing the application of LCA in corporate sustainability and aligning it with business activities/functions and business priorities (value creation). This alignment can mainstream and advance the implementation of LCM in business.

In order to fulfill the above mentioned objective, the authors explain the context of corporations, the opportunities for value creation, and the role of different functions in integrating sustainability in day-to-day business. Case studies show how LCA can be contextualized in business and connected to value creation opportunities. Based on case studies, the authors offer an iterative procedure to conduct LCAs and create sustainable value.

2 Background and Literature Review

2.1 *Context of Corporations and Products*

Corporations are among the main actors which can profoundly influence sustainability through their products and services that span across different locations through their supply chains and markets. There are broadly five forces requiring corporations to improve their sustainability performance more than ever before. These are megatrends (environmental, social, demographic), regulatory pressure, stakeholder pressure, supply chain risks and competitive pressures (Manda 2014). The developments underlying the megatrends are population growth and rising disposable income, increasing urbanization, growing share of elderly population, climate change, water scarcity, bio-diversity loss, resource scarcity, poverty and inequity (UN 2012; GSSD 2014; WWF 2012; UN DES 2013; Rockstrom et al. 2009). The regulations on corporate and product sustainability, emission standards and trading schemes are growing in many countries and regions (e.g. the USA, EU, China and India) (US-EPA 2014; World bank 2014; EC 2014; EDF and IETA 2013). The number and activity of global NGOs targeting the working standards among suppliers and the pollution they are causing is increasing year by year (O'Rourke 2005; Economist 2014; Jun 2014). Consequently, the interest of investors in sustainability aspects of corporations is growing. Companies are trying to reduce risks, reduce costs of scarce resources, and develop new products that can improve their sustainability performance and provide competitive advantage in the market.

Despite these pressures, managers in companies are still pressed to deliver value, and their performance is measured on how well they deliver the value. Therefore, managers often face the challenge of addressing stakeholder concerns in day-to-day business while simultaneously improving value and thereby financial performance of companies (Hart and Milstein 2003).

2.2 *Opportunities for Sustainable Value Creation*

It was found that the improved environmental and social performance of companies can have a positive impact on the financial performance through reduced costs, improved revenues, and avoidance of risks (Epstein 1996; Eccles et al. 2012; Hart and Milstein 2003). For example, process improvements could lower energy and water usage and save operational costs (Worrell et al. 2003); and improved raw material utilization not only decreases raw materials costs but also reduces costs for handling and disposal of waste while simultaneously reducing the environmental footprint. There are several risks that can be avoided by sustainability performance improvements (Koplin et al. 2007). Increased scarcity of raw inputs such as water can lead to disruption of operations, i.e. lost production activity, which will impact the revenue earning capacity. Companies have to increasingly pay higher fines for violations, they need to compensate wrongdoings, and need to earn the license to operate from the local communities by avoiding negative impacts. These are called regulatory and legal risks. There are possibilities for damaging corporate reputation, i.e. reputational risks, from media and NGO campaigns for not meeting stakeholder expectations such as workers' health and safety and labor practices, and safe living environments for local communities. Market and product risks can also occur when customers move to other products with better sustainability performance or when governments and organizations impose sustainable procurement policies.

On the other hand, there are several value creation opportunities for companies with superior sustainability performance for each risk category mentioned above. It is possible to obtain additional revenues from environmentally and socially superior products through a premium. Moreover, high sustainability performance of companies can positively influence the desire of customers to buy their products (brand image), the desire of employees to work for them (preferred employer) and the desire of investors for providing long-term capital (blue chip status or good rank in indexes such as Dow Jones Sustainability Index). Companies with superior sustainability performance can differentiate their products in the market against competitors to attract new customers and, consequently, create a competitive advantage. Business to Business (B2B) companies can help their customers, i.e. end-producers, to meet their sustainability goals by supplying superior intermediate products. In essence, the existence of a company or its profit making capacity can be affected by several ways described above through various risks and opportunities created by sustainability performance and stakeholder reactions. This shows the overlap between shareholder and stakeholder value which are interdependent and interrelated.

2.3 Business Functions/Activities

Businesses deliver products and services through the co-ordination of various activities. The main activities are innovation, business development, procurement/sourcing, marketing and sales, and production/operations. Figure 11.1 shows business functions and their contribution to sustainability in a company. Every activity of the organization has an influence on its sustainability performance. Therefore, integration of sustainability in business requires understanding of various business functions that deliver specific business activities in a company, and proper metrics should be developed to support decision-making.

The main activities of innovation are developing new products with novel, additional or improved functionality, improving existing products (e.g. by implementation of alternative process/manufacturing routes), and developing new processes to recycle waste. Business development is mainly involved in finding new applications or opportunities in different markets through collaboration with existing or prospective customers or by means of strategic alliances with other companies. It also defines the final product and positioning strategy and finds opportunities for the

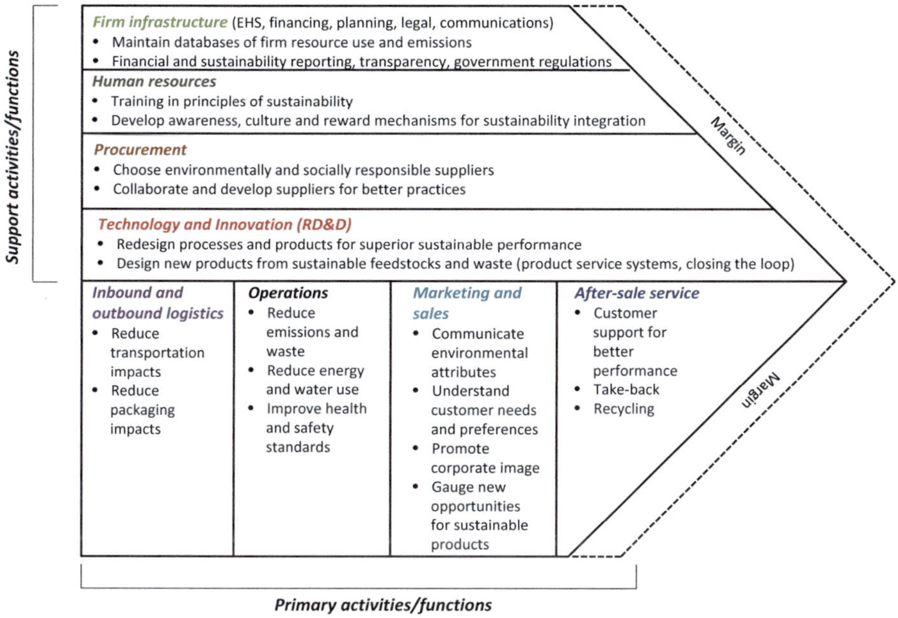

Fig. 11.1 Value chain of a company with business functions and relation to sustainability (Manda 2014) (Adapted from Porter (1985), Epstein and Roy (2001) and Porter and Kramer (2006))

development of new business models. In some companies business development is part of innovation.

Procurement/sourcing is involved in finding the right suppliers who can provide quality inputs at the right time. It also negotiates pricing and keeps track of the production timeline. Sourcing also conducts supplier environmental and social audits and deals with supplier certifications. Sourcing is responsible for avoiding any risks arising from supplier practices, including compliance with local laws and child labor. It can go further by not only looking at sustainability performance of suppliers but also at the influence of purchased items in the final product life cycle of customers. This would help close the information cycle between suppliers and final customers and may bring collaborative opportunities for future sustainable business development.

Marketing and sales is responsible for communicating the product attributes and differentiating their products against those of their competitors. It conducts market research to find opportunities for market growth and expansion to new markets. For Business to Consumers (B2C) companies, it also handles communication with consumers. Marketing conducts research to acquire the consumer intelligence related to how customers/consumers make trade-offs between price, environmental performance and other quality parameters.

To improve sustainability performance of processes and products, all these business functions need concrete guidance and insights on the product's impacts throughout its life cycle. This calls for sound tools to support business functions in decision-making.

3 Methodology

We use a framework which captures all the important elements presented in the above sections. It provides the applications of LCA in different contexts of companies for various purposes. This framework provides guidance for the description of case studies. A case study approach is recommended to investigate the value creation with sustainability (Reed 2001). Thus, the descriptive case study approach (Scapens 1990; Jupp 2006) was employed with two chemical companies. This enabled us to investigate the phenomenon of value creation with sustainability, especially LCA, in real-life contexts. The main elements of case studies are the drivers, the communication, embedding sustainability in business decisions, and sustainable value creation. Each case study focused on a specific example to study the similarities and differences of approaches between different companies with different examples. Multiple case study approach can confirm the reproducibility of the proposed approach (theory) and its practical implementation in real life situations. Therefore, a multiple case study approach is employed. This can provide stronger evidence than a single case study with regard to creation of value with LCA and its implementation in companies (Eisenhardt and Graebner 2007).

3.1 Description of Sustainable Value Framework for Chemical Industry

In order to integrate sustainability in the core business, functional managers, top management and investors need to be convinced of value creation from sustainability. The sustainable value framework, see Fig. 11.2, helps to link sustainability and value creation.

Figure 11.2 shows the multidimensional construct of the sustainable value framework that blends two dimensions. The vertical axis is about balancing the short term results such as financial improvements while making progress for the long term business growth and success in a competitive and disruptive technological landscape. The horizontal axis shows the firms' need to manage and grow internal organizational skills and capabilities while capturing the new perspectives, knowledge, and challenges posed by external stakeholders. Strategies on the left side are within the purview of a corporation, and the strategies on the right side are mostly driven by other stakeholders or changing environmental, social, and market circumstances. There are four sets of sustainability drivers that are also found in literature (Lozano 2013; Epstein and Roy 2001; UNEP/TU Delft 2007). The first set of drivers (lower-left quadrant) is pollution, waste and higher productivity, the second set of drivers (lower-right) is transparency and regulations, the third set of drivers (top-left) is growing environmental footprint, resource depletion, higher worker income and employability, the fourth set of drivers (top-right) is megatrends such as population, inequality and planetary boundaries. Addressing these drivers with a proper strategy

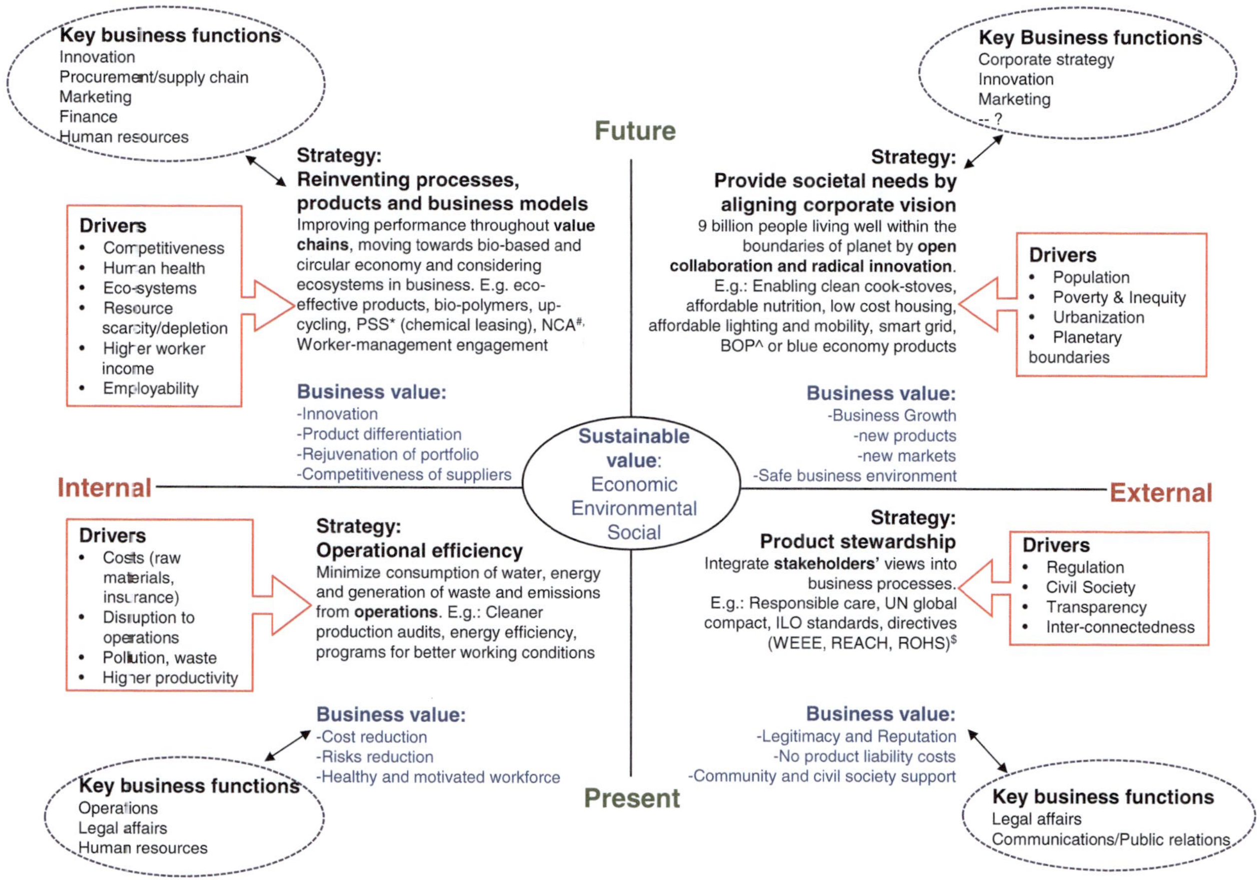

Fig. 11.2 Sustainable value framework (Adapted from Hart and Milstein (2003) (Manda 2014)). (*PSS is Product Service Systems, #NCA is Natural Capital Accounting, ^BOP is Base Of the Pyramid, $REACH is Registration, Evaluation, Authorization of Chemicals and ROHS is Restriction Of Hazardous Substances, WEEE is a directive on Waste Electrical and Electronic Equipment)

could lead to business value. For example, an operational efficiency strategy can address drivers such as raw material costs and pollution and can create business value such as reducing costs, avoiding risks and motivated workforce due to better working conditions.

Though the environmental drivers in top-left and top-right appear similar, their influence on businesses is different. In the top-left quadrant, businesses deal with the resource constraints and product impacts with a piecemeal approach. Whereas in the top-right quadrant businesses are aware of the limited capacity of our planet and its influence on their long term existence and, thus, openly collaborate with others to radically transform whole systems for planetary sustainability. Implementing strategies in each dimension needs the involvement of several specific corporate functions in order to reap the business value. In order to be successful in the short and long term, companies need to devise strategies in all quadrants to create sustainable value.

The ovals in Fig. 11.2 show various business functions that either take part in the implementation or benefit from the corresponding strategies. Depending on the structure of the company there might be differences in involvement by specific business functions in each quadrant. The environmental management or sustainability departments are not separately shown because they either implement the projects or they are involved in all quadrants depending on the type of project and they can also be part of operations, innovations or marketing in many companies. The role of various business functions is illustrated with real life case studies in Sect. 4 results and discussion. For a detailed description of framework with LCA examples from all quadrants, please refer to Manda (2014).

4 Results and Discussion

In this section, two case examples are described with two main aspects. First, the context of companies is provided in terms of different sustainability challenges, drivers and market situation such as customer needs and demands. Second, it is delineated how LCA insights and consequent business decisions are translated into value creation for the company. The two case studies mainly belong to the top-left quadrant of the framework shown in Fig. 11.2, i.e. reinventing processes, products and business models. The case study on vitamin C also has aspects from the lower-left quadrant, operational efficiency. Further, an implementation procedure is presented to translate LCA insights into value creation opportunities for business.

4.1 Case Study of Spun-Dyed Modal Fibers from Lenzing AG

The spin-dyeing case study is conducted for Lenzing, a bio-based fiber producer based in Austria. This company wants to further strengthen its position as a sustainable fiber producer by providing solutions to the challenges faced by textile industry and customers. Textile wet processing, i.e. dyeing and finishing, is known to cause environmental impacts such as water pollution. Spun-dyed fibers are developed against this background since these fibers avoid the conventional dyeing process of fabrics. A detailed LCA for the environmental impacts of conventionally dyed fabrics and spun-dyed fabrics made of wood-based modal fibers can be found in literature (Terinte et al. 2014).

4.1.1 Context and Drivers of Spun-Dyed Modal Fabrics

Figure 11.3 shows the relevant context and drivers of the spun-dyed fiber innovation. The most important driver is global population growth, which creates more demand for clothes with lower environmental impacts. The textile supply chains are mostly located in Asia where air pollution, water pollution and water scarcity problems are rampant. There is an increasing number of legislations targeting the textile industry in Asia (in order to raise the standards of practices), from the EU (e.g. REACH) to prohibit certain dyes and from selected European countries like Sweden that is planning to impose taxes on chemicals in textiles. International NGOs such as Greenpeace have criticized international apparel brands and retailers for the poor performance of their supply chain partners such as laundries and dyeing mills. The business context of the textile industry is also changing due to the initiatives such as Sustainable Apparel Coalition (SAC) that creates tools and awareness to reduce textile supply chain impacts. These initiatives and visibility to NGOs and consumers have created impetus for brands and retailers to be more sustainability oriented. Consequently, the sustainability agenda of B2B companies such as fiber producers are primarily driven by customer (brands and retailers) demands rather than consumers and NGOs. Increased scarcity raises the costs of water and power in Asian countries, increasing production costs. Since most of the textile industry has moved to Asia, EU producers need to find innovative ways to be competitive in the global market. All these drivers are important for the business functions of the fiber producer because they influence business decisions and help to contextualize the spin-dyeing innovation.

With these issues in mind, the LCA on spun-dyed fabrics can help business development and marketing.

4.1.2 Results of the Spun-Dyed Fabrics LCA Study

From the LCA it was found that, across all impact categories studied, spun-dyed fabrics cause only half to one third of the environmental impacts compared to conventionally dyed fabrics. Sensitivity analyses showed that the relative benefits of spin-dyeing, that nowadays takes places in Austria, are higher if it replaces conventional dyeing in countries like China or the USA where electricity grids and heat sources are predominantly coal-based. The liquor ratio and the number of washing cycles influence the results but do not alter the conclusions. Due to the comparatively very

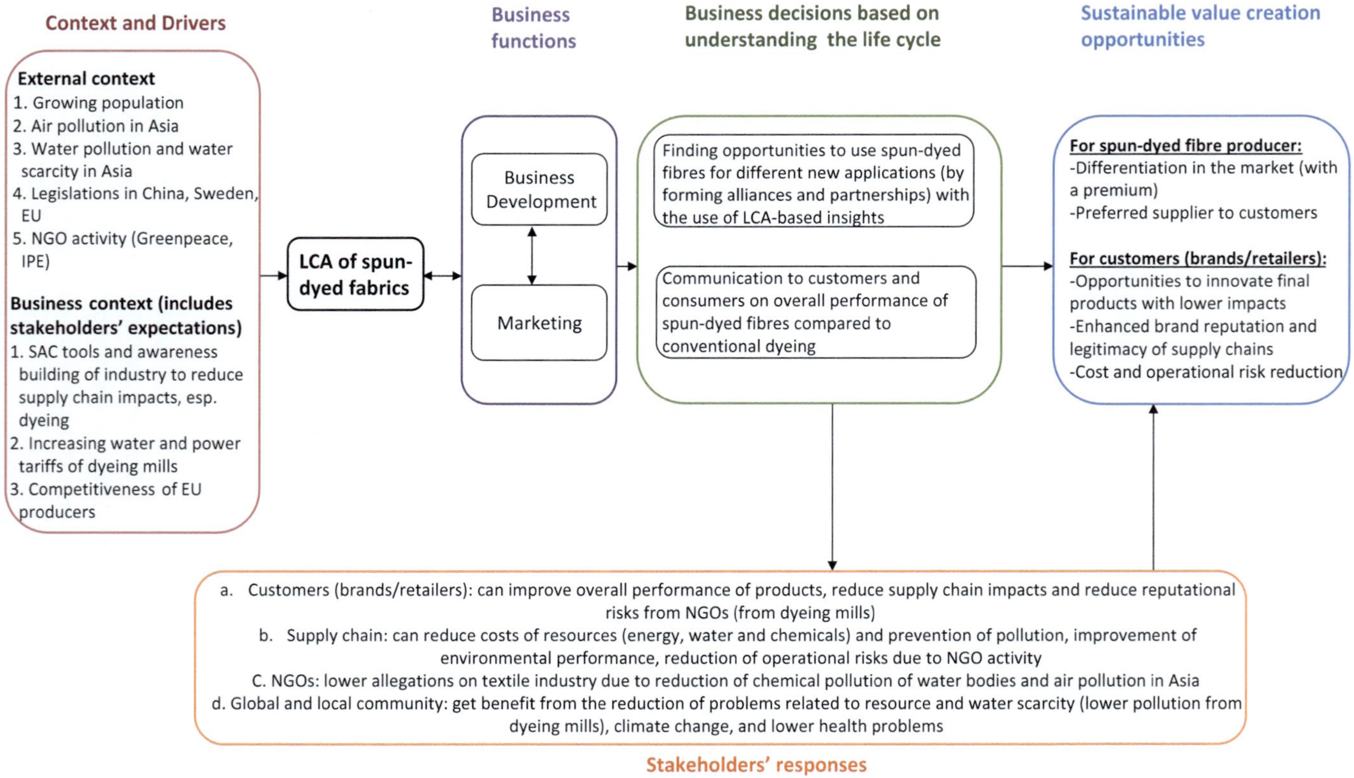

Fig. 11.3 LCA-based value creation approach for spun-dyed fabrics (relation between context/drivers and LCA of spun-dyed fibers and its support in business decision-making, key stakeholder responses to improved environmental performance and resulting value creation opportunities) (Manda 2014)

low amount of pigments required and entrapment of the pigment in the fiber structure, the spun-dyed fabric can be expected to cause substantially lower human and eco-toxicity impacts compared to conventionally dyed fabric. From a gate (fiber)-to-gate (dyed fabric) comparison, it was found that the spun-dyed fabrics need 60–90 % less water, chemicals, electricity and heat and generate 60 % lower wastewater emissions than conventionally dyed fabrics. These savings can reduce production costs. All technical details of the spun-dyed fabric LCA can be found in Terinte et al. (2014).

4.1.3 Business Decisions Supported by LCA

The business development department is reaching out to potential new customers to make new products/applications, such as fabrics for automobiles, based on the environmental performance of the spun-dyed fibers compared to conventional dyeing of different materials. The marketing department has been using the insights provided by LCA to communicate to textile brands and retailers. These business functions decided to use sustainability as a value proposition on par and without compromising other aspects like quality, functionality and price. By this means, sustainability is integrated into day-to-day business.

4.1.4 Stakeholders' Responses and Sustainable Value Creation

The innovative fiber production process triggered positive responses from various stakeholders. Due to the improved performance of the supply chains, there would be fewer allegations from NGOs against supply chain partners and thus lower operational risks. The customers (brand and retailers) can improve the overall performance of their products and they can reduce the reputational risks from supply chain partners such as dyeing mills. This would help brands and retailers to create value such as innovative products (garments) with lower impacts, enhanced brand reputation, and legitimacy (The Cambodia Daily 2014). The spin-dyeing needs fewer resources than conventional dyeing and thus reduces the variable production costs of final products. The local community can benefit from these improvements due to lower water pollution and reduced pressure on water resources. Owing to all the above aspects, the fiber producer could differentiate the spun-dyed fibers in the market and be a preferred supplier to brands and retailers.

This LCA was conducted during the business development stage. The innovation was accomplished before the LCA study was started. This LCA was used to quantify the environmental improvements which will be useful for developing new applications and attract new customers. Therefore the innovation department of fiber producer participated in the LCA to provide data and necessary information.

4.2 Case Study of Vitamin C Business Development by Royal DSM NV

Vitamin C, or L-ascorbic acid, was discovered as the nutritional factor lacking in the diets of sailors suffering from scurvy. It naturally occurs in fresh fruit and vegetables. The first industrial preparation of L-ascorbic acid was developed by Tadeus Reichstein for F. Hoffmann-La Roche in 1934. DSM acquired this business from Roche in 2003. Figure 11.4 shows the overall context, LCA-based insights, supported decisions and value creation opportunities of Vitamin C product.

4.2.1 Context and Drivers of Vitamin C

The world's population is still rapidly growing and expected to peak at nine billion people around 2050. Presently large groups of people in poor countries already suffer from malnutrition. Even if they have sufficient energy in their diets, these are deficient in micronutrients, just like sailors' diets did. Even in wealthy countries there are deficiencies of micronutrients, and people concerned about their health take precautionary supplements of micronutrients. Hence there is a large and growing demand for micronutrients, including Vitamin C. All over the world, even in developing countries, concerns about food quality and food safety are increasing, so food additives and supplements have to comply with high standards of quality and safety. Finally there are concerns about the environmental impact, including land use, of the activities required to feed the growing population. These concerns also lead to questions about the contribution of additives and supplements to food (environmental) footprint. There is strong competition and price pressure from China in the vitamins market. DSM has the only non-Chinese production site for Vitamin C in Dalry, Scotland.

The DSM Quality for Life™ program symbolizes quality, reliability and traceability. Quality for Life™ also means sustainability.

As part of the program DSM licenses the Quali-C® trademark to customers for positioning their consumer products. Quality for Life™ and the Quali-C® brand are integral parts of DSMs value propositions.

4.2.2 Results of the Vitamin C LCA Study

DSM carried out an LCA to determine the carbon footprint as a pilot in the UK carbon trust program. One of the key results was that energy contributes more than 50 % to the footprint of the product, even with the very low energy footprint, produced in a combined heat and power plant fired with natural gas. This high energy consumption was related to the production of the intermediate 2-ketogulonic acid (KGA) in an oxidation reaction. The oxidation can only be executed selectively if groups that should not be oxidized are protected by addition of acetone, which has

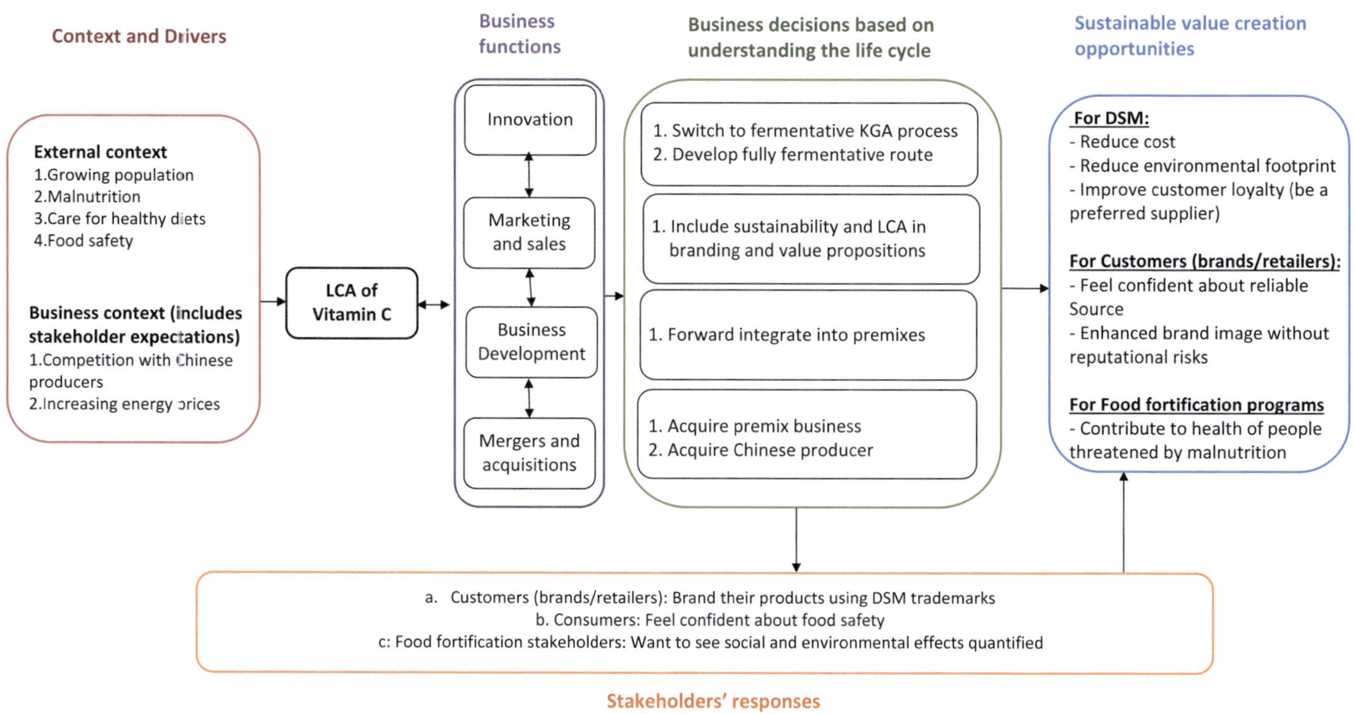

Fig. 11.4 LCA-based value creation approach for Vitamin C products (relation between context/drivers and LCA of Vitamin C and its support in business decision-making, key stakeholder responses to improved environmental performance and resulting value creation opportunities)

to be removed and recovered after the oxidation. In addition to the large energy requirement, this complicated procedure requires chemicals and produces waste.

DSM did not carry out a social LCA. The potential benefits to people taking vitamins as supplements or additives are very well known to customers and consumers, so there is no need to emphasize these. Production, starting from agriculture and fuel generation is completely European based, so there is little risk of social issues in the value chain.

4.2.3 Business Decisions Supported by LCA

Because of the cost and the high footprint of the KGA production DSM switched to the fermentative production of KGA. A complete fermentative route to further reduce cost and footprint is also considered. DSM has recently forwardly integrated into premix companies, who supply premixes of additives to the food and feed industry. This allows to actively influence the footprint of vitamin additives used. For strategic reasons DSM is acquiring a Chinese production site. The LCA was extended with an assessment of other environmental impacts and used to check the footprint of the Chinese product, and will be used to identify and assess initiatives to reduce the footprint.

4.2.4 Stakeholders' Responses and Sustainable Value Creation

Customers respond positively to Quality for Life™. They prefer products from a reputable and reliable source, with an undisputable track record in food safety, environmental performance and sustainability image. All these contribute to reducing their operational risks. This is valid in the western world, but certainly also in Asia.

Particularly in food fortification programs within the area of malnutrition, social LCA makes the benefits more tangible, and DSM is piloting this use. In addition, stakeholders in these programs are interested in environmental impacts, including those of small packaging sizes and distribution.

4.3 Implementation Procedure for Business Value Creation Based on Life Cycle Assessment in Companies

UNEP/SETAC has proposed the application of life cycle management (LCM) capability maturity model for developing the capacity of small and medium sized companies to achieve their sustainability goals (UNEP/SETAC 2009). The authors propose an implementation procedure for business value creation based on the insights gained from LCA studies and its integration in business functions (see also UNEP/TU Delft 2007; UNEP/SETAC 2009). This is an iterative procedure of various steps as shown in Fig. 11.5.

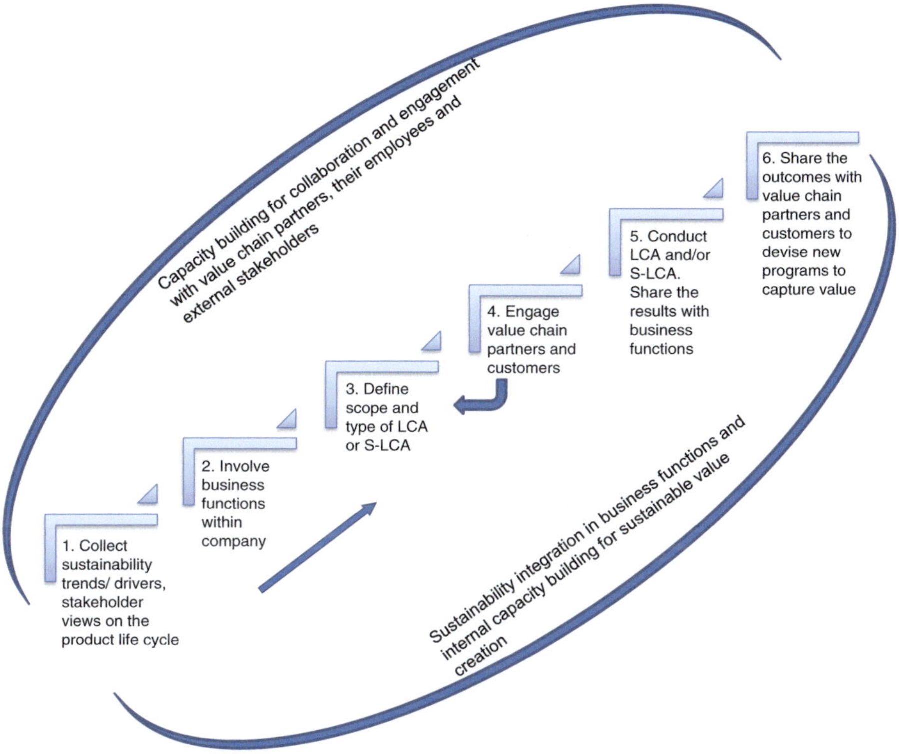

Fig. 11.5 Iterative procedure for translating insights from the LCA and S-LCA into value creation adapted from Manda et al. (2014)

The steps of the iterative procedure are:

1. Understand the context of the product by collecting information on various drivers and stakeholder views/concerns
2. Involve the relevant business functions within the company
3. Define the goal, scope and type of LCA and/or S-LCA
4. Engage the value chain companies and customers
5. Discuss the results of conducting LCA and/or S-LCA internally
6. Share the outcomes of the process with value chain companies and customers to capture value

4.4 Limitations of the Suggested Approach and Life Cycle Management

While implementing the suggested approach, practitioners would face some barriers and challenges.

We acknowledge the limitations of the LCA approach in terms of the development of methods for impacts such as biodiversity and toxicity, lack of inventory, uncertainties in data and methods, limited guidance on allocation for different product categories, etc. (Finnveden et al. 2009). Resource requirements are also acting as limitation for implementing LCA. Companies can also create value with customer experience, products with superior aesthetics, feel, etc. These aspects might not be captured by LCA, if its unit of analysis, i.e. the functional unit, cannot take these aspects into account. LCA is not easily able to capture the perception and preferences of people, taste, and emotional value attached to products, hence it cannot help companies to create value in terms of these aspects.

5 Conclusions

The main contribution of this chapter to the field of sustainability, especially corporate and product sustainability, is twofold. First, the authors have presented an approach to embed LCA in the corporate context and align environmental sustainability with business priorities. Second, they have shown how to synthesize and communicate all the different aspects in a coherent way, presenting LCA results to the managers of different business functions using an LCA-based value creation approach. This communication approach is a combination of "drivers-LCA metrics business decisions-stakeholder responses-value creation opportunities". There could be different barriers and challenges while implementing the presented approach. The barriers and challenges need further attention in future research. The integration of social aspects in the framework and in the case studies can be further improved. Matured approaches for social-LCA to address this issue are needed. The approach shown here is applicable to all other LCM tools such as S-LCA and LCC. Additional case studies can demonstrate the integration of the LCM tools using the approach described.

References

EC (2014) Non-financial reporting. http://ec.europa.eu/internal_market/accounting/nonfinancial_reporting/index_en.htm. Accessed 13 Aug 2014

Eccles RG, Ioannou I, Serafeim G (2012) The impact of a corporate culture of sustainability on corporate behavior and performance. Working paper, 12–035, 9 May 2012, Harvard Business School

Economist (2014) Enter the Chinese NGO: the Communist Party is giving more freedom to a revolutionary idea. http://www.economist.com/news/leaders/21600683-communist-partygiving-more-freedom-revolutionary-idea-enter-chinese-ngo. Accessed 11 June 2014

EDF and IETA (2013) The world's carbon markets: a case study guide to emissions trading. India case study. http://www.ieta.org/assets/Reports/EmissionsTradingAroundTheWorld/edf_ieta_india_case_study_may_2013.pdf

Eisenhardt KM, Graebner ME (2007) Theory building from cases: opportunities and challenges. Acad Manage J 50:25–32

Epstein MJ (1996) Measuring corporate environmental performance: best practices for costing and managing an effective environmental strategy. Institute of Management Accountants/Irwin Professional Publishing, Burr Ridge

Epstein MJ, Roy MJ (2001) Sustainability in action: identifying and measuring the key performance drivers. Long Range Plann 34(5):585–604

Finnveden G, Hauschild MZ, Ekvall T, Guinée JB, Heijungs R, Hellweg S, Koehler A, Pennington D, Suh S (2009) Recent developments in life cycle assessment. J Environ Manage 91:1–21

Gloria TP, Kohlsaat C, Bautil P, Wolf B, Early D, Ben-Zekry B (2014) A statistical approach to interpret relative environmental performance within product categories. Int J Life Cycle Assess 19:491–499

GSSD (2014) Global system for sustainable development at MIT. http://gssd.mit.edu/knowledge-system/guide-core-concepts/urbanization/urbanization-graphic. Accessed 12 June 2014

Hart SL, Milstein MB (2003) Creating sustainable value. Acad Manag Exec 17(2)

ILCD (2010) International reference life cycle data system (ILCD) handbook – general guide for life cycle assessment – detailed guidance. European Commission – Joint Research Centre – Institute for Environment and Sustainability, 1st edn. Publications Office of the European Union, Luxembourg

ISO 14040 (2006a) Environmental management – life cycle assessment – principles and framework. Geneva

ISO 14044 (2006b) Environmental management – life cycle assessment – requirements and guidelines. Geneva

Jun M (2014) Transparency and green choice. A presentation at sustainable apparel coalition half yearly meeting in Vietnam, May 2014

Jupp V (ed) (2006) The SAGE dictionary of social research methods. Sage Publications, London

Koplin J, Seuring S, Mesterharm M (2007) Incorporating sustainability into supply management in the automotive industry – the case of the Volkswagen AG. J Clean Prod 15:1053–1062

Lozano R (2013) A holistic perspective on corporate sustainability drivers. Corp Soc Responsib Environ Mgmt 2013. doi:10.1002/csr.1325

Manda BMK (2014) Application of life cycle assessment for corporate sustainability: integrating environmental sustainability in business for value creation. Dissertation, Utrecht University

O'Rourke D (2005) Market movements: nongovernmental organization strategies to influence global production and consumption. J Ind Ecol 9(1–2):115–128

Porter ME (1985) Competitive advantage: creating and sustaining superior performance. The Free Press, New York

Porter ME, Kramer MR (2006) Strategy and society: the link between competitive advantage and corporate social responsibility. Harv Bus Rev, Dec 2006

Rebitzer G, Buxmann K (2005) The role and implementation of LCA within life cycle management at Alcan. J Clean Prod 23(13):1327–1335

Reed DJ (2001) Stalking the elusive business case for corporate sustainability. Sustainable enterprise perspectives. World Resources Institute, 24 pp

Rockström J, Steffen W, Noone K, Persson Å, Stuart III Chapin F, Lambin E, Lenton TM, Scheffer M, Folke C, Schellnhuber HJ, Nykvist B, de Wit CA, Hughes T, van der Leeuw S, Rodhe H, Sörlin S, Snyder PK, Costanza R, Svedin U, Falkenmark M, Karlberg L, Corell RW, Fabry VJ, Hansen J, Walker B, Liverman D, Richardson K, Crutzen P, Foley J (2009) Planetary boundaries: exploring the safe operating space for humanity. Ecol Soc 14(2):32, http://www.ecologyandsociety.org/vol14/iss2/art32/

Sandin G, Clancy G, Heimersson S, Peters GM, Ten Svanstrom M, Hoeve M (2014) Making the most of LCA in technical inter-organisational R&D projects. J Clean Prod 70:97–104

Scapens RW (1990) Researching management accounting practice: the role of case study methods. Br Account Rev 22:259–281

Terinte N, Manda BMK, Taylor J, Schuster KC, Patel MK (2014) Environmental assessment of coloured fabrics and opportunities for value creation: spin-dyeing versus conventional dyeing of modal fabrics. J Clean Prod 72:127–138, http://dx.doi.org/10.1016/j.jclepro.2014.02.002

The Cambodia Daily (2014) Garment workers begin boycott of overtime. http://www.cambodiadaily.com/archives/garment-workers-begin-boycott-of-overtime-53140/. Accessed 12 June 2014

UN (2012) World population prospects: the 2012 revision. http://esa.un.org/unpd/wpp/unpp/panel_population.htm. Accessed 12 June 2014

UN DES (2013) World economic and social survey 2013: sustainable development challenges. United Nations, Department of Economic and Social Affairs, New York

UNEP/SETAC (2009) Life cycle management: how business uses it to decrease footprint, create opportunities and make value chains more sustainable, UNEP/SETAC 2009. http://www.unep.fr/shared/publications/pdf/DTIx1208xPALifeCycleApproach-Howbusinessusesit.pdf

UNEP/TU Delft (2007) Design for sustainability: a practical approach for developing economies. http://apps.unep.org/publications/pmtdocuments/Design_for_sustainability.pdf

US EPA (2014) Carbon pollution standards. http://www2.epa.gov/carbon-pollution-standards. Accessed 1 June 2014

World Bank (2014) State and trends of carbon pricing. http://documents.worldbank.org/curated/en/2014/05/19572833/state-trends-carbon-pricing-2014

Worrell E, Laitner JA, Ruth M, Finman H (2003) Productivity benefits of industrial energy efficiency measures. Energy 1081

WWF (2012) Living planet report 2012

Part III
Life Cycle Management as Part of Sustainable Consumption and Production Strategies and Policies

This part broadens the scope of life cycle management and presents it as part of sustainable consumption and production, discussing strategic opportunities for policy action and related responsibilities of consumers and policy makers, among other stakeholders, along the value chain.

Chapter 12
Hotspots Analysis: Providing the Focus for Action

Mark Barthel, James A. Fava, Curtis A. Harnanan, Philip Strothmann, Sofia Khan, and Simon Miller

Abstract The technique 'hotspots analysis' belongs to the toolbox life cycle management. 'Hotspotting' or 'hotspots analysis' is an emergent technique being used in a growing number of different analytical disciplines, so research disciplines and functions within organizations (e.g., R&D, new product development, procurement), and in diverse geographies, in support of the green economy and the United Nations post-2015 Sustainable Development Goals. It can be used to inform government policy priorities, drive growth and innovation in business and empower citizens.

Due to the growing interest of various stakeholders in applying hotspot analysis methodologies, the UNEP/SETAC Life Cycle Initiative initiated the project "Global Principles and Practices for Hotspot Analysis". During the first phase of the project, a study was conducted to map existing hotspots analysis methodologies and studies world-wide, which culminated in the report: *Hotspots Analysis: mapping of existing methodologies, tools and guidance and initial recommendations for the development of global guidance*. This chapter draws on knowledge derived from this report and presents a number of relevant findings.

M. Barthel (✉)
Waste & Resources Action Programme (WRAP), Banbury, Oxon, UK
e-mail: Mark.Barthel@wrap.org.uk

J.A. Fava
thinkstep, West Chester, PA, USA

C.A. Harnanan
thinkstep, Ottawa, ON, Canada

P. Strothmann
FSLCI, Berlin, Germany

S. Khan
UNEP, Albany, NY, USA

S. Miller
3keel LLP, Oxford, UK

G. Sonnemann, M. Margni (eds.), *Life Cycle Management*,
LCA Compendium – The Complete World of Life Cycle Assessment,
DOI 10.1007/978-94-017-7221-1_12

Keywords Hotspotting • Hotspots analysis • Life cycle assessment • Life cycle management • Sustainability • UNEP/SETAC life cycle initiative

1 Introduction

The information-age has led to a proliferation of content, ranging from the assimilation and analytical challenges associated with 'big data' through to ever-increasing publication lists of research and innovation findings. The major challenge for businesses, policy-makers, academic researchers and consumers is deciding where and how to act to have the maximum impact. For any action a balance must be struck between speed of response and pragmatism and the need to be informed by reliable and trustworthy science-based evidence.

This prioritization method is called 'hotspotting' or 'hotspots analysis' and is an emergent technique being used in a growing number of different analytical disciplines.

This technique belongs to the toolbox life cycle management (LCM). LCM is the *application* of life cycle approach including LCA and related methods as SLCA and LCSA in business to drive business improvement. Life cycle management essentially embraces many applications of life cycle approaches, including product – as well as company-related approaches – to ensure that the full range of risks and opportunities are known and actions taken to reduce impacts across the value chain.

We believe there is now an opportunity to collate existing techniques from around the world to develop a proven methodology for hotspots analysis. This can be adopted by multiple research disciplines and functions within organizations (e.g., R&D, new product development, procurement) and in diverse geographies, in support of the green economy and the United Nations post-2015 Sustainable Development Goals (SDGs). It can be used to inform government policy priorities, drive growth and innovation in business and empower citizens.

Due to the growing interest of various stakeholders in applying hotspot analysis methodologies, the UNEP/SETAC Life Cycle Initiative decided in 2012 to focus its Flagship Project 3a on the topic under the title of "Global Principles and Practices for Hotspot Analysis". During the first phase of the project, a study was conducted to map existing hotspots analysis methodologies and studies world-wide, which culminated in the report: *Hotspots Analysis: mapping of existing methodologies, tools and guidance and initial recommendations for the development of global guidance*[1] (Barthel et al. 2014).

This chapter draws on knowledge derived from this report and presents a number of relevant findings.

---

[1] The report can be accessed at: http://lifecycleinitiative.org/wp-content/uploads/2014/10/Flagship3a-Hotspots-Mapping.pdf

2 What Is Hotspots Analysis?

Over the past few years, hotspots analysis has become a helpful and effective tool that assists in the identification of areas to be prioritized for action.

Hotspots analysis (HSA) is defined as a methodological framework that allows for the rapid assimilation and analysis of a range of information sources, including life cycle based studies, market, and scientific research, expert opinion and stakeholder concerns. The outputs from this analysis can then be used to identify potential solutions and prioritize actions around the most significant economic, environmental, ethical and social sustainability impacts or benefits associated with a specific country, city, industry sector, organization, product portfolio, product category or individual product or service. Hotspots analysis is often used as a pre-cursor to developing more detailed or granular sustainability information (Barthel et al. 2014).

The findings from hotspots analysis provide a comprehensive understanding of impacts. They also allow for the prioritization of resources and actions in countries, cities, industry sectors, product portfolios, product categories or individual products that really matter by virtue of their environmental, social and ethical impact profile and/or their physical trading volumes and economic value in the economy. In addition to streamlining research and analysis, a common feature of hotspots analysis is the presentation of information and findings in accessible formats, including for non-technical audiences, who are often the key decision-makers in policy and business settings.

The benefits of hotspots analysis include the following factors (Barthel et al. 2014):

- The rapid assimilation and analysis of multiple evidence threads leading to accessible outputs and a clearer understanding of the actions required to eliminate, reduce or mitigate identified hotspots
- A highly cost-effective approach to life cycle thinking and management across multiple impact categories and issues, sectors or product categories that is perhaps more suited to developing countries, emerging economies and SMEs trying to find an evidence-based focus for their actions
- The provision of both technical and non-technical information to decision-makers in government, business and civil society

2.1 *Typical Steps to Conduct Hotspots Analysis*

Hotspots analysis employs a materiality-focused prioritization approach to identify sustainability impacts across a range of attributes such as economic, environment, social and governance. The results from hotspots analysis typically allow decision-makers to identify sustainability impact improvement opportunities and prioritize impact reduction actions. Hotspots analysis methodologies often use research and expert inputs and stakeholder views to develop criteria and a mechanism for prioritization and interpretation of the outputs according to the scope and scale of the study.

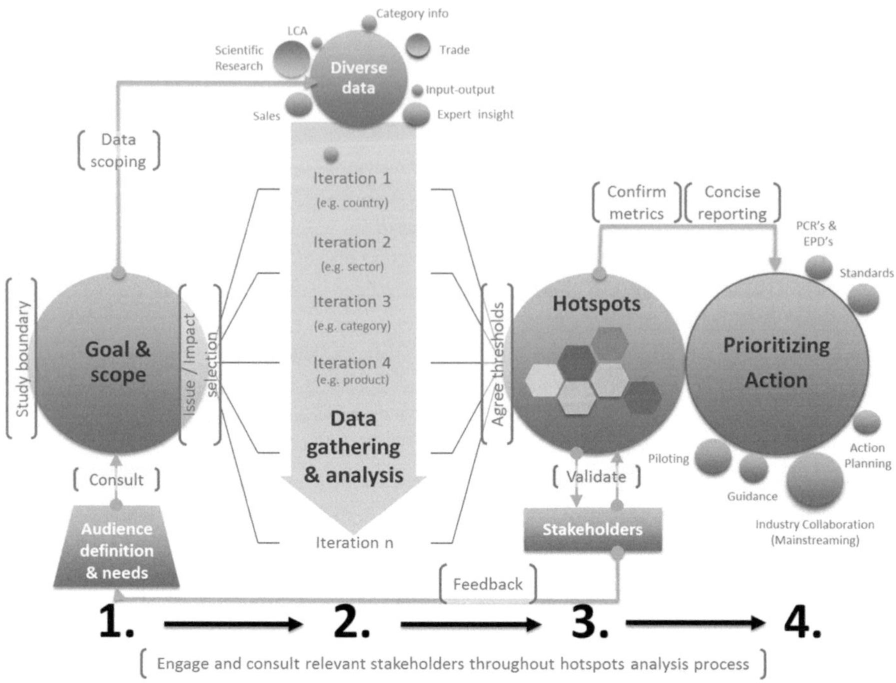

Fig. 12.1 Typical steps in a hotspots analysis methodology (Barthel et al. 2014)

The following steps illustrate what a typical hotspots analysis involves (see also Fig. 12.1).

1. *Goal and Scope definition*: The first step involves defining goal and scope by understanding the requirements of the hotspots analysis study. The next step involves stakeholder mapping and engagement to identify and gain consent on the study boundary, attributes, impact categories, identifying the ground rules of data mining and analysis and the need to use proxy data where this is required. In essence this step would include defining goal; scope and agreement on broad materiality and prioritization approach for the study.
2. *Data gathering, expert insights and analysis*: This step would include knowledge building through data mining, data analysis, data validation, expert interviews and stakeholder consultations. This step typically involves drawing together different evidence threads such as the findings from life cycle studies, input/output analysis data, scientific research studies, product information, sales volumes/economic value and trade information, depending on the scale and scope of the study (e.g. whether the analysis is being done at the national, city, sector or product-category-level).
3. *Hotspots identification and validation*: Once all the required data and knowledge is gathered, it is time to identify, develop, discuss and obtain agreement on materiality thresholds (i.e., when does an impact become a hotspot) and the

criteria to be used for ranking and prioritising hotspots for action, including a stakeholder validation process relating to identified hotspots. The next step includes the identification and prioritization of impact reduction opportunities, reviewing and validation of the identified hotspots by a larger audience and identifying implementation gaps and recommendations required to achieve impact reduction opportunities.

4. *Prioritising action*: The last step involves using the outputs from the study to achieve sustainability improvements. Typically, this step would involve action planning, development of industry guidance and standards, piloting or road-testing of potential solutions, industry collaborations and voluntary agreements, etc.; and further working with relevant stakeholders to disseminate and mainstream proven or effective solutions based on feedback from piloting activities.

2.2 Approaches of Hotspots Analysis

Most hotspots analysis use quantitative, qualitative or a combination of quantitative and qualitative approaches.

Quantitative approach typically uses quantitative data such as traditional life cycle assessment data, product data, sales and trade data, input–output data or material flows analysis or contextual market data for the study. Most hotspots analysis use some form of quantitative approach with stakeholders' involvement in the development process according to the study by (Barthel et al. 2014). Examples of hotspots analysis methodologies that use a quantitative approach include the *GHG Protocol's Value Chain (Scope 3) Accounting and Reporting Standard*[2]; and the *GHG Protocol's Product Life Cycle Accounting and Reporting Standard.*[3]

Qualitative approach typically involves securing access to extensive expert knowledge and professional judgment, gaining a better understanding of stakeholder's concerns and deeper stakeholder engagement as required. This approach has more emphases on qualitative elements. The Sustainability Accounting Standards Board (SASB) Materiality Map[4] is an example of hotspots analysis that uses a qualitative approach.

Combination approach would use both quantitative and qualitative measures to identify the hotspots. Such studies use quantitative data, expert opinions and stakeholder engagement to conduct hotspots analysis. Many of the existing hotspots analysis methodologies use a combination approach to identify sustainability impacts and improvement action as observed in a recent study (Barthel et al. 2014).

---

[2]A copy of the *GHG Protocol's Value Chain (Scope 3) Accounting and Reporting Standard* may be accessed at: http://www.ghgprotocol.org/standards/scope-3-standard

[3]A copy of the *GHG Protocol's Product Life Cycle Accounting and Reporting Standard* may be accessed at: http://www.ghgprotocol.org/standards/product-standard

[4]The Sustainability Accounting Standards Board (SASB) Materiality Map may be accessed at: http://www.sasb.org/materiality/sasb-materiality-map/

Examples of such approach include The Sustainability Consortium (TSC),[5] Public Gardens Sustainability Index[6] hotspots analysis and AHAM hotspots analysis[7] and WRAP's Product Sustainability Forum (PSF)[8] in the UK (WRAP: Waste & Resources Action Programme).

3 Applications/Use of Hotspots Analysis

The purpose of hotspots analysis is to help policy-makers, businesses and other stakeholders to collate, analyze and visualize sustainability hotspots information and insight drawn from a range of different data and information sources in order to move more swiftly from research and analysis towards tangible, practical actions (Barthel et al. 2014).

Typical application of hotspots analysis include:

- Product- and sector-level sustainability standards
- Government or trade association sponsored voluntary agreements with industry
- Policy, research and innovation activities to drive more sustainable forms of production and consumption
- Strategic prioritization of areas for impact management in global value chains
- Information to support consumer-facing campaigns or business-to-business communications and messages on key sustainability themes
- Pilots, value chain and stakeholder collaborations and partnerships to address key sustainability hotspots

A range of different stakeholders, such as industry, government, trade associations and collaborative groups, are using hotspots analysis methodologies to identify sustainability improvement opportunities and develop action plans to address identified hotspots. Table 12.1 shows some of the existing hotspots analysis methodologies developed by different stakeholders.

---

[5] More information on The Sustainability Consortium may be accessed at: http://www.sustainabilityconsortium.org/

[6] A copy of the Public Gardens Sustainability Index may be accessed at: http://www.publicgardens.org/files/files/Longwood%20Gardens%20-%20Sustainability%20Index%20for%20North%20American%20Public%20Gardens%20v%201_0%20-%20Final%2020130514.pdf

[7] More details on the Association of Home Appliance Manufacturers (AHAM) hotspots analysis used in the development of the *AHAM 7001-2014/CSA SPE-7001-14/UL 7001, Sustainability Standard for Household Refrigeration Appliances* may be found at: http://lcacenter.org/lcaxii/final-presentations/513.pdf

[8] More information on the Waste Resources Action Programme's (WRAP) Product sustainability Forum (PSF) may be accessed at: http://www.wrap.org.uk/content/product-sustainability-forum-psf

Table 12.1 Examples of different stakeholders using hotspots analysis

Led by	Examples
Industry	Textiles and clothing (WRAP, UK)
	Home improvement products (WRAP, UK)
	Detergents
	Electrical and electronic products (The Sustainability Consortium, USA)
	Drinking water filtration systems (WQA, North America)
	Home appliances (AHAM, North America)
Government	Grenelle I and II Laws (France)
	EU Product/Organizational Environmental Footprint
Collaborative groups	The Sustainability Consortium (USA)
	Product Sustainability Forum (WRAP, UK)
	Product Category Rules (PCR) Guidance
	Water Footprint Network (Netherlands)
	Ellen MacArthur Foundation (UK)
UNEP 10 year framework of programmes on SCP	Consumer Information Programme
	Sustainable Lifestyles and Education Programme
	Sustainable Tourism, including Ecotourism Programme
	Sustainable Food Systems Programme

3.1 Scale of Application

Hotspots analysis methodologies are flexible and can be adapted based on the scope and scale of application required by those commissioning or undertaking studies. In general hotspots analysis can be applied at:

National and city: Countries and cities use hotspots analysis to help government policy-makers to focus on voluntary agreements or action plans with industry, citizens and communities in areas where sustainability hotspots have been identified. Examples include: the Water Footprint Network's analysis of water scarcity hotspots in major river catchments, World Resources Institute's work to quantify cities carbon emissions, etc.

Sector/industry and product category/product: Businesses use hotspots analysis to identify improvement opportunities and action plans, particularly in areas such as future resource availability and management, global supply chain risks and volatility, waste prevention and management, etc. Examples of sector initiatives based on hotspots analysis include: the UK grocery retailer – Tesco, tackling the food losses and food waste associated with the international sourcing of its products and their use by consumers; and The Sustainability Consortium building consensus around the key sustainability hotspots to be addressed in consumer goods value chains.

3.2 Attributes

Hotspots analysis studies tend to cover a wide range of attributes depending on the scope and scale of application. Broadly hotspots analysis can cover single or multiple attributes.

Single attributes would cover only one attribute, such as environmental or economic or social sustainability or the governance issues and challenges associated with the use of natural resources (e.g., water, biodiversity). Most of the existing single attribute hotspots analysis studies cover environmental issues, with some focusing on single impact categories, like carbon management or greenhouse gas (GHG) emissions. Examples of such studies include: *GHG Protocol's Product Life Cycle Accounting and Reporting Standard*, Japan Environmental Management Association for Industry (JEMAI)'s Carbon Footprint Program[9] and the Water Footprint Assessment Methodology.[10]

Multiple attribute studies would cover more than one attribute such as environmental-social or economic-environmental-social or economic-environmental-social-governance, etc. Many of the existing hotpots analysis cover multiple attributes. Examples of multiple attributes hotspots analysis study include: SASB Materiality Map, The Sustainability Consortium (TSC), WRAP's Product Sustainability Forum (PSF), AHAM hotspots analysis, etc.

3.3 Commonalities and Differences Among Methodologies

The UNEP/SETAC Life Cycle Initiative Flagship Project 3a report documents some general commonalities and differences among the 21 hotspots analysis methodologies analyzed during the project's initial phase (Barthel et al. 2014). The findings are summarized below:

Commonalities

- All of the key methodologies engage several stakeholders in their development.
- All (with the exception of two) methodologies include environmental impacts.
- All methodologies, at a minimum, utilize a quantitative life cycle approach. Some exclusively, others also incorporate qualitative elements.
- The majority of approaches address multiple impacts.

Differences

- National-level methodologies all exclusively utilize a quantitative approach which addresses environmental impacts, while sectoral- and product-level

[9] More information on Japan Environmental Management Association for Industry (JEMAI)'s Carbon Footprint Program, may be accessed at: http://www.cfp-japan.jp/english/

[10] More information on the Water Footprint Assessment Methodology may be accessed at: http://www.waterfootprint.org/downloads/TheWaterFootprintAssessmentManual.pdf

methodologies tend to be more diverse in the impacts and issues they address, as well as in the use of more qualitative elements and inputs.
- National-level methods tend to focus on the use of input/output analysis or materials flows analysis; whereas sector- or product-level methodologies tend to focus on life cycle approaches or "beyond LCA[11]" approaches.
- While the hybrid funding (i.e., a combination of both public and private funding) appears to be dominant among the methodologies, there is no common model that can be attributed to any of the three methodology levels.

3.4 Case Studies

In order to provide some perspective on all of the attributes of hotspots analysis (HSA) covered so far in this chapter, a few HSA case studies have been selected to illustrate and elaborate the different types of hotspots analysis methodologies being used. In the examples given below, we will provide a brief description of the HSA methodology that was/is being applied and the salient features of the methodology. Table 12.2 (see Sect. 3.4.4) provides additional information on various aspects of HSA from development and application to stakeholder engagement. These example hotspots analysis methodologies were chosen to help visualize the range and variability/diversity in scope and scale of hotspots analysis from approach to application. It is not the authors' intention to convey that these HSA methodologies are superior or preferred to any other HSA methodologies available.

3.4.1 Association of Home Appliances Manufacturers (AHAM)

AHAM utilizes hotspots analysis as the main tool to identify and prioritize life cycle sustainability impacts that would be addressed in its resultant product sustainability standards. Since 2010, AHAM has been utilizing hotspots analysis on several products including: refrigeration appliances, clothes washers, cooking ranges, dishwashers, as well as a range of portable and floor care appliances.

AHAM's hotspots analysis methodology seeks to identify the most significant environmental, social and governance impacts across the life cycles of these products. This process is overseen by a task force comprising: AHAM, its standards development partners (UL Environment and the CSA Group), a range of appliance manufacturers, experts with significant product or relevant industry experience, as well as its sustainability consultant – PE INTERNATIONAL (now re-branded as thinkstep).

[11] In the use of the term "beyond LCA" the authors mean that hotspots analysis, as a complementary tool, is able to expand upon the scope and range of impacts that may be identified via life cycle assessment (as encompassed by environmental life cycle assessment, social life cycle assessment and life cycle costing). "Beyond LCA" should not be interpreted as better than or superior to life cycle assessment. LCA and hotspots analysis are in fact complementary tools with their own strengths and limitations.

Table 12.2 Summary of hotspots analysis methodologies and application (Barthel et al. 2014)

Methodology	Approach	Breadth of impacts	Stakeholder involvement	Target audience	Application
AHAM Hotspots Analysis	Qualitative and quantitative (Combination)	Environmental, Social, Governance	Stakeholder review of methodology and pilot testing of standards	AHAM's sustainability standards Task Force members as well as key value chain stakeholders	Home appliance product sector
WRAP's Product Sustainability Forum	Qualitative and quantitative (Combination)	Environmental, Economic	Multiple stakeholders involved	Major retailing and manufacturing businesses, NGOs, academics and governments	Food, DIY (home improvement) products, electrical and electronic products, clothing and textiles, retail sectors
Sustainability Accounting Standards Board (SASB) Materiality Map	Qualitative	Environmental, Social, Economic and Governance	Corporations, market participants, and public interest and intermediaries	SASB standard development working groups and other relevant stakeholders	Several sectors such as Architecture & Engineering Consumption Energy Financial Food & Beverage Forestry and Paper Health Care Home & Office Non-renewable Resources Personal Care Products etc.
Global Protocol for Community-scale GHG emissions	Quantitative	Environmental	Multiple stakeholders	Anyone assessing the GHG emissions of a geographically defined area	Transportation, Waste, Industrial Processes, Agriculture, Forestry and Other Land Use sector

Prior to commencing the hotspots analysis, the task force creates a template value chain heat map to identify sustainability attributes for the resultant Standard. This involves the review and condensation of the typical life cycle assessment (LCA) impact categories into a more simplified format that would facilitate harmonizing, categorizing, and analysing environmental/sustainability issues (i.e., hotspots) with broader stakeholder pressures (i.e., hot buttons).

The first step of the hotspots analysis involves a review of available literature to identify significant product life cycle impacts. This review includes: life cycle assessment studies on the candidate product or product components, manufacturer product life cycle data, academic studies, environmental product declarations (EPDs) and existing standards.

Another key contribution to the hotspots analysis involves interviews with selected manufacturers. Each manufacturer is asked to rank the level of importance (i.e., high, medium, or low) of addressing each of environmental impacts across the five life-cycle stages of the product category. The results of the literature review, stakeholder interviews, as well as review of other existing standards were aggregated into a heat map to graphically provide a preliminary view into priority environmental impacts and stakeholder concerns.

Throughout the process, an extensive stakeholder engagement is conducted to review the results of the hotspots analysis and solicit feedback. In these engagements, AHAM typically include representatives from the following sectors: manufacturers, suppliers, retailers, government agencies, consumer groups, and non-governmental organizations.

As an additional level of rigor, prior to translating these hotspots into criteria within its sustainability standards, AHAM often conducts a screening-level life cycle assessment using primary (where available) and proxy data from manufacturers and industrial databases to verify and validate the result of the hotspots analysis.

Pilot-testing of the resultant sustainability standards by appliance manufacturers is also a core component of the overall process.

3.4.2 Global Protocol for Community Scaled Greenhouse Gases Emission Inventories (GPC)

GPC (global protocol for community scaled greenhouse gases emission inventories) conducted hotspots analysis to identify requirements and provide guidance for calculating and reporting city-scale GHG[12] inventories, consistent with the 2006 IPCC[13] Guidelines for National GHG Inventories. The goal is to allow for more credible reporting, meaningful benchmarking and aggregation of climate data and greater consistency in GHG accounting. The Global Protocol for Community-Scale GHG Emissions (GPC) is the result of a collaborative effort between the World Resources Institute (WRI), C40 Cities Climate Leadership Group (C40), and ICLEI – Local Governments for Sustainability (ICLEI).

[12] GHG: greenhouse gases.

[13] IPCC: Intergovernmental Panel on Climate Change.

The GPC specifies the principles and rules for compiling a city-level GHG emissions inventory; it does not require specific methodologies to be used to produce emissions data. It provides guidance on calculation methodologies (i.e., defining boundaries, defining emission sources, calculation guidance) for individual emission sources including stationary energy, transportation, waste, industrial processes and product use emissions and agriculture, forestry and other land use.

3.4.3 WRAP's Product Sustainability Forum (PSF)

WRAP's PSF (Product Sustainability Forum) was established in late 2010 in response to a request from the UK governments and major retailers and manufacturing companies to establish a pre-competitive space for collaboration between governments, business, NGOs, academia and other key stakeholders to come together to build the evidence to help quantify, reduce and communicate the whole lifecycle environmental impacts and hotspots associated with consumer products in the UK economy. Since its creation in 2010, WRAP's PSF and the organizations that support it have been working together to achieve these objectives. Following almost 3 years of research WRAP's PSF is now beginning to apply its work in a growing number of international supply chains through Pathfinder demonstration projects; mainstreaming projects that seek to embed lifecycle and sustainability thinking at the heart of organizations; and the growing membership and geographical coverage of the International Network of Product Sustainability Initiatives[14] (INPSI), which WRAP's PSF was instrumental in establishing in 2012.

WRAP's PSF uses a similar methodology to identify hotspots as identified in Sect. 2; more information on the specific methodology and application, including links to its free, on-line knowledge base is available.[15] In the last year, WRAP's PSF has moved away from a primary focus on hotspots-related research to more of an action-orientated approach, focusing more of its resources on driving change within its supporting organizations through Pathfinder demonstration projects and mainstreaming and embedding activities. Feedback from all of these collaborative activities is now being used to improve and update the data and information gathered during the initial hotspots analysis phases based on production-specific and value chain performance information.

3.4.4 Other Attributes

Table 14.1 provides further information on different attributes of HSA methodologies discussed in this chapter such as approach, breadth of impacts covered, stakeholders' involvement, target audience and application for selected existing methodologies.

---

[14] More information on the International Network of Product Sustainability Initiatives (INPSI): http://www.product-sustainability.net/

[15] http://www.wrap.org.uk/content/psf-knowledge-base-0

The purpose of Table 12.2 is to allow visualization of the flexibility and diversity of HSA methodologies development and application.

4 Other Tools Used to Identify Sustainability Impacts

A growing number and diversity of tools, resources and methodologies are being developed and used to identify sustainability impacts. This often leads to confusion among stakeholders with regards to the best tools for the job and the best way to use them to generate science-based and actionable outputs and information. Sometimes this confusion – and the range of tools and methodologies available to users – leads to erroneous conclusions, such as HSA being superior to, or replacing life cycle assessment studies (LCA). This section would discuss the roles of HSA and other methodologies as tools to identify and take action on sustainability impacts.

Life cycle thinking and a value chain perspective is imperative to achieve robust sustainability assessment results. ISO 14040 (2006) is the *de facto* globally accepted standard for conducting life cycle assessments (LCA) to identify the environmental impacts of the product or system being analyzed. The precepts and process steps contained in ISO 14040 have also been adapted to inform the development of other life cycle-based approaches and analytical tools, including those that focus on economic impacts (i.e., life cycle or whole life costing) and social impacts (i.e., social life cycle assessment). These latter applications are not as well developed, nor are they used as widely as traditional environmental LCA.

The ISO 14040 series provides a technically rigorous framework for conducting life cycle assessments (LCA), which is widely used for measuring the environmental dimensions of sustainability. Life cycle costing (LCC) and social life cycle assessments (S-LCA) also use the ISO 14040 framework with some adaptations. The UNEP/SETAC Life Cycle Initiative's publication "Towards a Life Cycle Sustainability Assessment, 2011" elaborates the use of ISO 14040 framework to conduct environmental LCA and further adapted to conduct LCC and S-LCA studies in detail (UNEP/SETAC Life Cycle Initiative 2011).

4.1 Hotspots Analysis Versus Other Tools

Environmental LCA, S-LCA and LCC each quantify specific impacts related to their respective attributes of sustainability as shown in Fig. 12.2. Hotspots analysis is more flexible and can incorporate several attributes of sustainability, such as economic, environment, social, ethical and governance, depending on the scope and scale of application, as shown in Table 12.1. Figure 12.2 shows the scope and impacts covered by LCA, LCC, S-LCA and hotspots analysis respectively.

Hotspots analysis offers a complementary approach to more traditional life cycle-based methods; it is not a replacement or competitor to LCA, LCC and

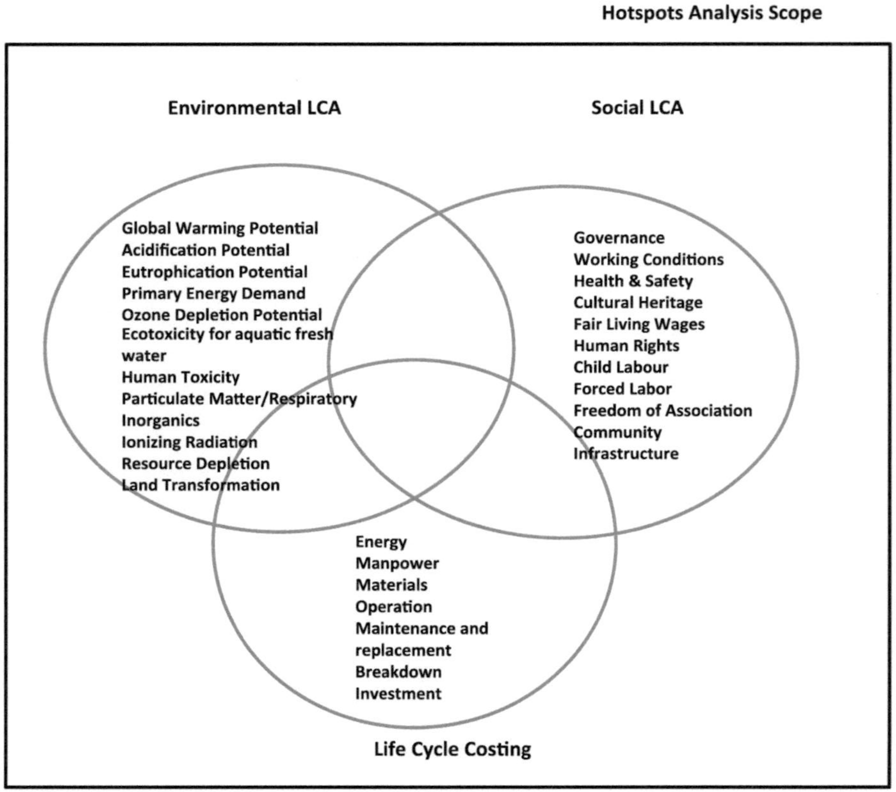

Fig. 12.2 Scope of representative impacts covered by LCA, LCC, S-LCA and hotspots analysis methodologies

Table 12.3 Comparison of hotspots analysis with LCA, LCC and S-LCA

Methodology	Stakeholder engagement	Potential impact coverage	Ease of use	Approach	
				Qual.	Quant.
Hotspots analysis	■■■	O$✠◆	++	✓	✓
Life cycle assessment (i.e., ISO 14040)	■■	O	+		✓
Social life cycle assessment (Adapted from ISO 14040 series)	■■	✠	+	✓	✓
Life cycle costing studies	■	$	+		✓

■■■Full engagement and pilot testing; ■■Moderate engagement throughout the process; ■Limited phases of engagement; $Economic; O Environment; ◆Governance; ✠ Social; +++ Easy; ++ Moderate; + Difficult

S-LCA. In fact, in most cases hotspots analysis can be the initial step to identify and prioritize sustainability impacts at pre-competitive/pre cursor level, support decision-making and also obtain stakeholder engagement and validation. In some cases, where available, existing LCAs (both data and information) are used as input into the hotspots analyses. The second step could be to conduct LCA, S-LCA, LCC (depending on the scope) to quantify impacts and further support decision-making, comparison and improvement at greater accuracy based on primary/secondary data and models. HSA allows initial scoping, identification and prioritization, while LCA, LCC and S-LCA may be used to engage in a more in-depth analysis and perhaps to validate the outcomes of HSA.

Table 12.3 compares hotspots analysis with ISO 14040 series based-environmental LCA, S-LCA and LCC across various indicators such as: level of stakeholder engagement, types of sustainability impacts covered, ease of use and type of approach (i.e., qualitative vs. quantitative).

5 Key Observations

Within the UNEP/SETAC Flagship Project 3a report (Barthel et al. 2014) some general observations were made with regard to the hotspots analysis methodologies analyzed therein. Several of these observations are discussed below.

5.1 Audience and Application

Existing hotspots analysis methodologies are being developed with a number of audiences and sustainability-based applications in mind. Some studies are being used to help government policy-makers to focus voluntary agreements or action plans with industry in areas where sustainability hotspots have been identified. For example, as is the case with WRAP's Product Sustainability Forum's work in the UK food chain, the French Government's work to provide more sustainability information to consumers, or the Water Footprint Network's analysis of water scarcity hotspots in major river catchments.

Businesses are using hotspots analysis to focus their resources, drawing up action plans and practical programs of work to eliminate, reduce or mitigate hotspots in their global value chains; and tackling major societal and commercial issues like food waste, food and resource security (future supply risk and resilience issues); and water use in agriculture. For example, the work of UK grocery retailer, Tesco, to tackle the food losses and food waste associated with the international sourcing of its products and their use by consumers; and the work of The Sustainability Consortium in building consensus around the key sustainability hotspots to address in consumer goods value chains. Other stakeholders are using the findings from hotspots analysis to inform their thinking. For example, the Oxford Martin School at Oxford University is working alongside WRAP in the UK to use hotspots analysis to inform its thinking on the research, policy and business drivers to facilitate a mass movement over time to healthier, more sustainable eating patterns or diets.

5.2 Beyond LCA

In some cases, the scope of hotspots analysis methodologies and studies are broadening beyond consideration of one or more environmental impact categories and including "beyond LCA" approaches and wider sustainability topics like biodiversity management, animal welfare, fair trading arrangements, land use and land use change and governance issues around raw materials or water resources.[16] This development would suggest that both methodology developers and users see the value in securing a more holistic view of hotspots, allowing them to identify where trade-offs may need to be considered (e.g., between traditional intensive agricultural practices and the potential impact on the agri-ecosystems that support them). The importance of taking a "beyond LCA" approach to the development of hotspots analysis methodology was also highlighted by stakeholders as important.

While there is still a clearly defined niche for traditional LCA approaches that solely utilize quantitative data and exclusively address environmental impacts, there are a growing number of hotspots analysis methodologies that move beyond traditional LCA and include either additional quantified data and information (e.g., trade, market and sales data; contextual sector or product category information; sup-

[16] In the use of the term "beyond LCA" the authors mean that hotspots analysis, as a complementary tool, is able to expand upon the scope and range of impacts that may be identified via life cycle assessment (as encompassed by environmental life cycle assessment, social life cycle assessment and life cycle costing). "Beyond LCA" should not be interpreted as better than or superior to life cycle assessment. LCA and hotspots analysis are in fact complementary tools with their own strengths and limitations.

porting scientific research and innovation; materiality studies); and/or qualitative inputs, such as expert opinions, stakeholder concerns, consumer attitudinal and behavioral insights, etc. This trend appears to be most prominent among product- and sector-level hotspots analysis methodologies.

This observation does not preclude the fact that the majority of methodologies share a common foundation in that they utilize a life cycle approach to hotspots analysis. Most of the methodologies reviewed by the authors also follow a pragmatic approach that includes the identification of all life cycle aspects and impacts within a study boundary before applying materiality criteria or significance thresholds in order to define which ones are "hot". In some circumstances (not necessarily for the key methodologies identified) a methodology may not consider the whole life cycle at the start, since there may be sufficient existing studies for the same sector or product category suggesting that the hotspots always lie in one or more specific life cycle stages.

The initiation of methodology development stems from a variety of different organisations, and is often linked to a specific objective. Governments may act in relation to policy objectives or priorities, whereas the private sector may act based on a recognition of a business case for action. NGOs may be informed by recognition that a methodology can help in articulating the need for action in line with their objectives.

5.3 *Commonalities*

Common features of all the methodologies identified by the authors are their engagement with a wide stakeholder base in development and their quantitative nature, though some methodologies also incorporate qualitative information drawn from a range of sources. The majority are focused on multiple impacts and issues, with most covering a core set of environmental issues, though issue-specific methods also exist. The national-level methodologies reviewed exclusively utilize a quantitative approach which addresses environmental impacts, while sectoral- and product-level methodologies tend to be more diverse in the impacts and issues they address, as well as utilising qualitative elements.

While the hybrid funding of methodology development and application appears to be dominant among the methodologies, there is no common model applied at a national, sector, or product level.

5.4 *Ease of Use*

Another interesting observation that was alluded to in the analysis was that none of the hotspots analysis methodologies listed was "easy" to use. More than half were considered to be "difficult" requiring some expert knowledge or experience; and the remainder were considered as "moderate" and may require some expert guidance in order to use. As a tool that is used to facilitate decision-making as a precursor to (or in lieu of) a more detailed analysis, hotspots analysis still seems to require at least some expert input.

5.5 Gaps

In terms of gaps, few methods appear to incorporate financial data, in particular on the costs and benefits of addressing hotspots. The methods are generally linked to quantification activity. Links to identification of a range of associated opportunities or solutions to reduce the impact of hotspots identified are often sparse, with notable exceptions in the methodologies developed by The Sustainability Consortium and WRAP's Product Sustainability Forum. Whilst the need for action is recognized in principle, its incorporation into methods is generally limited. In particular, there is a lack of guidance on how to assess the potential for reducing a hotspot.

6 Challenges and Next Steps

Currently there is no common global approach to hotspots analysis; nor has there been any effort to bring together or share best practice amongst those organizations or initiatives currently developing and using these methods. There is also no accepted guidance on how to translate and apply the results of hotspots analysis into meaningful sustainability information and insight for use by industry, governments and other stakeholders.

Recognizing that this situation may result in a range of negative impacts, including a lack of consistency in the methodological approach, difficulties in comparing the results of hotspots studies and the potential for conflicting sustainability information in the marketplace; the UNEP/SETAC Life Cycle Initiative established Flagship Project 3a to address these and other issues as noted in the introduction to this chapter.

The primary focus of this project is to identify existing methodologies, tools and resources that can or could be applied at three scales or levels of detail, namely at the national, sector or product category-level. A secondary research objective is to seek to determine the potential use, adoption or adaptation of these methodologies by developing countries, emerging economies, SMEs or for use at the city-scale. During the first phase of the project, the UNEP/SETAC Life Cycle Initiative conducted a study to map existing hotspots analysis methodologies and studies world-wide (Barthel et al. 2014).

Phase 2 of this flagship project is seeking to respond to the urgent timeframes and requirements of a number of implementation programs within the UNEP's 10 Year Framework of Programmes on SCP (10YFP), and the 10YFP Secretariat's need to provide national-level hotspots analysis guidance to National Focal Points to help them prioritize their SCP policies and programs.

The second phase will involve a series of rapid-prototyping workshops in one designated geographical location, supplemented with regional webinars to solicit input and feedback from key experts and stakeholders from around the world that were identified during Phase 1 of the project. These workshops and webinars are intended to meet the ultimate objectives of this project, to facilitate the rapid development and refinement of:

1. A common methodological framework and global principles and guidance for hotspots analysis that is capable of being used at different scales or levels of detail (national, sector, product category and city)
2. Best practice guidance for the appropriate use and communication of sustainability information derived from hotspots analysis and other life cycle approaches

These workshops will also be used to test with stakeholders the feasibility of a number of options to bring together the findings, data and information from existing hotspots analysis studies to provide a richer, more global picture of the sustainability hotspots in society and the economy.

With its Flagship 3a, the UNEP/SETAC Life Cycle Initiative continues to address these issues on its way towards developing "Global principles and guidance for hotspots analysis" in the second Phase of the project.

References

Barthel M, Fava J, Harnanan C, Valdivia S, Khan S, James K, Smerek A (2014) Hotspots Analysis: mapping of existing methodologies, tools and guidance and initial recommendations for the development of global guidance. UNEP/SETAC Life Cycle Initiative-Flagship Project 3a (Phase 1) v.5.2. UNEP/SETAC Life Cycle Initiative website. http://lifecycleinitiative.org/wp-content/uploads/2014/10/Flagship3a-Hotspots-Mapping.pdf

ISO 14040 (2006) Environmental management – life cycle assessment – principles and framework. A copy of this standard can be purchased. http://www.iso.org/iso/catalogue_detail?csnumber=37456

UNEP/SETAC Life Cycle Initiative (2011) Towards a life cycle sustainability assessment: making informed choices on products. United Nations Environment Programme website. http://www.unep.org/pdf/UNEP_LifecycleInit_Dec_FINAL.pdf

Chapter 13
From Sustainable Production to Sustainable Consumption

Marc-Andree Wolf and Kirana Chomkhamsri

Abstract The objective of this chapter is to explain which secondary environmental consequences (often called rebound effects) life cycle assessment (LCA) and life cycle management (LCM) of products need to consider in addition to the conventional product LCA, and which roles different actors in society have in the context of environmentally sustainable consumption. The key issue is that any consumption decision affects the consumer's household resources of available income, time, and space (volume, area), what leads to additional or reduced overall consumption, within the limits of further consumption constraints and cross-category effects. Exactly how any additional resources are used by the consumer strongly affects the overall consumption. Moreover, this chapter considers the consumption on person, on national and global level, with some focus on sustainable lifestyles, and concludes with recommendations on next steps towards better measurement and management of the environmental secondary consequences of consumption.

Keywords Consumption constraints • Life cycle assessment • Life cycle management • Rebound effects • Secondary consequences of consumption • Sustainability • Sustainable production • Sustainable consumption

1 Introduction

Sustainable consumption has been defined already very early in explicit relation to the life cycle perspective: "*The use of services and related products which respond to basic needs and bring a better quality of life while minimizing the use of natural*

M.-A. Wolf (✉)
maki Consulting GmbH, Sepp-Zehentner-Str. 33, 83071 Stephanskirchen, Germany
e-mail: Marc-Andree.Wolf@maki-consulting.com

K. Chomkhamsri
P.P.P. Intertrader,
109/202 M. 9 Leaungmung-Pakkred Rd., Bangpood T., Pakkred A., Nonthaburi 11120, Thailand

G. Sonnemann, M. Margni (eds.), *Life Cycle Management*,
LCA Compendium – The Complete World of Life Cycle Assessment,
DOI 10.1007/978-94-017-7221-1_13

resources and toxic materials as well as emissions of waste and pollutants over the life cycle of the service or product so as not to jeopardize the needs of future generations." (Norwegian Ministry of the Environment 1994, adapted). Beyond the product perspective to life cycle management (LCM), this definition highlights the needs fulfilment, or utility, of the products. Indeed, but there are other aspects that differ when taking a consumption perspective on products: "*Sustainable consumption is often seen as a concept mirroring the production side. While sustainable production has a production and upstream perspective searching opportunities to alter production processes and related activities in a more sustainable direction, sustainable consumption is directed to how and why goods and services are demanded, used and consumed.*" (Thidell 2011). This indicates why there is a controversy about the "right" way to assess the environmental performance of products, which the authors try to explain in the following, before explaining the details:

Life cycle assessment (LCA) of products guided by ISO 14040 (2006a) and 14044 (2006b) is a well-established framework for analysing, improving and comparing the environmental performance of products. ISO LCA is also the core basis for life cycle management (LCM), while further standards that are based on ISO 14040 and 14044 support it. With the current framework ISO 14044:2006, the interim product and waste flows as well as the interventions with the environment are taken into account as the inputs and outputs of the analyzed system. The effects on other, not functionally connected products and infrastructures, are, however, not addressed in the framework and also not implied.[1] In other words, the current framework of LCA is taking a product perspective, not a consumption perspective.

This product perspective and ISO LCA itself has been criticized (e.g. by Girod et al. 2010) for lacking to capture the various secondary consequences of consumption decisions, including due to changes in the available income, time, and other household resources.[2]

In this chapter, instead of asking for changing product LCA, the authors follow a different approach, namely that of proposing differentiated life cycle modelling methods for the two fundamentally different perspectives: the product perspective and the consumption perspective, resulting in dedicated approaches for different actors and applications (Lundie et al. 2007; European Commission 2010; Wolf et al. 2012): different applications imply a different purpose and scope of supported decisions and hence require a differentiated guidance. The Product Environmental Footprint (PEF) guide (European Commission 2013), that largely builds on the product-decision support ("Situation A") guidance variant of the ILCD Handbook (European Commission 2010), is an example of such application-specific guidance.

The authors argue that the product perspective is a very useful logic for product developers: The decisions by the consumers beyond the use and end-of-life man-

[1] The system expansion to include functions within the system boundary of the analysed product serves exclusively to render two products of an only partly identical set of functions comparable.

[2] Such rebound effects and economy-wide consequences can also be environmentally positive, i.e., result in a negative environmental impacts, why we adopt here the more inclusive term "secondary consequences" of the ILCD Handbook, instead of "rebound effects" that implies a negative effect.

agement of the purchased product itself is essentially beyond the influence of the producer of the analyzed products. The authors derive that there is no requirement to integrate such consequences outside the analyzed product into its analysis. To provide data on the environmental performance of a product over its life cycle is the essential step for developing more sustainable products. It is also a core contribution, when looking at products from the consumption perspective. However, if the consumption of products is the viewpoint of the analysis or the evaluation of lifestyle concepts, additional components should be added to the analysis: This methodological extension to the consumption side and the secondary consequences that consumption decisions cause is the main scope of this chapter.

There is, however, another aspect where the consumption perspective goes beyond the product-perspective: expanding from the product's functional unit to the contribution to human needs fulfilment.[3] This aspect will at least briefly be addressed in this chapter.

This chapter hence aims at providing an overview of the secondary consequences of consumption and the products' contribution to the human's needs fulfilment. This draws on the works of Hofstetter et al. (2006), Weidema (2008), Girod et al. (2010) and others, and ends in an outlook of recommended next steps to further develop, differentiate, and deploy LCA as a tool that equally well supports efforts towards sustainable consumption as it already supports decisions on sustainable production and products.

Still, both perspectives "product" and "consumption" initially serve exclusively the concept of relative sustainability, i.e. of better efficiency. To expand the view to absolute sustainability, one needs to bring in further elements, e.g. breaking down the planetary boundaries (Rockström et al. 2009) to the individual citizen's environmental impact budgets. This next step is not part of this chapter.

2 Secondary Consequences of Consumption Decisions

2.1 Overview

This section provides an overview of the secondary consequences that purchase, use and end-of-life of a product can induce outside the actual product system itself. These consequences go beyond functional relationships with other products, i.e. beyond part-system and system-system relationships that are part of regular product LCA, and which are described in the ILCD Handbook, Chap. 7.2.2.

The following consumption-induced secondary consequences and aspects that modify them can be differentiated (compiled from Becker 1976; Eyerer and Wolf 2000; Hofstetter et al. 2006; Weidema 2008; Girod et al. 2010, with additions):

[3] The authors build here on the concept of human needs developed by Maslow (1954) and expanded by Max-Neef (1991) and others over the following decades.

- Changes in the available household resources income, time, and space, as well as further constraints to consumption, particularly food calories uptake capacity, drink intake capacity, skills and information availability, and access to products
- Use of the freed household resources (or restrictions in case of reduced household resources)
- Cross category effects
- Mental secondary consequences

The above listed secondary consequences cause additional or reduced consumption.

Beyond these, other secondary consequences occur on local, regional, national or international scales that are not addressed in this chapter.

In three additional sub-sections the authors look briefly into higher order consequences due to economic transactions, present a new measure of the environmental life cycle performance of products from consumption perspective and reflect on possible harm due to needs over-fulfilment.

2.2 Changes in Available Household Resources and Consumption Constraints

The following household resources have been considered for studies on sustainable consumption (compiled from Becker 1976; Eyerer and Wolf 2000; Hofstetter et al. 2006) Weidema 2008; Girod et al. 2010, with own additions and examples):

- Available income[4]
- Time
- Space (volume, area)

Moreover the following elements, which are better understood as constraints to consumption, are to be considered (Hofstetter et al. 2006):

- Food calories uptake capability
- Drink intake capability
- Skills and information
- Access to products and technologies

2.2.1 Available Income

Changes in available income – always a decrease – occur with any purchase decision, while decisions during use (e.g. more efficient use, shared use) and when selling a used product, the available income can increase compared to the default case. This additionally available income allows for additional consumption.

[4] *Income is the sum of all the wages, salaries, profits, interests payments, rents and other forms of earnings received… in a given period of time*. (Case et al. 2014)

2.2.2 Time

While each purchased product initially reduces the consumer's time budget, due to the purchase process, some products have relative time saving advantages compared to the average product and others enable to actually increase the available time budget of the consumer on a net basis: car navigation systems or apps with traffic avoidance, integrated washing and drying machines, crease-free shirts, faster internet connection, etc. all save time. This time is made available for additional consumption but also for other activities (e.g. economic activity to generate extra income, or resting, i.e. (near) non-consumption).

2.2.3 Space

Space (volume and area) to store or use goods is a physical limitation. Examples are the living area that limits the amount of furniture that can sensibly be put, storage space to keep clothes, parking space in cities that may relevantly limit the option to have a second car, the consumer's skin surface that can only so many times per day be treated with crèmes or lotions, or storage space on storage media in a computer.

However, many products offer the possibility to increase or better use the available space, either as a secondary product property (e.g. a Smart car may be an option as second car even in city centres with severe parking space limits), or as a primary property (e.g. vacuum bags for storing bed clothes, shelves, external hard disks, etc.). While each Euro can only be spent once, and the number of things one can do at the same time is limited, spatial limitations are arguably less absolute, while at least an asymptotic saturation of the available space can be observed in reality.

2.2.4 Food Calories, Drink Intake

The amount of calories that we can digest is also limited. The market growth potential of the food industry therefore lies in selling further processed food with higher value added as well as in convenience food in smaller packages at a higher per calorie price. Diet food is another key option to sell more food without surpassing the individual's overall calorie uptake limit. Eating more calories is formally one way to expand this limit, although again not limitless and with possibly harmful consequences (see Sect. 2.8). Still, certain limits exist in the volume that people can or at least want to eat.

Similar limitations exist for drinks, while our body is better able to put through water than carbohydrates, proteins, and fat.

2.2.5 Skills and Information

Skills and available information can be an important limitation to consumption, but we argue that they are of a different nature than the previously listed ones: with the purchase of a good or service the consumer is not negatively affecting his or her skills and available information, in contrast to the situation for money, time, and space. If anything, they would be increased. Skills and available information are however a constraint to consumption of those goods that require a higher level of skills or information.

Next to absolute limitations of personal aptitude, the necessary money and time needed to acquire the specific skills can be a relevant obstacle to consumption. Examples are leisure activities such as operating a small sailing boat, constructing the own furniture, or playing the piano. However, the idea of a knowledge based society and the growing offers of online courses for virtually everything and instructions and courses offered by do-it-yourself markets are however eroding the limitation of information.

At the same time, the process of learning the required skill may be a key part of the activity and the success to have mastered the skill can be an important contribution to the individual's needs fulfilment. Legend are however the households that have a piano with nobody in the family being able to play it.

2.2.6 Access to Products

Similarly, limitations to access products are an obstacle, but they are not affected by the consumption of individual other products, while the wider consumption pattern can strongly affect them, particularly if consumption thresholds need to be surpassed to make them economically viable: A good example of such limitations is the availability of car sharing outside the larger cities' centre. Such access restrictions are partly a matter of relative demand limitations – while in a hen-and-egg situation, where limited demand means that the necessary threshold is not achieved to make the product available – but partly also absolute, where a frequent public transport service would not be economically or even environmentally sensible for very remote places with virtually no population.

2.2.7 Interchange Ability of Household Resources

While these household resources and other constraints have their own budget, we note here that compensation across some of these is possible to a large extent, firstly time and money: a consumer can free additional time by hiring another person for cleaning or other household work, or by buying time-saving equipment. Similarly, space and money: the available space can be extended by using some of the available household income for renting a larger flat or extra storage. And we have given already the example of using time (and potentially money) to acquire new skills or information. Finally, investment of time can mean to take up extra economic activities to increase the available household income.

However, where money transactions are involved – such as in the example of contracting a cleaning service – the service provider received additional household income, hence can consume more. We will come back to this characteristic of money being conserved, and what this implies for sustainable consumption.

2.3 *Use of Freed Household Resources*

2.3.1 Overview

Where the availability of any of the listed household resources is reduced by a consumption decision, other products that require the same resource(s) may be affected: most prominently, money can be spent only once. Reduced time availability can be compensated only in certain cases (e.g. doing two things in parallel – see chapter "Cross Category Effects"). Reduced space means space has to be freed by another product, while with the above-mentioned individual flexibility to expand the available space or accept a further cramping of the available space. Eating food calories means that less other food (with calories) will generally be eaten, with the above-mentioned, limited flexibility for compensation.

It is important to note that for consumption studies, the individual options how to react to reduced household flexibility will be of less interest, but the average situation and patterns of effects will be the focus of analysis. Individual flexibility however adds to the variance of the average situation and provides options for scenario definition on different systematic ways how to react to reduced household resources availability.

Inversely, increased availability of any of the household resources allows the consumer to use it for additional consumption: more available income can be spent e.g. on a further away holiday destination (as Eyerer and Wolf (2000) have exemplified), a larger TV set, or any other good or service.

It is relevant for quantifying the environmental impacts of the changed availability of household resources, which products are quantitatively affected by the changed consumption. Expanding on the proposal by Girod et al. (2010), we see four distinctions:

- More of the same product
- More of the same function or need fulfilment
- Marginal shift to better fulfil the less well fulfilled needs
- General increase of average consumption

This first variant of using the freed resource is – if the household resource is income – also termed direct rebound effect, substitution effect, or pure price effect (Greening et al. 2000). The direct rebound effect for energy-efficiency increase has first been postulated already by W.S. Jevons in context of increased coal efficiency (Jevons 1875), cited in Gillingham et al. 2013), and in the more recent discussion on energy-efficiency policies again by Khazzoom (1980). The last three variants are also called indirect rebound effect, income effect, or secondary effect (Greening et al. 2000).

2.3.2 More of the Same Product

"More of the same product" means that the consumer will e.g. drive more kilometres, if the new car is faster (saved time) or more energy efficient (saved income). Or he/she will eat two servings of the low calorie desert, etc.

2.3.3 More of the Same Function or Need Fulfilment

"More of the same function or need fulfilment" (originally termed "More of similar" by Girod et al. (2010)) means that e.g. saved time due to a faster car will be used to generally increase travelling, also with other transport means. In support of this effect, Schafer and Victor (2000) have compiled data from a survey that shows that the average time per day and person spent on travelling across a wide range of cultures worldwide and the entire scale of city/villages sizes and over several decades essentially does not differ and ranges between 50 and 90 min per day. While such average values have to be interpreted with care, we could derive that the reduced travel time (e.g. due to home office) will be used 1:1 for other travelling. Kitou and Horvath (2008) have shown such an effect of e.g. home-office staff joining colleagues for lunch, while in that case interpreted largely due to less congested roads, as the study looked at wider adoption of home office work. While in this example the need that is fulfilled is a different one, the same function of personal transport is affected.

On the level of the same need, an example would be that eating the reduced calorie desert would result in eating other food in addition, whether at the same meal or at another time. It should be noted that in this last example, the calorie uptake is both a human need and also limit to consumption.

2.3.4 Marginal Shift to Mix of Less Well Fulfilled Needs

Thiesen et al. (2008) have assumed for their calculations that freed household resources – in their case for additional available income – will be used for the delta between the consumption profile of the analysed income level and that of the next higher level. Example if some money is saved in the lowest income level, the money was assumed to be spent to a larger-than-average share for dwelling use and maintenance (based on spending statistics), and to a lower degree on a range of other products.

In a more general perspective, we argue that it makes sense to assume that on average the individual would spent any saved resources on those needs that he/she feels are least well fulfilled. Example if time is saved by a single mother, she may spent it primarily on playing more with her child, while a stressed single manager might spend it on mental relaxation exercises or seeing friends. The logic behind this concept is that the use of the available household resources is optimised by the individual, plus that more basic needs are fulfilled first. Only when these are fulfilled, higher needs are increasingly met. While this variant of how additionally available household resources are used is the least well defined one, we argue it to be the most plausible one, if looking at the individual consumer.

2.3.5 General Increase of Average Consumption

Finally, as a default option on the other end of the range, the consumer may just increase its average consumption. While this logic may be less accurate for the individual, particularly if the individual is barely able to meet his or her most basic needs, it may be a robust and quite accurate approach when looking at the average consumer in society.

2.4 Cross Category Effects

Cross category effects (Hofstetter et al. 2006), also termed technology rebounds by Weidema (2008)), relate to technology changes that affect the availability of other technologies or alter their effect on the available household resources. An example is the parallel use of a product A, enabled by a new product B, while not relevantly impacting on each other's functions. An example is the use of a laptop during a train travel (enabled by the portability of the computer and by the trains power outlet and Wi-Fi access). This situation can be argued to free time, as the work on the laptop is working time, if assuming that total working time is not increased.

Other effects are more indirect and can interact with other mechanisms on society level, affecting e.g. infrastructure availability.

2.5 Mental Secondary Consequences

The knowledge (or sometimes only belief) that a product X is more environmentally friendly may lead to an additional consumption "*because the product X has less impacts*", as Girod et al. (2010) argue. Examples are the more fuel-efficient car or more energy-efficient lighting that lead to driving further or having more lamps, respectively.

Similar to the use of freed household resources, also the mental consequences can lead to more use of the same product, as in the above examples, of products that fulfil the same function or meet the same need, or of other products or activities (e.g. "*because I separate my waste, it is ok that I …*"). These mental consequences can also be interpreted as having a mental budget for environmental impacts, as Girod et al. (2010) suggest. However, this mental budget is less accurate than the income or time budget and we can easily be misled as to the actual environmental benefit of a product.

Next to such negative secondary consequences that were in focus in previous work, the authors argue here that these can also be positive: the individual may like the good feeling that e.g. the decision of "being a vegetarian for a week" trial gives and he/she becomes fully vegetarian. Or the positive feeling of knowing to do something good for the environment leads to the decision to adopt a more sustainable lifestyle, i.e. consumption decisions in other product categories and meeting other human needs.

2.6 Higher Order Consequences of Economic Transactions

In addition, spending the saved money on other products means that this money is made available to individuals in a different product's supply chain. The net effect of the individual's available household income is hence not only depending on what he/she spends it, but also on the net change in impacts due to changed consumption depending on where the money that is being spent is going: buying a banana from Gran Canaria will – next to the local retailer – bring income to the wholesale/importer and the Spanish farmer. Buying it from Costa Rica will bring the income to people in different countries and cultures. The spending of additional income can be expected to differ between cultures, age classes, education levels, and between different income groups. If we however assume for simplification that the consumption profile of the different supply-chains do not differ from each other, the secondary consequence in the supply-chain is zero and the net effect is exclusively the extra consumption by the consumer.[5]

2.7 Towards a New Measure of the Environmental Life Cycle Performance of Products from Consumption Perspective

The fact that "money is conserved", as e.g. Dragulescu and Yakovenko (2000) highlight, makes it distinct from the other household resources. Saved time, as example, is actually net extra time available for activities and not handed over to the producers of the purchased products, in contrast to the situation for money that is merely transferred when purchasing a product.[6]

---

[5] This does not yet consider that personal and corporate taxes modify the available income for consumption in the supply-chains. Still, if we assume that the taxes are used for purchase or investment by the governments, the money is still used for consumption, albeit with an again different consumption profile.

[6] Note that also saving money in a bank account means consumption, as it allows other economic actors to take a loan and invest, same as buying on credit by the consumer. The only way to avoid that available household income is available for consumption, is to keep it at home (while that may mean that it marginally affects inflation).

Eco-efficiency, i.e. the quotient of price per environmental impact of a product therefore is a useful indicator: in its most simple form, a twice as expensive product of the same impact effectively reduces the ability of the consumer to spend the money on consuming other goods (Huppes and Ishikawa 2005).

However, we argue here that the concept of eco-efficiency is not considering the effect of additional consumption if the product is only produced cheaper, but a higher profit is kept by the producer: this profit is used for investment by the producing company or distributed to the company owners, e.g. shareholders, and hence available for additional consumption, the same way as it would be available to the consumer of that product if instead the product price would be reduced. More precisely, all profits along the supply-chain (and of consumables during product use and end-of-life services) need to be excluded from the economic component of eco-efficiency, to avoid this distortion.

This insight clarifies from a different perspective that the limiting factor in global consumption is global production, which is obviously limited by the output of the active labour force: If a product is produced with less workforce along its supply-chain, the not anymore required workforce is available for producing more of this product (or other products), hence increasing global production and consumption and hence environmental impacts. And the more qualified this not anymore required workforce is, the more overall production is increased, given the on average higher productivity of the higher qualified workforce. In short: The higher the quotient of the qualification-weighted amount of human working time q*t of a product and overall (i.e. normalized and weighted) environmental impact over the life cycle of a product, the less impacting the product, including considering the secondary effects of freed human work productivity (what is structurally equivalent to the effect by the enabled additional consumption by the product's consumer due to additionally available income).[7] This is called "Environmental work productivity" WP_{ENV} (Eq. 13.1):

$$WP_{ENV} = \frac{\sum_{i=1}^{n} q_i \cdot t_i}{\sum_{j=1}^{m} N_j \cdot W_j \cdot LCIA_j} \quad (13.1)$$

With N being the normalization factor, W the weighting factor, and $LCIA$ the LCIA result of the product, per impact category j.

The price of a product of the eco-efficiency concept is hence replaced by the work productivity, avoiding the distortion due to profits that are part of the consumer price of a product.

If we inverse this quotient, we get a measure for the environmental intensity of human work productivity WI_{ENV} (Eq. 13.2):

[7] Note that the other secondary consequences that were addressed above are, however, not yet included.

$$WI_{ENV} = \frac{\sum_{j=1}^{m} N_j \cdot W_j \cdot LCIA_j}{\sum_{i=1}^{n} q_i \cdot t_i} \quad (13.2)$$

If using the global human productivity and global environmental impact, this is the global average environmental intensity of qualification-weighted human work $WI_{ENV,G}$.

We can use this measure to integrate the effect of different work intensity of a product to correct the life cycle wide environmental impact of the product. By forming the quotient of the product-specific $WI_{ENV,P}$ and the global average $WI_{ENV,G}$, we obtain a normalized factor that expresses the potential net change of environmental impacts due to the amount of human productivity our product binds. Applying this factor to the normalized and weighted LCIA results of the analyzed product yields it's actual impact IMP_{net}, including considering the approximated secondary consequences due to the specific environmental intensity of its production (Eq. 13.3):

$$IMP_{net,P} = \frac{WI_{ENV,P}}{WI_{ENV,G}} \cdot \sum_{j=1}^{m} N_j \cdot W_j \cdot LCIA_{j,P} \quad (13.3)$$

It is important to highlight that this formula does not capture other secondary consequences and that it takes a product perspective.

2.8 *Happiness or Harm Due to Need (Over-) Fulfilment*

> The whole economy of human society is based on one general and simple principle: I want to be happy…. Denis Diderot (1713–1784), as cited in Elchardus (1991) (see Eckersley et al. (2001)

Hofstetter et al. (2006) propose an explicit approach to measure semi-quantitatively the contribution of a product to the fulfilment of the various needs, i.e. to the consumer's happiness. We can use this idea to expand on the new measure that we have proposed in the preceding subsection by integrating the utility of the product to the consumer, i.e. how much it is contributing to the consumer's needs fulfilment, i.e. happiness. The approach by Hofstetter et al. (2006) needs further refinement and testing, as the authors make clear. Among others, we see as one main aspect for improvement the way of how the different kind of information is aggregated (see also the recommendations in Wolf and Chomkhasmri (2012) on substitutability/orthogonality of criteria). It should also be considered to exclude limitations due to required skills and information, as it can be argued that for the individual, who will make a consumption decision, only those products for which he/she has the necessary skills and information will be considered anyway. Finally and as a general limitation, the degree of needs fulfilment for the 'softer' needs, such as identity, participation, and so on, will much depend on the individual, and will likely escape a general agreement when trying to quantify a specific product's fulfilment of such needs.

Combining the net environmental impact of the analysed product $IMP_{net,P}$ that we have proposed above with a quantitative happiness-utility indicator of the analysed product HU_P (that we do not further work out here), we obtain a measure for the environmental intensity of needs fulfilment (aka of the product's happiness-utility) HUI_P. Note that this one is yet excluding yet other social and socio-economic secondary consequences other than those captured by the creation of qualified work and income in the supply-chain (Eq. 13.4):

$$HUI_P = \frac{IMP_{net,P}}{HU_P} \tag{13.4}$$

In contrast to the average consumer, to which we refer with this formula, for any specific consumer any specific consumption decision will of course look at activities and products that enable these that best meet that consumer's currently least well fulfilled needs. The indicator would still be the same, but the consumer would only consider those products that contribute to fulfilling his/her specific, most pressing current needs.

Finally, it needs to be highlighted that an over-fulfilment of some of our needs is leading to physical and mental health issues, be it overweight, dependence on alcohol, nicotine and other drugs (including on medication, gaming). Also information overload and the limited ability to keep abreast with new technologies can be understood to potentially counteract needs fulfilment and happiness.

Hence, maximizing needs fulfilment in the sense of summing up the happiness-utility results has limits for some of the needs. Also the linearity and the balance across the needs should be observed when looking at the overall needs fulfilment of a person.

Further work is needed here.

3 Sustainable Consumption on Different Levels

3.1 Product Level: From Functional Unit to Needs Fulfilment

Comparative product LCA studies analyse the life cycle wide impact products per functional unit of each product; i.e. in relation to "which function(s)" each product provides, "how much" of the function, "how well" and for "how long". This basis serves to compare alternative products.

In a consumption perspective, and particularly for consumer products, it makes sense to expand this functional unit also to the human needs fulfilment: The direct function of a product ultimately serves to meet a range of human needs. While its primary, technical function of many or most products typically relate to only one of the physical basic needs, e.g. mobility, housing, or food, it always contributes also to meet other, psychological basic needs, such as for example affection, participation, and identity (Max-Neef 1991). The relevance for consumption decisions to meet also needs such as "identity" can be illustrated by the relevance that brands have in clothes consumption decisions. This example also illustrates that it will be an individual judgment how well a product meets these "soft" needs.

3.2 Person or Household Level: Sustainable Lifestyles

Moreover, it should be highlighted that needs fulfilment is often done rather by complex activities, that involve different products in a specific combination that create a new quality, rather than by simply consuming each of them: a simple walk in the park may involve a combination of outdoor clothing, maybe an umbrella, a bus trip to reach the park and for the way back, using the restaurant service to have a cake or ice-cream etc. and – important in the context of secondary consequences – involve an individually decided period of time. While the distance walked and the life time of the shoes have some relevant causal relation, otherwise the duration of many activities can be largely independent from the actual consumption of goods. Particularly the fulfilment of higher needs are less directly related to product consumption, other than more basic needs such as food and shelter.

In view of efforts to a more sustainable consumption and lifestyles, it is important to consider that very different activities – using possibly also the same amount of the household resources, but with a hugely different environmental impact – may still contribute to the same degree to the needs fulfilment and happiness for the same individual person. Using leisure time and money for meetings our "soft" needs can differ as much as taking a longer motorbike ride, playing a game on a smartphone, or practicing yoga, depending on the person's preference. Also meeting our physical basic needs can be done in different ways, while again using the same amount of the other household resources. One of the possibly most widely discussed component of sustainable consumption is eating vegetarian versus a meat-rich diet. As another example, for the need shelter/housing, zero-energy houses have much lower overall life cycle impacts than less well designed and insulated houses, possibly at the same total cost of ownership.

On the next more complete level, we look at the entire consumption of a person or it's household. We agree with the literature that the individual will aim at optimizing the use of his or her household resources to achieve a maximum fulfilment of the needs. Which needs are considered how relevant and how well the individual understands which products best contribute to fulfil these needs, is obviously different for each individual.

The quantification of the impact of consuming a product, i.e. including the many secondary consequences, carries a very high uncertainty, as illustrated in the preceding chapters. Somewhat surprisingly, a much more accurate guidance can be given to individuals if looking at the entirety of consumption: The sum of all consumption – e.g. in form of lifestyle scenarios – has no secondary consequences across the person's available household resources, as they are all covered in the total by definition. This allows to build scenarios of different lifestyles and calculate and compare their overall environmental impacts.

Some limitations will still reduce the accuracy and precision of the results of lifestyle-level studies:

- Accurate LCI data are not available for many specific products yet, respectively approximations are less precise, and available data from different countries is not widely interoperable (see e.g. the findings of a recent survey among National LCA databases globally in Wolf (2014c))
- Secondary consequences on society level are not covered or including them adds a relevant uncertainty, e.g. changes in road congestion if the individual uses public transport instead of a car.
- Effects on changing consumption patterns upstream of the supply-chain, i.e. at those individuals that earn extra income by contributing to the production of the purchased goods.
- Finally, the calculation of how well the specific lifestyle fulfils any specific person's needs will have a high uncertainty.

However, defining alternative lifestyles and assessing their overall environmental impact and utility, using the approach proposed in Sects. 2.7 and 2.8, will allow individuals to reflect on his/her own lifestyle and allow to adopt or adapt a more sustainable one. In summary, sustainable consumption decisions mean to meet the same needs in a less impacting way without overly triggering secondary consequences by changing the available household resources.

3.3 National Level: From Territorial Inventory to Including Burdens of Imported and Exported Products

Quantifying the environmental impact of different lifestyles on national level would have to look at different adoption-levels, as e.g. sparsely distributed electro-charging stations means additional travel to recharge the vehicle and additional transport means to come to these stations and back home or to the office. On a national level of consumption, we can hence capture such effects on infrastructure within the country. The only secondary consequences that escape the analysis are changes in international infrastructure, such as e.g. airports, and via changed amount of imports from those countries.

On the national level, past studies on the nations "footprint" have often looked at the territorial level only. However, since a number of years, more and more studies also consider the import and export of goods and delivery of international services, and the upstream burdens associated with their life cycles.

An advanced approach to this idea has been piloted in a study commission by the European Commission in 2008 on the consumption-based national resource efficiency (European Commission 2012): Territorial data, mostly based on official statistics, were combined with full process-based life cycle data for the 15 most important traded product groups. These product groups were represented by representative products (e.g. "passenger car" for the product group "road vehicles" or "methanol" for "organic chemicals") and the inventories were scaled up to the amount of goods traded in each product group. The rest of trade was approximated by the mix of those that were explicitly modelled. It was moreover possible to model the inventories of the traded goods for the two or three most important source countries. Despite some weaknesses, particularly in the territorial data, the study could show for many impact categories that a shifting of burdens occurred from Europe to other countries, i.e. while territorial impacts were slowly reducing, due to an increased import or higher processed products, the overall EU consumption-based environmental impact is increasing with time.

The main sources of lack of accuracy and of uncertainty in such approaches – next to the mentioned territorial data that is weak in several impact categories – are limitations in life cycle data on specific products for a range of product groups, particularly more complex consumer products and services. Also, the approximation of a product group by one representative product carries a relevant uncertainty, which can be overcome only by increasing the number of products to approximate a product group. The recent increase in availability of Environmental Product Declarations and Footprints for all kinds of products is a promising development, which can be expected to substantially ease such calculations.

Such studies are valuable to inform policy makers about true consumption-based trends in environmental impact, and to identify the main product groups and trade partners from and to where such a shifting of burdens happens and inform related policies. One key advantage of this approach is that these studies can be tailored and further developed to be very specific on traded products to address specific policy questions. In contrast, Environmentally Extended Input Output (EEIO) studies are limited by the very broad range of whole industry sectors, which cannot well differentiate below industry sectors. Moreover, EEIO is based on economic relations across the economy, hence its life cycle data is closely correlated with money that is an important limiting factor to consumption as discussed above, hence will lead to only rough and possibly distorted results.

3.4 Global Level: The Sum of All Consumption Versus the Planetary Boundaries

Studies on global level necessarily take a more comprehensive perspective, including all human activities. Monitoring the overall environmental impact is a very high-level indicator that can be used also to evaluate in how far we surpass the planetary boundaries. It shows on the largest of scales and only slowly over years whether all the measures by individuals up to national governments and international agreements show success in terms of slowing or reversing the trend of increased environmental impact on a global level.

One important topic under discussion in the context of sustainable consumption is shifting production to low income countries with potentially lower environmental standards: Cheap products have often been criticized as increasing the environmental impact, as they allow for more consumption. If we moreover assume that the most cheap products are so because the staff in their supply-chain are poorly paid (next to general productivity increase), the cheaper products mean also a shifting of parts of the income and hence consumption from middle or higher income countries to low income countries, but also to richer people (in both low and high income countries), since the cheaper products mean they have additional household income available. As statistics show, the allocation of the available household income considerably varies depending on income level. The net effect of this consumption shift still needs to be quantified.

4 Actors in Sustainable Consumption and Their Possible Roles

4.1 Companies: Sustainable Products

As argued initially, the main sphere of influence for companies is the development and production of goods and the operation of services. It is essential that the environmental performance of these goods and services is improved, based on their functional unit. If these products change the available household resources, it is a consumer choice what to do with any additional resources.

Still, companies might want to better understand the secondary consequences of their products and how well the products meet the individual needs beyond the mere function. This should help companies to be prepared for the discussion on sustainable consumption. In fact, many companies do parts of this analysis already: fashion and lifestyle, time and space saving are key selling points in many product segments. The environmentally negative secondary consequences are however out of the scope of the analysis. This also means that conflicts are to be expected between offering an environmentally more efficient product with time-saving properties to the consumer, while at the same time trying to avoid that this leads to secondary consequences that party or fully compensate the environmental advantages of the product.

A design for sustainable consumption would then be an extension of conventional Ecodesign by considering consumption-related secondary consequences, and – as Hofstetter et al. (2006) argue – basic needs that require satisfaction. A first step would be to provide quantitative information on the effect on the consumer's household resources, particularly time and space saving, and expand on the information on the cost of ownership, that is currently legally to be provided for some consumer goods only. Regarding food calories, this information is already standard information on food products.

On lifestyle level, companies across sectors may start working together to develop and promote a portfolio of products for sustainable lifestyles and/or ecosystems.

4.2 Citizen: Sustainable Consumption Decisions and Lifestyles

All final consumption is decided on by the consumer, while marketing, the role models that media personalities play, and the media in general, but also family and friends are influencing consumption decisions. For implementing sustainable consumption to reach a sustainable life style, consumers require awareness of sustainability and need to receive sufficient and correct information to support their consumption decisions (Wolf 2014a).

Taking more sustainable consumption decisions means purchasing, using, and end-of-life managing products that – while fulfilling the consumer's human needs in at least the same degree as alternative products do – have a lower environmental life cycle impact, including to quantitatively consider the secondary consequences, including the higher order consequences in the society and due to possibly freeing human working time. In Sects. 2.7 and 2.8 we have already sketched a respective quantitative measure on this last named aspect "Sustainable consumption" has been defined quite early already, such as in (Norwegian Ministry of the Environment 1994).[8]

The entirety of consumption of a citizen is facilitating the person's lifestyle. A lifestyle is – in its broadest sense – "Ways of life, encapsulating representations, values and beliefs, behaviors and habits, institutions, economic and social systems." (UNEP 2011). In context of this chapter however, we refer mainly to the consumption of goods and services that enables the individual to create and live his or her lifestyle, similar to the definition promoted in context of the Marrakech process: "*Sustainable lifestyles are patterns of action and consumption, used by people to affiliate and differentiate themselves from others, which: meet basic needs, provide a better quality of life, minimise the use of natural resources and emissions of waste and pollutants over the lifecycle, and do not jeopardise the needs of future generations.*" (Thidell 2011, adopted from CSD 2004).

The individual will aim at maximising the utility of his or her household resources, i.e. optimize the needs fulfilment. In Sect. 3.2 it was explained why it will be more accurate to calculate the environmental impacts of the entire consumption profile of an individual's lifestyle, than of individual products. We therefore argue that it makes sense to define a range of lifestyles and calculate their overall

[8] Unfortunately, in one of the most prominent and recent global efforts to "Develop recommendations for effective policies on Sustainable Lifestyles" (UNEP 2011), the reference to life cycle approaches is essentially limited to the glossary.

environmental profile. Individuals can then learn which lifestyles have which environmental consequences and see if they want to shift their own lifestyle into the direction of one of the less impacting ones.

However, such "model lifestyles" needs variation, not only because of differences in the individual taste and belief, but already because the available household resources vary (e.g. different times and income bound for commuting distances, with/without children, different health, other long-term obligations, etc.).

Moreover, when defining one's lifestyle, people often refer to approaches, rather than individual products or to concepts that would capture the entirety of the lifestyle. "I am vegetarian", "I separate waste", and "I buy local" are a few examples of such approaches. These approaches help consumers to group specific decisions and to communicate them, as well as combining a lifestyle. The challenge is that not all decisions that are taken in line with such approaches are actually environmentally beneficial (already if not considering secondary consequences). Some approaches that are perceived as environmentally advantageous can even be more impacting, see e.g. examples for misconceptions about polymers in Wolf et al. (2010). It will be important to analyse which of these approaches are actually environmentally beneficial, again including considering the secondary consequences on available household resources.

If any such approaches are fully followed, they can also lead to infrastructural changes at the consumer. For example, "I prefer public transport" may lead to the decision to not have a private car anymore.

We would like to add that the above refers implicitly to middle and upper consumers, while families of low and lowest income classes will have less choices to shift to more sustainable lifestyles, in their struggle to meet at least their most basic needs. Moreover, given their low income, they typically have a lower per person environmental impact than better-off families.

4.3 Governments: Facilitating Sustainable Consumption and Lifestyles

Several past and current policies and initiatives have supported sustainable consumption (and production). Starting on the international level, the Sustainable Consumption and Production (SCP) program by the United Nations Environment Programme (UNEP) is based on the achievements of the 1992 United Nations Conference on Environment and Development in Rio de Janeiro (the Earth Summit), and the 2002 World Summit on Sustainable Development (WSSD) in Johannesburg (UNEP 2012). The European Union launched the Beyond GDP initiative, aiming at developing indicators that are as clear and appealing as GDP, but intend to be more inclusive of environmental and social aspects of progress, and the Action Plan on Sustainable Consumption and Production (SCP) that has life cycle thinking in its core, to name a few. Similar programs have been started in many other countries worldwide.

The government itself is a big consumer, with governmental spending in the range of 1/6 of nation-wide spending (e.g. in the EU 2002: 16 %). Green Public Procurement (GPP) is therefore a means that can have a key steering effect for more sustainable products. Continuous efforts are made for better informed GPP with comprehensive, life cycle based indicators (e.g. in the recently started project EURECA for GPP of data centre services (NN 2015)).

Beyond this product-perspective and next to creating markets for less impacting products, the scale of government procurement can also facilitate the creation of infrastructures in support of green procurement by consumers: Governments set the rules of the society and establishes or steers the development of key infrastructures, which can be favouring more sustainable consumption. Government can hence also provide options for less environmentally impacting consumption, e.g. public transport.[9] Similar to the situation of companies that offer cost and time saving aspects of their products for the direct benefit of the consumers, also governments generally follow the approach of saving costs and time for the citizen. Therefore, only by understanding the society-wide implications including due to secondary consequences of their projects and policies, the governments can fully take their role of steering consumption towards a long term stable, i.e. sustainable one. This includes to steer or counteract the transformational effect (Greening et al. 2000).

Identifying or developing elements that make up sustainable lifestyles and facilitating their adoption by implementing the required infrastructure are key tasks. Promoting sustainable consumption and sustainable lifestyles, as well as facilitating them by financial measures are other, main leverages of governments, on the way to a sustainable society. R&D investments into sustainable products and lifestyles and their infrastructure, Green Public Procurement, and education courses and campaigns for schools and university courses, are further examples for suitable governmental activities. On international level, the coordination with other national governments will help improving the common understanding of sustainable consumption and measures. This list above illustrates the crucial role that governments have in steering the society to sustainability.

4.4 *Others*

A range of other actors play a role in society and also in efforts to a more sustainable consumption:

[9] Such measures can have relevant negative environmental secondary consequences, if they free household resources at the consumer, i.e. if they are cheaper, save time (or allow to do two things at the same time, e.g., working during commuting), as already mentioned. Gillingham et al. (2013) however have found from the analysis of studies that negative environmental secondary consequences of energy-efficiency improvements are typically in the range of 5–30 % and hence less than is sometimes feared and warn that paying too much attention to single cases where the effect is higher may be used as excuse to not take action.

Industry associations – similar to companies – are important sources of high quality life cycle data for products that best represent the industrial reality. They can moreover disseminate information about sustainable products and sustainable consumption aspects to their members and bring in the voice of the represented industry into the public discussion on sustainable consumption.

Green and consumer NGOs can support bringing understanding and knowledge on sustainable consumption to consumers and contribute to the public discussions on sustainable lifestyles.

Research bodies and consultants are essential to help increasing the understanding and knowledge base on secondary consequences of consumption decisions, by developing better methods and models, and by offering software tools and data to support the analysis.

5 Conclusions and Next Steps

Life cycle management as a process has a history that reaches back more than 25 years, while as a term it has risen to prominence only about 15 years ago. Big progress has been made in collecting and analyzing data and information on environmental interventions along the supply-chains and working together to improve the environmental performance of many thousands of products and many tens of thousands of processes, worldwide. The life cycle data availability for this kind of analysis has constantly grown and now allows – while less so for countries with a shorter history in life cycle approaches – to get reliable results also for complex products, particularly, if the producing industry is actively involved and experienced experts support the analysis.

Product life cycle analysis and management was and still is the core also for the slowly developing field of sustainable consumption analysis, which needs to employ in addition to LCA complementary methods and data to also capture the secondary consequences outside the analysed product and directly connected products. While first life cycle based studies on the secondary consequences go back to the late 1990s, given the much more complex effects and higher effort, a much smaller share of studies has looked into it and the body of evidence is growing only slowly.

With increasingly better availability of process-based life cycle data and more and more companies publishing Environmental Product Declarations and Footprints, the evidence and process-based life cycle data basis for consumption and lifestyle studies is now further expanding.

If we want to achieve a reduced global environmental impact, we need to approach this from both the production and the consumption side. Unfortunately, the reduced environmental impact of many products and per functional unit is so far overcompensated by increased overall consumption – more products with a larger function per average person and an overall growing world population.

As recommendations, we consider the most important steps for public and private organisations to support consumers in their consumption and lifestyle choices to be the following ones:

- Increase the availability of interoperable life cycle inventory and impact data, including Product Environmental Footprints and other Environmental Product Declarations, so they can serve for better consumer information and can be combined into lifestyle studies. Agreements on interoperability across industry and governments – ideally on a global level – would be needed for this, in coordination with the software and data developers in consulting and research that support such work.
- Improve the data and evidence base for consumer choices on how freed household resources are reallocated. As interim step, working with scenarios as described in Sect. 3.2 can serve. Governmental research efforts should be directed at this task.
- Develop robust methodologies for capturing the secondary consequences of consumption, expanding on proposed approaches e.g. of Hofstetter et al. (2006) and of the environmental intensity of needs fulfilment, e.g. advancing the approach that we have sketched in Sects. 2.7 and 2.8. Government research funding or dedicated method development calls – as multi-stakeholder projects rather than a research exercise – in support of government analysis would be essential here. In this, it will be important to bring together experts from the fields of economy and life cycle experts: differences in terminology and approaches need to be overcome in interdisciplinary work.
- Stepwise develop a wide set of more environmentally sustainable lifestyles as archetypes for consumers to adopt and adapt from. As argued in Sect. 3.2, it is more accurate to assess the overall environmental impact of a whole archetype lifestyle than for many individual products, because the important secondary consequences on available household resources are automatically covered and zero. Scenarios on lifestyle approaches (e.g. "buy local", "eat vegetarian") should be analyzed and used to support communication with consumers to separate the more efficient and effective approaches from those that only apparently reduce the environmental impact.
- Consider secondary consequences when developing public infrastructure, financial instruments and other legislation directed at more sustainable consumption and lifestyles, e.g. in public transport, internet bandwidth[10] and others.
- Continue and strengthen the efforts for developing and distributing education and information materials for industry, government officials and citizen – the understanding of what makes up a "sustainable products" and sustainable lifestyles is argued here to be very limited outside a few dedicated expert groups.

[10] A recent expert workshop on environmentally sound data centers, organized by the European Commission's DG CONNECT and composed of data centre developers, operators and users, has warned that "*bandwidth growth needs to be better linked to the ability of technological developments to cope with it in terms of the related energy consumption and environmental impacts*" (Wolf 2014b).

Expand this with information on the secondary consequences of consumption – to enable those that are interested to adopt a more sustainable lifestyle have the necessary information to do so and avoid negative secondary consequences and hence largely or fully useless efforts. Edutainment TV programs might be one suitable format to this aim, particularly to reach out to consumers.

While we could not address such topics in this chapter, we would like to point out that the social and socio-economic impacts of consumption must not be forgotten but need to be integrated into the analysis, next to the environmental impacts that were the focus of this chapter.

The challenge ahead for humanity is truly one that needs the combined effort of all actors, globally: The otherwise wanted and fostered increase in the eco-efficiency and energy-efficiency counteracts sustainability, what Hofstetter et al. (2006) call the "efficiency-trap". In consequence and further amplified by continued population growth, the absolute pressure on the environment is increasing, even though the products' environmental impact per functional unit is generally decreasing.

However, it depends on the choice of the individual consumer, i.e. each and any of us, which lifestyle we adopt and how we meet our true needs with our available household resources.

References

Becker GS (1976) The economic approach to human behavior. The University of Chicago Press, Chicago

Case KE, Fair RC, Oster SM (2014) Principles of economics, 11th edn. Prentice Hall, Upper Saddle River, 816 pp

CSD (2004) "Every little bit helps…" Overcoming the challenges to researching, promoting and implementing sustainable lifestyles, vol 48. Centre for Sustainable Development, University of Westminster

Dragulescu A, Yakovenko VM (2000) Statistical mechanics of money. Eur Phys J B 17(4):723–729

Eckersley R, Dixon J, Douglas B (2001) The social origins of health and well-being. Cambridge University Press, Cambridge

European Commission (2012) Life cycle indicators for resources: development of life cycle based macro-level monitoring indicators for resources, products and waste for the EU-27. European Commission, Joint Research Centre, Institute for Environment and Sustainability. EUR 25517 EN. Publications Office of the European Union, Luxembourg. ISBN 978-92-79-26423-8

European Commission (2013) Recommendation (2013/179/EU) on the use of common methods to measure and communicate the life cycle environmental performance of products and organisations, Annex II (Product Environmental Footprint (PEF) Guide

European Commission – Joint Research Centre – Institute for Environment and Sustainability (2010) International reference life cycle data system (ILCD) handbook – general guide for life cycle assessment – detailed guidance. First edn March 2010. EUR 24708 EN. Publications Office of the European Union, Luxembourg

Eyerer P, Wolf MA (2000) Zero emission – was geht und was nicht. In: Altner G, Mettler-von Meibom B, Simonis UE, von Weizsäcker EU (eds) Jahrbuch Ökologie 2001. C.H. Beck, München, pp 136–148

Gillingham K, Kotchen M, Rapson D, Wagner G (2013) The rebound effect is over-played. Nature 493:475–476, http://www.nature.com/nature/journal/v493/n7433/full/493475a.html. Accessed 9 Mar 2015

Girod B, de Haan P, Scholz RW (2010) Consumption-as-usual instead of ceteris paribus assumption for demand – integration of potential rebound effects into LCA. Int J Life Cycle Assess 16:3–11. doi:10.1007/s11367-010-0240-z

Greening LA, Greene DL, Difiglio C (2000) Energy efficiency and consumption – the rebound effect – a survey. Energy Policy 28:389–401

Hofstetter P, Madjar M, Ozawa T (2006) Happiness and sustainable consumption. Psychological and physical rebound effects at work in a tool for sustainable design. Int J LCA 11(1):105–115. doi:10.1065/lca2006.04.018

Huppes G, Ishikawa M (2005) A framework for quantified eco-efficiency analysis. J Ind Ecol 9(4):25–41

ISO 14040:2006 (2006a) International standard – environmental management – life cycle assessment – principles and framework. International Organization for Standardization, Geneva

ISO 14044:2006 (2006b) International standard – environmental management – life cycle assessment – requirements and guidelines. International Organization for Standardization, Geneva

Jevons WS (1875) The coal question. McMillan, London

Khazzoom JD (1980) Economic implications of mandated efficiency in standards for household appliances. Energy J 1:21–40

Kitou E, Horvath A (2008) External air pollution costs of telework. Int J LCA 13(2):155–165

Lundie S, Ciroth A, Huppes G (2007) Inventory methods in LCA: towards consistency and improvement. Final report. June 2007. In: UNEP/SETAC Life Cycle Initiative, Life Cycle Inventory programme, First phase 2001–2005. Task force 3

Maslow A (1954) Motivation and personality. Harper & Row, New York

Max-Neef M (1991) Human scale development. Routledge, New York

NN (2015) Datacenter EURECA project: Green DC EU public sector procurement coordination and support action. Horizon 2020. Project ID 649972. 2015 – ongoing

Norwegian Ministry of the Environment (1994) Oslo roundtable on sustainable production and consumption. http://www.iisd.ca/consume/oslo004.html

Rockström J, Steffen W, Noone K, Persson Å, Chapin FS III, Lambin E, Lenton TM, Scheffer M, Folke C, Schellnhuber H, Nykvist B, De Wit CA, Hughes T, van der Leeuw S, Rodhe H, Sörlin S, Snyder PK, Costanza P, Svedin U, Falkenmark M, Karlberg L, Corell RW, Fabry VJ, Hansen J, Walker B, Liverman D, Richardson K, Crutzen P, Foley J (2009) Planetary boundaries: exploring the safe operating space for humanity. Ecol Soc 14(2):32, http://www.ecologyandsociety.org/vol14/iss2/art32/. Accessed 3 Mar 2015

Schafer A, Victor DG (2000) The future mobility of the world population. Transp Res A Policy Pract 34(3):171–205

Thidell Å (2011) Sustainable consumption and green lifestyles – definitions and concepts. Background material for the Workshop on sustainable consumption and green lifestyles of the Nordic Council of Ministers. http://esa.un.org/marrakechprocess/pdf/Issues_Sus_Lifestyles.pdf. Accessed 3 Mar 2015

Thiesen J, Christensen T, Kristensen T, Andersen R, Brunoe B, Gregersen T, Thrane M, Weidema B (2008) Rebound effects of price differences. Int J Life Cycle Ass 13(2):104–114

UNEP (2011) Visions for change – recommendations for effective policies on sustainable lifestyles – global survey for sustainable lifestyles. ISBN 978-92-807-3116-3

UNEP (2012) Global outlook on SCP policies: taking action together. ISBN 978-92-807-3250-4

Weidema BP (2008) Rebound effects of sustainable production. Presentation to the "sustainable consumption and production" session of the conference "bridging the gap; responding to environmental change – from words to deeds", Portorož, 14–16 May 2008

Wolf K (2014a) Life cycle sustainability assessment in support of sustainable life styles. 11th Asia Pacific roundtable for sustainable consumption and production, Bangkok, 19–20 May 2014

Wolf MA (2014b) Environmentally sound data centres: policy measures, metrics, and methodologies. Summary report on a workshop organised by the European Commission's DG CONNECT on 1 Apr 2014, 9 July 2014

Wolf MA (2014c) National LCA databases – status and ways towards interoperability. www.maki-consulting.com. Accessed 5 Mar 2015

Wolf MA, Chomkhamsri K (2012) Selecting the environmental indicator for decoupling indicators – how evaluation approaches influence the results. Oral presentation and short paper at: the 10th international conference on EcoBalance 2012, Yokohama, 20–23 Nov 2012

Wolf MA, Baitz M, Kreissig J (2010) Assessing the sustainability of polymer products. In: The handbook of environmental chemistry – polymers – opportunities and risks II. pp 1–53. doi:10.1007/698_2009_10

Wolf MA, Pant R, Chomkhamsri K, Sala S, Pennington D (2012) International reference life cycle data system (ILCD) handbook – towards more sustainable production and consumption for a resource-efficient Europe. JRC reference report, EUR 24982 EN. European Commission, Joint Research Centre. Publications Office of the European Union, Luxembourg

Chapter 14
Life Cycle Management Responsibilities and Procedures in the Value Chain

Fritz Balkau, Eskinder Demisse Gemechu, and Guido Sonnemann

Abstract Product life cycles and companies' value chain dynamics now extend to far-away countries, linking a multitude of end-users with numerous upstream suppliers and manufacturers. The breadth of the sustainability issues of popular concern, together with the complex nature of supply chains from which they arise, leads to serious management challenges. These challenges have been met in different ways depending on the interests and the institutional context of the actors. Corporations are strongly focused on optimizing product performance through a reliance on life cycle assessment based procedures. Commodity sectors are often seeking harmonized sustainability performance across a broad geographical range. Management institutions and business associations are providing life cycle management frameworks for corporations, followed up with training, and further research into improved metrics. At regional level some efforts have been made to introduce life cycle approaches, e.g. sustainable procurement, but the formal application of structured life cycle management is not yet widespread. The different approaches taken by the above actors reflects not only their different situations, but also the lack of a clear universal framework for life cycle management and a more generalized toolbox that will support their sustainability ambitions throughout the value chain. Limitations of current life cycle assessment methodologies imply that not all sustainability challenges are addressed in a consistent manner.

Keywords Life cycle assessment • Life cycle management • Responsibilities • Supply chain • Value chain

F. Balkau (✉)
Sustainable Solutions, 4 rue le Regrattier, 75004 Paris, France
e-mail: fbalkau@gmail.com

E.D. Gemechu • G. Sonnemann
University of Bordeaux, ISM, UMR 5255, 33400 Talence, France

CNRS, ISM, UMR 5255, 33400 Talence, France

G. Sonnemann, M. Margni (eds.), *Life Cycle Management*,
LCA Compendium – The Complete World of Life Cycle Assessment,
DOI 10.1007/978-94-017-7221-1_14

1 Introduction

Product life cycles and companies' value chain dynamics are always complex. Value chains now extend to far-away countries, linking a multitude of end-users with numerous upstream suppliers and manufacturers. Production inefficiencies at each step rapidly cascade through these chains, leading to unwanted products and wastes accumulating along the way:

> ...only about 1 % of all the materials mobilized to serve America are actually made into products still in use six months after sale, ... (Hawken et al. 1999)

Social conditions have gained greater visibility. The popular press throws a regular spotlight on deplorable labor practices and human rights conditions before a product even reaches the consumer. With these expanded notions, the environmental impacts have become more difficult to understand by the general public and many decision-makers. Our concerns have progressed from simple notions of waste and resource depletion to encompass also subtle effects of trace chemicals (e.g. endocrine disruptors) and the massive degradation of our life-support systems. There is much to do in response to these pressures, from improving the performance of individual products to a general overhaul of our entire production and consumption system. In both cases, we need a life-cycle approach to our management intervention if we are to do more than just tinker with the problem.

2 Looking into the Life Cycle Management Agenda

When referring to procedures within life cycle management (LCM), it is important to be clear about our propositions. There is an evolution in language that includes mention of life cycle, value chains, supply chains, materials cycles, circular flows and so on. In part this reflects the growth of life cycle type studies, processes and evaluations, but it soon spills over also into operational management concepts. Even the notion of 'life' of a material or product is not fixed. And if we are concerned with both upstream and downstream aspects of company activities we can no longer refer simply to 'supply chain', in which only the upstream focus is considered. In this paper we have chosen to refer to the 'value chain' since it covers multiple product life cycles of a company. Value chain can better include non-material and non-environmental values such as social appreciation and cost as well as management dimensions that are directly related to company activities and not to the engineering aspects of a product. All these are important considerations as the notion of sustainable consumption and production becomes more and more a question of operationalization in a business context.

So how can we manage the value chain from a sustainability perspective? The breadth of the issues of popular concern, together with the complex nature of the materials chains from which they arise, leads to a serious management issues (Balkau and Sonnemann 2011). As well as ensuring economic viability, managers need to take into account a host of externalities, many of them from outside the corporation. Attempts to link economic self-interest (i.e. efficiency gains) with these externalities has only been partially successful as some issues clearly represent a non-recoverable cost. It is easy to speak of the 'triple bottom line'; operationalizing it for extended value chains remains a challenge for individuals and enterprises alike.

A part of LCM is concerned with improving profitability of corporations. However, many LCM objectives are directly drawn from the sustainability agenda, i.e. building a better future for us all. While the main components – social, environmental and economic – are now universally accepted, the specific goals to be achieved within them are subject to wide interpretation. This is especially the case for the social components of labor practices, human rights and cultural and intellectual property. Even environmental standards vary widely.

The term life cycle management is currently applied to a wide variety of initiatives that show major differences in approach (Remmen et al. 2007). Objectives vary widely, from optimizing a product to saving the planet. Much of LCM can be seen as a fusion of supply chain management (SCM) and extended producer responsibility (EPR), using life cycle assessment (LCA) and other assessments as input. But while LCA and a variety of corporate management tools have been codified, LCM itself is still subject to varying definitions, interpretations and practices according to the viewpoints and objectives of the principal actors (Seuring 2004). It does not help that the notion of sustainable development is itself evolving, resulting in shifting targets for any management endeavor.

In fact, we are seeing several approaches to LCM evolving side by side. Thus companies unilaterally managing the upstream dimension of their supply lines work in parallel with sector-wide initiatives that operate under an agreed chain-of-custody framework that includes codes of conduct, independent review and sanctions for non-performance. Some progressive companies do both simultaneously, albeit not always for the same products.

While we observe that LCM has already been extensively used, closer examination reveals that many of these applications lack the rigor that is a feature of LCA itself. Supply chains are kept deliberately short to enable more effective management control, and objectives are reduced to a few parameters that reflect current corporate viewpoints rather than the aggregate needs of the environment. A common deficiency of many LCM exercises is that they often overlook the importance of the downstream consumer where much of the sustainability impact actually occurs. In short, for practical reasons much LCM lacks both depth and breadth.

3 Corporate Practice in Life Cycle Management

Much effort by corporations and supporting institutions has gone into translating the results of LCA into actual management interventions of their value chains (Remmen et al. 2007). Classic examples include the LCM practice of major manufacturing companies like 3M (2015) as well as of global retailers such as Walmart (2015). Their approaches show many similarities, but also some differences in the way they are managed. While the retailers see LCM predominantly through the prism of supply chain management, the manufacturers are often more focused on using LCM internally to improve product development and reduce production costs. In such cases internal co-ordination and collaboration between different parts of the organization becomes important. The breadth of this collaboration will depend on the objectives of the LCM exercise, i.e. whether it is primarily intended to influence product development or more aimed at addressing external sustainability issues in the supply chain. Like most management exercises, a mainstreaming of LCM in a corporation under the direction of the CEO (Chief Executive Officer) will be more successful than creating a special 'add-on' LCM service. The latter faces too many obstacles of acceptance by the existing mainstream departments.

For the retailers, it is vital to bring the entire upstream supply chain into LCM exercise as this is where many of the major environmental and social impacts need to be addressed. Thus IKEA's Code of Conduct (IKEA 2012) specifies minimum requirements on its 1600 suppliers covering social, environmental and labor-related conditions. While the code of conduct provides a clear statement of objectives, compliance is unlikely to be achieved automatically; much work is needed with supply chain actors to inform, explain, educate and train key partners along the chain. Some 'recognition' instruments can be used to identify reliable suppliers e.g. certification of ISO 14001. While major companies do work directly with their suppliers to facilitate compliance with company requirements, it is not well documented how far up the supply chain the influence is actually exerted.

TRUCOST has undertaken surveys of how many companies publicly report on their suppliers' impacts. TRUCOST found that of the environmental damage caused by the world's largest 3,000 companies annually, 49 % comes from within supply chains (Salo 2015).

> Some companies, such as Puma, are actively collecting supply chain environmental performance information for use in their business decision-making. Puma, part of the PPR Group, conducted a detailed analysis of the environmental impacts of its operations and supply chain. Only 6 % of the impacts come from Puma's offices, warehouses, stores and logistics. The rest come from its supply chain, more than half from the production of raw materials for manufacture. The findings were used by Puma to review where its raw materials came from, and what materials to source at all.

Direct suppliers are the most visible. But managing the suppliers of the suppliers remains a complicated exercise for individual companies. Traditionally corporations simply relied on contracts with its supply chain. An earlier study by Seuring and Goldbach (2002) showed that the collective negotiation model for sustainable supply chain management is more successful than a command and control approach.

In some cases, the direction of the coordination may also need to be reversed. Suppliers with a number of different clients may find that the latters' demands are not compatible, or are inconsistent in other ways. This has led to some collective arrangements where several clients and their different suppliers agree on a common agenda, and perhaps also a common communication and certification system.

LCM involving sub-contracting and global supply chain management may face serious political hurdles and trade barriers. Sustainability requirements on suppliers may not be well accepted by foreign governments who see this as interference in their national affairs and an unwanted application of western environmental standards. The political disputes over products manufactured by child labor, and/or subject to lax or unenforced safety and environmental standards have been rumbling on in global trade negotiations for many years. Fair-trade labels are not seen the same way from opposite ends of the supply chain, and LCM has to be sensitive to such issues. While company requirements on their suppliers may appear to be simple contract arrangements between companies, the issues easily spill over into political rancor and trade reprisals. As well, the WTO has a general policy to avoid environmental conditionality in trade arrangements.

Management techniques will depend on which parts of the value chain are included in the LCM exercise. There is a gradual movement, often spurred by legislation, to consider downstream issues of consumer protection, efficiency in use and end-of-life disposal. Managing the downstream parts of the life cycle requires different procedures and skills because the consumer needs to be persuaded rather than commanded into conformity with the LCM objectives. While shaping consumer behavior remains a delicate marketing issue, some corporations are already reaching out to their clients about the appropriate use of their products. The example of Unilever is shown below (Unilever 2015). Unilever's sustainability strategy addresses environmental impacts across the value chain.

"Our commitment to reduced environmental impact extends right across our value chain – i.e. from the sourcing of raw materials through our own production and distribution to consumer use and eventual disposal of residual packaging. Consumer use accounts for around 70 % of our greenhouse gas footprint. Engaging consumers …. will be key to achieving our vision. Metrics for our four priority environmental impact areas across the value chain include greenhouse gas (GHG) emissions, water, waste, and sustainable sourcing. These metrics are designed to measure the impacts of our products when used by consumers, such as grams of greenhouse gas per single usage occasion. During 2009 around 1 500 products were assessed to allow us to understand their water, waste and GHG impacts in 14 of our largest markets. In 2009 we also started to develop a set of metrics covering social impacts. For .. brands with social missions, the metrics seek to measure the benefits they bring to society. In 2010, Lifebuoy used the new metrics, helping track the impact of Lifebuoy programmes on hand washing behaviours over a five-year period".

We are still a long way from a universal application of his concept, even if some examples are well documented. The notion of downstream LCM is linked also to the idea of Product Service Systems where the manufacturer has replaced a product by a service, as for example in renting products instead of selling them (the story of Interface, SafetyKleen and others).

4 Collective Model for Life Cycle Management

An alternative LCM model with good success is based on a formal collaboration that binds the principal value chain stakeholders through a common, agreed code. The Code specifies objectives, governance, partner responsibilities and a monitoring and reporting mechanism. This model is especially suited for sector-wide LCM programs that address system issues, with useful examples found in forest, fisheries, food and chemicals stewardship. While similar in most aspects in addressing supply chain issues, many of these programs do not necessarily label themselves as LCM. A particular feature is that many of them are not even based on formal LCA as a precursor; they address directly well-understood concerns of sustainability or of safety.

Several initiatives in the natural resource exploitation area have developed sophisticated LCM procedures based on agreed objectives. The FSC (Forest Stewardship Council) (https://ic.fsc.org/) and MSC (Marine Stewardship Council) (www.msc.org/) pioneered this approach using a formal code of operation and on-going monitoring and review. Company operations are independently verified and certified as code compliant, with public reporting on key performance indicators. Stakeholders have extensive opportunities to provide input via electronic means, in workshops and in governance meetings. The supply chain extends from producer through distributors to retailers and eventually down to the consumer who is able to exercise a choice based on sustainability criteria specified on the product label. 'Management' thus occurs at several decentralized points in this chain in line with an agreed collaborative charter. It remains that while the 'production' parts of the supply chain are well controlled, the downstream consumer end is weakly addressed, being reduced to a simple label on the product (fish, timber). Like much of LCM, there is no advice concerning the use of the product and how to deal with consumer waste at end of product life.

In addition to the above, a range of other initiatives can be found concerning consumer products aimed at fair trade, organic produce and the like. Many rely on a product label that identifies a particular environmental or social quality; however the management of the quality assurance behind the label is an unknown quantity. Third party certification is sometimes used, but much organic produce, for example, seems to have little rigorous LCM behind it (or LCA ahead of it).

In terms of life cycle reach the most complete LCM instrument and the most rigorous structure is without doubt the International Cyanide Management Code (www.cyanidecode.org) that aims to ensure safe and environmentally sound handling and use of this toxic chemical in the gold mining industry as shown below.

The International Cyanide Management Code was developed to help gold producers operate in ways that prevent risks to workers, the public and the environment at all points of the product's life. The Code requires signatory companies to apply strict rules, both technical and procedural, for handling cyanide along its life cycle. Signatory companies are regularly third-party audited and certified, with mandatory incident reporting. Reports are publicly available. An independent institute administers the code. A particular feature of the Code is its total value chain reach: even the upstream chemical manufacturers, suppliers and transporters must comply with the Code requirements before the end-user mining company can purchase the substance. The Code is sharply focused; it deliberately addresses only a limited number of sustainability objectives concerned with cyanide toxicity; it is not a 'do everything' code. Code membership brings operational benefits and public relations and financing advantages. A number of financing institutions and planning bodies now require code compliance before they deal with gold mining companies, while companies regularly mention the code in their public communications.

The code concept has also been used in related sectors, such as by diamond (the so-called Kimberley process (www.kimberleyprocess.com/) and jewelry suppliers (www.responsiblejewellery.com/) who have put in place a 'chain of custody' control of operations by their members. In each case, the entire supply chain is subject to the sustainability requirements of the end-user – pollution, risks, social conditions, etc. The selection of sustainability criteria varies among the different codes, usually incorporating a strong emphasis on social issues (e.g. Kimberley) as well as pollution-type factors (e.g. cyanide code). It may be criticized that these voluntary codes only bind their members. In the case of the cyanide code, over half of world gold production is now code compliant, an achievement that few other activity sectors can claim. In any case, a non-regulatory character is a feature of nearly all LCM exercises.

The usefulness of sector-wide agreements such as the above becomes clear when we consider LCM from a supplier perspective. Suppliers often have many different clients. It would be overly burdensome if each client were to impose its own sustainability or quality standard on the supplier who is then not only faced with additional administration but also a fragmentation of product quality requirements. Several initiatives of sector-wide LCM are know in, for example, the international textile trade, to coordinate diverse LCM requirements. These initiatives often focus on social and labor issues, less so on environmental agenda although examples of this are known also.

In some instances, government participation is necessary to ensure sufficient management buy-in. The UK's Defra, for example, sponsored an LCM initiative in the clothing sector to ensure better alignment with social and environmental expectations.

The initiative has voluntary participation from all major stakeholder groups implicated in the sustainable management of clothing, including at the supply end in Asia. The structure is less rigid than the codes of ICMI or FSC; however, it has the advantage of handling a broader range of sustainability issues (Fig. 14.1).

Another useful example comes from the electronics industry. The GeSI initiative (GeSI 2015) brings telecoms, appliance manufacturers and service companies together in an effort to improve traceability of materials used in manufacture of appliances (as well as actual improved environmental and social performance along

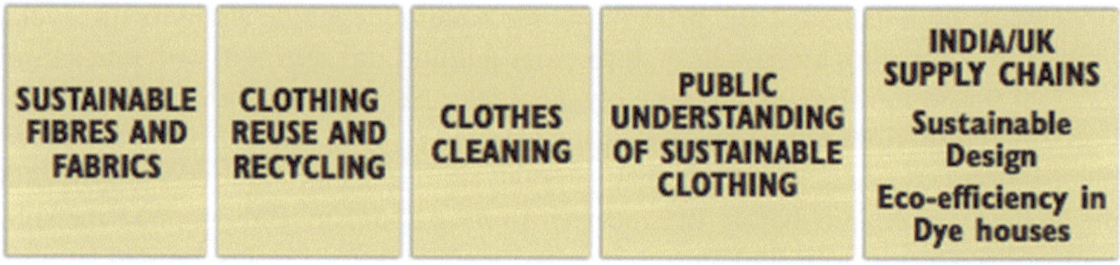

Fig. 14.1 The LCM initiative in the clothing sector (Source: Defra 2011)

the supply chain). Among other objectives it aims to improve environmental performance and resource efficiency and support sustainable supply chain practices. Of direct relevance to LCM is the conflict-free smelter initiative incorporating a labor code, certification, inspections and compliance listing. Further along the chain, manufacturers are encouraged to prefer purchasing from conflict-free smelters.

Table 14.1 Management instruments used in collaborative schemes

Stakeholder dialogue/conference
Agreement on criteria, objectives, targets
Code of conduct for members
Technical or procedural guidelines/codes
Chain of custody linkages along the value chain
Training for compliance initiatives
Third party audits
Certification of compliance
Non-compliance reporting
Transparent reporting
Product labels
Customer use advice
Membership-based structure
Secretariat to administer, monitor, review
Compliance may be required by planning/financial institutions
Compatible with supply chain management practices, eco-labels, environmental audits, sustainable procurement

The GeSI initiative is noteworthy for three aspects: (i) it is sector-wide and engages a large number of companies, (ii) it spans most of the value chain from mining of resources to consumer use and end-of-life recycling, and (iii) it includes multiple sustainability criteria from labor/social to energy to waste. GeSI has partnered with non-industry stakeholders like business groups, international institutions and NGOs.

We can summarize some of the management instruments used at various times in these collaborative schemes as per Table 14.1.

Each of the above collaborative initiatives has its particular origins. Each has taken an independent path to arrive at similar end-point where the collaboration is subject to an agreed framework and action plan. There is, however, no standardization of this approach that gives guidance to future initiatives elsewhere and in other sectors. An attempt to provide such a framework through ISEAL attempted to federate individual initiatives and to facilitate experience sharing. So far ISEAL regroups over 20 members (Fig. 14.2) mostly ones concerned with ethical and social issues surrounding resource exploitation (ISEAL 2015).

While obviously compatible with conventional LCM ideas, some of the collaborative initiatives may operate under other names such as SSCM, SMM, material management and so on. The diversity of nomenclature is perhaps one of the consequences of a lack of clear, formal and universally accepted framework of LCM.

5 Life Cycle Management and Business Organizations

Due to their influence over corporate management practice, it is useful to review also the role of management institutions and business associations in improving value chain performance. The peak business body WBCSD (www.wbcsd.org), while broadly endorsing a life cycle management approach, is concentrating on

Fig. 14.2 ISEAL members (ISEAL 2015)

improving the underlying metrics as, for example, practical materials accounting procedures.

This concern with metrics is also a preoccupation of other business groups, many of which have put substantial resources into developing appropriate measurement and assessment techniques for common use. ICMM (International Council for Mining and Metals) (www.icmm.com), for example, is a long-standing partner in the UNEP/SETAC Life Cycle Initiative. ICMM has been advocating a deeper use of 'sustainable materials management' (SMM) thinking by minerals companies.

This direction is echoed by the Australian government and the Minerals Council of Australian (MCA) in their advice to companies (Commonwealth of Australia 2006). The reach of the approach (up-stream suppliers to end-users), the breadth of sustainability objectives, its basis in scientific assessment, the range of management instruments to employ and the upstream/downstream multi-stakeholder involvement make these recommendations true LCM under another name (see Table 14.2, source ICMM (2006)). The value of these advisory initiatives is substantial because the advocated management framework ensures a coherent approach to LCM across

Table 14.2 Elements of successful implementation and integration – from ICMM (2006)

Strategy	Strong leadership for the initiative – champions in key business units, clear direction from the CEO and corporate action plan on sustainable development
	Clear understanding and awareness throughout the company of the business value
	A well-articulated policy or vision statement that addresses or include materials stewardship
Systems	Link to existing management systems, such as ISO 14001, as a means for issue identification, impact assessment, target setting and continual improvement
	Integrate materials stewardship into appropriate business planning processes and reward or incentive systems
Programs	Training and awareness raising programs for business units and departments where skills need to be developed – for instance on life cycle thinking, risk assessment, material flows, knowledge of environmental issues across the life cycle and in end-use markets, regulatory environmental trends
	Engagement programs – with suppliers, downstream manufacturers, users, recyclers and other stakeholders to identify opportunities for improving production, manufacture, use and end-of-life management of materials
	Measuring and reporting programs – develop targets and indicators for materials stewardship and report publicly to enhance reputation and accountability
	Commodity specific stewardship plans – develop in conjunction with other actors in the value chain
Tools	Assessment and decision support tools – eco-efficiency, design-for-environment, life cycle assessment and thinking, risk assessment and management, materials flow analysis
Data and other information	Life cycle inventories, risk assessments (environmental, health and safety), material flow analysis (common templates for compiling and communicating resource flows), recycling rates data, use/application data
	Organizations and contacts along the value chain

an entire industry. A sector-wide approach to LCM ensures a level playing field for companies and gradually moves to a more widespread adoption of the procedures.

Other sector associations are also recommending LCM approaches to their members.

The chemicals sector has also acknowledged the importance of managing value chain issues. ICCA's 2006 Responsible Care Global Charter (www.icca-chem.org/) states that "… member companies and associations commit to promoting the Responsible Care ethic, principles and practices along their own value chains … and to …develop and share best practices through mutual assistance, and to … work in partnership with upstream suppliers and downstream chemical users to collaborate on improved processes for the safe and effective uses of chemicals". ICCA organizes workshops and produces guidance publications to assist companies to implement these principles.

The International Solid Waste Association (ISWA) (www.iswa.org) published already in 2002 a forward-looking 10-year Perspective paper (ISWA 2002) advocating a holistic approach to waste management based on more complete consideration of the life cycle of materials that ultimately become waste. According to ISWA "… there is a need for a new approach. The focus must be upon the whole product life cycle and not only on the waste phase of products, substances and materials. Important decisions in relation to the amount of waste generated are taken both at the concept and design stage and further on during the production process. There is a need for dematerialization of the economic process…" Some of ISWA's workshops and documents are aimed at helping members move in this direction.

Some sector programs run in parallel with company initiatives. Thus AISE is encouraging a better use of its members' products (detergents) across their life cycle (AISE 2015), while Unilever is pursuing its own initiative to encourage clients to use less water and energy in the wash cycle (Fig. 14.3).

The frameworks developed by sector organizations are often the basis for LCM implementation by individual companies. Corporate members of FSC, MSC, ICCA, etc. frequently make reference in their stewardship and CSR (Corporate Social Responsibility) programs of their membership of these bodies, and the

Fig. 14.3 Cleanright best use tips (Source: AISE 2015)

compliance with the relevant codes, as well as conformity with independent initiatives such as GRI and Global Compact (see below).

Independent institutions are also supporting LCM through advice, training and sometimes through formal procedures. A good example of the latter is the Global Reporting Initiative (GRI) (www.globalreporting.org). The GRI Board of Directors identified disclosure on sustainability issues across the supply chain as a revision priority based on feedback from its committees and the wider GRI network. While GRI is not itself undertaking LCM, the increased transparency of individual company activities leads to greater management attention to the corporation's impacts along the value chain, eventually resulting in actual LCM oriented management interventions to improve value chain performance.

The Global Compact (www.unglobalcompact.org/) brings major corporations around a number of agreed sustainability principles. The GC Toolkit contains relevant instruments required for an application of life cycle management approaches, from assessment techniques to value chain interventions such as SCM and EPR.

There is also a role for policy intervention, beyond simply encouraging information and training. An in-depth review by the OECD titled "Sustainable Materials Management – Making Better Use of Resources" (OECD 2012) explores a set of policy principles for SMM application, examines how to set and use targets and analyses case studies from member countries. In this publication, SMM is closely aligned with LCM and explores the subsidiary instruments that make it function.

6 Life Cycle Management in Regional Development

LCM is not limited to corporate activity; it is an important concept at regional level where we have substantial resource and product flows, major infrastructure developments and important issues of social and human development. Many regional administrations have formally adopted sustainability objectives and are already undertaking footprint studies and ecological assessments. Life cycle ideas are creeping into individual programs and operations. But there is as yet little visibility of formal LCM approaches in most administrations, especially when confronted with long-term issues and complex material flows. This is not to say that LCM in not relevant, rather its use has not yet filtered down to regional administrators in a major way.

Nevertheless, some initiatives are already in this direction. Sustainable public procurement has become more evident in municipalities, although mostly without the technical backup in assessment. Nor do these programs actively "manage" the products they are purchasing – they are more of a 'buy/not buy' nature that sends relatively weak signals through the supply chain.

Thus while the Province of Nova Scotia has a sustainable procurement requirement under legislation, there is not a strong administrative capability to make it function effectively. A recent study in Canada found that supplier engagement was weak in most sustainable procurement programs at provincial level (Reeve 2012).

The OECD, UNEP and the World Bank are now undertaking a programme of studies, workshops and publications to improve the quality of public procurement to boost national and regional economic progress. The sustainability dimension is included in this work.

Some states regulate ostensibly for life cycle objectives, as for example in California.

> California's Green Chemistry Initiative 2012 (GCI) is a new law to regulate toxic chemicals in consumer products. The law incorporates a life cycle "alternatives analysis" to evaluate alternatives during their manufacture, in use and at disposal. Based on the evaluation, the state may restrict or ban the use of chemicals of concern.

Materials flow studies are now of increasing interest to public authorities, as is the use of footprint studies to better understand the regional sustainability dynamics. The OECD Environment strategy for the first decade of the twenty-first century (OECD 2001) outlined the need for governments to look for integrated management solutions that link resource use and prevention of waste into a coherent policy approach, such as the one embodied in the sustainable materials management (SMM) paradigm. The recent OECD report on SMM (OECD 2012) is directly aimed at clarifying the policy measures that improve the efficiency of materials flows at national and regional levels. But while useful in identifying national case studies, the policy recommendations sometimes seem rather remote from on the ground life cycle considerations.

The recent series of summer schools held by the University of Salento was directly aimed at exploring the application of life cycle approaches to sustainable regional development (www.lcss.unisalento.it/). This initiative has now been expanded into a publications project to illustrate how LCM techniques can be applied to regional planning, resource management, infrastructure, industrial development and environment protection.

But there is also the question of how LCA and LCM techniques can be further adapted to serve the special needs of these institutions and the administrators that would use these techniques.

7 Link with Other Sustainability Management Initiatives

LCM has to be seen also in the context of other resource management concepts as most of these have their roots in some form of life cycle thinking, and many of them advocate a particular management approach. Concepts of circular economy, industrial ecology, 3R, cradle to grave, etc. have been around for many years (Fig. 14.4).

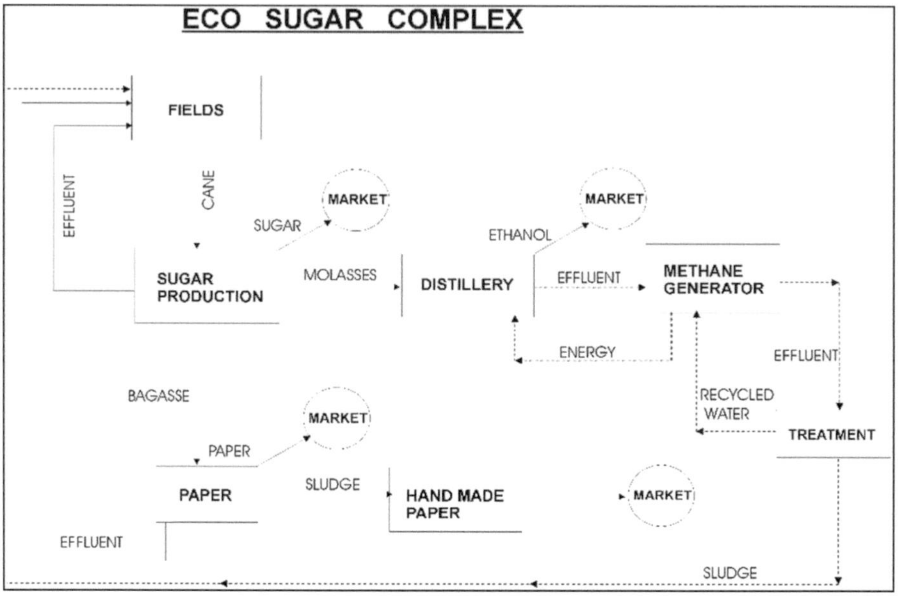

Fig. 14.4 Opportunities for LCM in a circular, life cycle, industrial ecology, 3R model for industrial production (Source: UNEP)

Life cycle in their conception, they are usually presented in materials flow rather than management terms and have their main impact on LCM objectives rather than on the management processes used. Conversely, the various programs on human rights and social conditions are not especially focused on a value chain approach but their objectives can still be taken into account in LCM applications. Finally, it is useful to try to link the downstream dimension of LCM with various 'consumption side' programs in the UN system, with green purchasing initiatives, PSS ideas and EPR for producers and retailers. Many of the solutions they propose can be more easily operationalized through LCM than on a stand-alone basis.

The extension of LCM into the downstream consumption area presents particular challenges, however this is precisely where much of the impact, and hence the real need, often lies. For many appliances and for buildings, for example, over 80 % of energy use, and much materials consumption occurs during use. In the food cycle, end-point cooking is the most energy-demanding step. We already saw above the attempts by the detergent industry to address this point. Overall, LCM's reach so far into the downstream parts of the value chain can best be described as timid. Effective management action will no doubt need new partnerships with public policy institutions, governments and consumer organizations, but it cannot deny that this is an important area to address.

On the manufacturing side the concepts of cleaner production, eco-efficiency, green productivity and 'pollution prevention pays' have been adopted in many places, often incorporating recommendations for management that can be incorporated into LCM procedures (audits, materials flows, partnership-building …). Many useful solutions have come from these concepts especially in major corporations that have the management strength to operationalize them.

It is in organizational frameworks that LCM finds its closest 'management' allies. The EMS framework of ISO 14001 is ideally suited to incorporate a value chain version of sustainability management. Corporate Social Responsibility (CSR), although not itself standardized, can also provide a framework for those LCM objectives related to social, human rights and labor issues. Management constructs like SCM and EPR underlie much of the life cycle 'management' methodology. And it hardly needs pointing out that the panoply of standardized (or not) assessment tools based around ISO 14040 series (among others) is what underpins LCM in the first place.

Many of the above management tools have tended to be used individually. LCM incorporates them into larger framework, adding, where needed, procedures such as product service systems (PSS), eco-design, eco-labeling, risk management and so on.

It is also evident that LCM in its attempt to influence the value chain of materials and products needs to find an accommodation with other management regimes, ranging from the WTO position on environmental condition in trade, the demands of future climate agreements, to the EU and OECD directives on chemicals management (e.g. REACH). Some of these have a certain value chain influence themselves that can be 'captured' by LCM.

At the policy level we have seen how SMM adopts the same value chain perspective as LCM.

As the building of stakeholder relations is at the core of successful LCM (whether internal or with external partners), the management of key player identification, communication, transparency and trust building is at the heart of the exercise. Coercive management models can have only limited application in a value chain. LCM can usefully learn from certain business sectors that have extensive experience in using such partnership techniques. Thus the toolkit developed by ICMM for successful community stakeholder building would be useful also in guiding LCM in other sectors.

Inevitably LCM, like many other areas of decision-making, will rely on third-party information, criteria and standards as a basis of its action. Its sustainability objectives are commonly based on international codes of practice and a variety of standards, whether ISO or other. In some cases work with, or endorsement by, major respected international NGOs will give credibility to the LCM exercise. But major corporations also adopt their own independent objectives on, for example, embedded energy or water use, recyclability and chemical composition. The use of third party standards and certification puts an extra dimension on the LC management exercise with which practitioners may be unfamiliar.

LCM is not the only sustainability game in town. In the end, LCM has to interface with the operational level to parallel initiatives and concepts. There can be no single universally valid technique because circumstances and objectives can be so diverse. But it does suggest that LCM should at certain moments be seen as a concept rather than a fixed technique, to be in turn the umbrella or the tool, as circumstances dictate. LCM's contribution will always be to propose an intervention strategy across the value chain reach of whatever is under consideration, whether it be a product (e.g. a detergent) or a procedure. (e.g. procurement).

8 Summary and Conclusions

The complex world of product life cycles and companies' value chain dynamics in our material world is mostly hidden from view. The complexity of addressing an extensive network of value chain stakeholders, coupled with the multiple criteria in the sustainable development agenda, means that a rigorous 'total life-cycle' management is almost impossible. A number most LCM initiatives are selective, focusing on only a limited number of steps and a selected few sustainability criteria. Realistic scoping is thus an important prerequisite to success. Within these limitations a number of different approaches to LCM are being pursued.

The success of any LCM exercise depends greatly on its objectives, how it is conceived and on who and what is included. For corporations focused on their products, the objectives will not be the same as for public institutions interested in optimizing the system as a whole. It follows that the methodologies and tools will not be identical. Thus collective code-based LCM is often interested in a global improvement in certain common issues such as social and labor conditions. Their methods include a big dose of multi-stakeholder building, often reaching down to their consumer base, plus transparency and communication. Individual manufacturing companies are more focused on using LCM to improve environmental performance of their products through reducing wastes, enhancing technologies, creating new markets and reducing liabilities. Much of their methodology is applied in-house although supplier engagement is necessarily a part of the exercise. Big retailers are interested in a positive profile for their products with consumers, and thus reach deep into their supply chain to try to achieve this. Few of the above have yet put downstream consumers, product effectiveness and end-of-life issues as mainstream components in their LCM exercises to the same extent as their focus on materials and social content. As a general statement, we can say that LCM has found more application in product enhancement than in systems optimization.

The public policy interface with LCM is still relatively undeveloped. There are many reasons for this, including political and short-term economic factors. But a major barrier is lack of appreciation by regional and national administrators of how LCM techniques could improve their program delivery. The currently inefficient use of sustainable public procurement is one example of this. This problem is exacerbated by the fact that many life cycle instruments are poorly suited for use at this level, either being too complex to use, or by not sufficiently incorporating some of the sustainability criteria that preoccupy the regions. It is also true that regulatory procedures often lack the flexibility to incorporate life cycle instruments due to the compartmentalized way, in which regulations are often developed. Further work by OECD and other institutions may help to break down these barriers and encourage a more comprehensive and less sectoral approach to sustainable materials management (SMM) and the like.

The varying interests of the major players mean that we can see several parallel strands of LCM being pursued, sometimes masquerading under other names. This parallelism arises from independent but converging programs. The visible UNEP SETAC Initiative with its past work on LCM was heavily influenced by the experience and viewpoints of the LCA community. The sector programs arose out of a need to address controversial sustainability issues that could not be resolved unilaterally such as overfishing, forest depletion and cyanide accidents. The SMM work sprang from the understanding by sector associations and governments that the issue of advanced waste management through resource stewardship had to be tackled at a level higher than single companies. Even earlier, the use of sustainable supply chain management arose from a need to reduce corporate liability for actions of environmentally unscrupulous suppliers and contractors. Sustainable procurement was seen as one of the immediate concrete actions that could be taken by institutions to implement Agenda 21. The various management operations under the above have some common elements but also some differences. The cooperative mechanisms may for example not rely on LCA as a basis of action. But it is also notable that the language and definitions of the theoretical constructs vary significantly, and there are a few human contact points where exchange of experience takes place. The LCM family is still very diverse, and its management approaches show significant differences.

In the above context, a comment can also be made about the adequacy of current assessment techniques as a basis for LCM. We noted above that LCA, while useful for resource flows, environmental impacts and cost evaluations, does not provide adequate input on management related parameters along the whole value chain of a company. Thus it gives too little information about the key (management) players actually found along the value chain who will have their part to play in effective LCM coalitions. Nor does LCA yet comprehensively evaluate a number of important sustainability criteria such as biodiversity, social conditions or trade impacts. Without such enhancements in the assessment phase, the management operations are handicapped in their ability to deal with a significant number of sustainability issues. Without such enhancements with regard to considering these criteria, LCM will continue to rely on a particular view with regard to sustainable consumption and production as well as some level of intuition and individuals' subjective understandings as much as hard data.

Overall, there remain significant challenges to a more rigorous approach to LCM at all levels. For the moment LCM remains more a management art than science, trying to bridge the gap between hard data and intuitive decision-making. Promising work in bringing more science in this field has been done for instance by Morel (2014). This type of work needs to continue and to include also management responsibilities in the products' life cycles and companies' value chains.

References

3 M (2015) 3 M sustainability. Retrieved March 15, 2015, from http://www.3m.com/3M/en_US/sustainability-us/industry-solutions/

AISE (2015) Consumer activities. Retrieved March 10, 2015, from http://www.aise.eu/our-activities/information-to-end-users/consumer-activities.aspx

Balkau F, Sonnemann G (2011) Addressing sustainability issues through enhanced supply chain management (chapter 18). In: Renko S (ed) Supply chain management – new perspectives, Industrial engineering and management. InTech, Rijeka. doi:10.5772/799

Commonwealth of Australia (2006) Stewardship – leading practice sustainable development program for the mining industry. Retrieved from www.minerals.org.au/file_upload/files/resources/enduring_value/stewardship.pdf

Defra (2011) Sustainable clothing roadmap: progress report 2011. Retrieved from https://www.gov.uk/government/publications/sustainable-clothing-roadmap-progress-report-2011

GeSI (2015) Global e-sustainability initiative. http://gesi.org/

Hawken P, Lovins AB, Lovins LH (1999) Natural capitalism: the next industrial revolution. Boston: Little, Brown & Co.

ICMM (2006) Maximizing value – guidance on implementing materials stewardship in the minerals and metals value chain. International Council on Mining & Metals (ICMM), London, Retrieved from www.icmm.com

IKEA (2012) The IKEA our code of conduct – IWAY. Retrieved March 17, 2015, from http://www.ikea.com/ms/en_IE/about_ikea/our_responsibility/iway/

ISEAL (2015) ISEAL members. Retrieved March 10, 2015, from http://www.isealalliance.org/

ISWA (2002) ISWA's 10 years perspective on waste management. http://www.iswa.org/en/home/search/

Morel S (2014) L'empreinte environnementale à l'ère de la société collaborative: de l'Analyse du Cycle de Vie comme outil expert à une instrumentation collaborative pour conduire une transition organisationnelle, École Nationale Supérieure des Mines de Paris. https://pastel.archives-ouvertes.fr/tel-01152159/document

OECD (2001) OECD environmental strategy for the first decade of the 21st century. Retrieved from http://www.oecd.org/environment/indicators-modelling-outlooks/1863539.pdf

OECD (2012) Sustainable materials management – making better use of resources. http://www.oecd.org/env/waste/smm-makingbetteruseofresources.htm

Reeve (2012) The state of municipal sustainable procurement in Canada – best practices & current trends. Retrieved from https://reeveconsulting.files.wordpress.com/2011/05/2012-state-of-munic-sust-proc-canada.pdf

Remmen A, Jensen AA, Frydendal J (2007) Life cycle management – a buisness guide to sustainability. UNEP, http://vbn.aau.dk/files/12154798/_Final__version_for_internal_use_20070509.pdf

Salo J (2015) Beyond the brand: leaders in supply chain environmental sustainability. Retrieved March 17, 2015, from http://www.trucost.com/blog/104/beyond-the-brand-leaders-in-supply-chain-environmental-sustainability

Seuring S (2004) Industrial ecology, life cycles, supply chains: differences and interrelations. Bus Strateg Environ 13(5):306–319. doi:10.1002/bse.418

Seuring S, Goldbach M (eds) (2002) Cost management in supply chains. Physica-Verlag, Heidelberg, pp 15–30

Unilever (2015) Sustainable living. Retrieved March 14, 2015, from http://www.unilever.com/sustainable-living-2014/our-approach-to-sustainability/our-strategy/

Walmart (2015) Responsible sourcing. Retrieved March 17, 2015, from http://corporate.walmart.com/global-responsibility/ethical-sourcing/

Chapter 15
Policy Options for Life Cycle Assessment Deployment in Legislation

Annekatrin Lehmann, Matthias Finkbeiner, Clare Broadbent, and Russ T. Balzer

Abstract Life cycle thinking is on the political agenda and widely used in practice. Moreover, numerous industries have actively been developing life cycle assessment (LCA) approaches for many years. As the authors think that it is in substance "right" to base environmental legislation on LCA, they started to explore and to develop policy options for integrating LCA into legislation. Commissioned by WorldAutoSteel, the authors focused on CO_2 legislation in the automotive industry, but the options developed based on this example can be used for other industries and other environmental impacts as well. It was found that theoretically a broad range of policy options exists, and that practically some of them are already implemented in real world legislation and that there is no clear scientific overall preference for one single option. It was also shown that solutions for most technical requirements are already available, but that a consensus on proper setting of these requirements is missing.

Keywords Automotive sector • CO_2 • Legislation • Life cycle assessment • Life cycle management • Life cycle thinking • Policy initiatives

1 Introduction

Life cycle thinking (LCT) and life cycle assessment (LCA) (ISO 14044 2006) (Baitz et al. 2012; Finkbeiner 2012) gain increasing importance in policy (Reimann et al. 2010; Inaba et al. 2003) and the authors think that to base environmental legislation on LCA is in substance "right". In fact, LCT is already considered in

A. Lehmann (✉) • M. Finkbeiner
Technische Universität Berlin, Strasse des 17. Juni 135, 10623 Berlin, Germany
e-mail: annekatrin.lehmann@tu-berlin.de

C. Broadbent
World Steel Association, Brussels, Belgium

R.T. Balzer
WorldAutoSteel, Springboro, OH, USA

G. Sonnemann, M. Margni (eds.), *Life Cycle Management*,
LCA Compendium – The Complete World of Life Cycle Assessment,
DOI 10.1007/978-94-017-7221-1_15

some current legislations and investigated in ongoing policy initiatives, e.g. the Product Environmental Footprint method (PEF) (European Union 2013), which is currently widely discussed amongst various stakeholders (Finkbeiner 2014; Galatola and Pant 2014; Lehmann et al. 2015). Besides this, LCA has been widely applied in practice for many years, and several industries, companies and associations are actively developing LCA approaches (Finkbeiner et al. 2000). Examples are the World Steel Association and WorldAutoSteel, its automotive group, which published their position to plea for LC based regulation, e.g. in automotive CO_2 legislation and proposes to consider LCT in post 2020 legislation (World Steel Association 2013).

Against this background, in 2013 WorldAutoSteel commissioned Technische Universität Berlin (TUB) to explore and develop policy options for integrating LCA into (automotive) legislation. The research project is still ongoing.

Automotive CO_2 legislation is a relevant example for illustrating the necessity for considering a LC perspective. The reason is that it is shown that the current focus on tailpipe or exhaust emissions (the use phase) is not sustainable anymore because CO_2 reductions in the use phase can come along with increasing CO_2 emissions in other LC phases like the production phase (Daimler 2014; Krinke 2009; Kendall and Price 2012; PE International 2013), which are typically referred to as the embodied energy or environmental footprint.

The first idea in the research project was to "simply" move from tailpipe based CO_2 limits to LC based CO_2 limits. But the second thought was to explore alternative policy options as well. The underlying methodology and the results are presented in Sects. 2 and 3 respectively. A discussion of the key findings and an outlook is provided in Sect. 4.

Though the policy options are developed and described for the example of CO_2 legislation in the automotive sector, they generally can be transferred to other sectors and other environmental impacts as well.

2 Methodology

The methodology of the research process to explore and describe different policy options is presented below. It comprises the identification of various policy options including prioritization and the identification and description of characteristics of the prioritized policy options, including e.g. technical requirements as well as potential strengths, opportunities, and threats.

2.1 Identification of Policy Options

To identify the policy options, the authors chose both a theoretical and practical approach. In the theoretical approach, they used four different structural elements, which were combined to develop policy options. In the practical approach, they

Table 15.1 Structural elements to define policy options, their features and description

Structural elements	Features of policy option	Description of the policy option features
Type of enforcement	Mandatory	Policy is legally binding; defined requirements (e.g. limit/target values) have to be fulfilled
	Voluntary	Policy is not legally binding, but intends to have indirect effects
Levers	Performance	Policy defines product requirements; if they are not met, product needs to be re-designed
	Process	Policy defines requirements on company level for process improvement
Use of LCA	Direct (full LCA)	Policy directly defines LC based limit/target values and/or requires communication of full LCA results
	Indirect (LCT)	"Backoffice" use of LCA or LCA data during policy development such as target value setting processes
Market role	Market access	Policy provides minimum requirements (threshold values) for products to be able to enter the market
	Market incentive	Policy provides framework (e.g. criteria) for promoting environmentally preferred products

analyzed existing legislations, which are related to CO_2 and/or have implemented LCT, i.e., include an LC perspective. A first focus was put on Europe (EU).

The structural elements are summarized in Table 15.1 and describe individual features of possible policy options. The features reflect different types of enforcement (mandatory or voluntary), different levers (on a process- or a product level), different stringencies on the use of LCA (consideration of full LCA or LCT) and different market roles (access or incentive). The policy options defined based on the structural elements as well as theoretical and practical examples (where available) are presented in Sect. 3.

Regarding the market role of policy options: "market access" is used to exclude products (and services or processes) that have a low performance, e.g. environmental performance – from the market (typically ca. 10–20 % of products). An example of how to implement this market role is the Conformité Européenne (CE)-mark (EC 2015). "Market incentive" usually aims at promoting the 10–20 % "best in class" products (and services or processes). It can for example promote environmentally friendly products by using the Eco label type I (ISO 14024 1999) labelling system in the market. Therefore the majority of products (ca. 70–80 %) are set in between these two market roles.[1] General effects of "market access" and "market incentive" policy options on the share of environmentally preferred products on the market are shown in Fig. 15.1.

After having identified various policy options, they were prioritized based on three criteria, namely rigor of implementation (referring to the type of enforcement

[1] These two roles can be implemented by several policy tools e.g. type approval, limit values, permits, labels, taxes, subsidies, and procurement rules (e.g., Green Public Procurement (GPP).

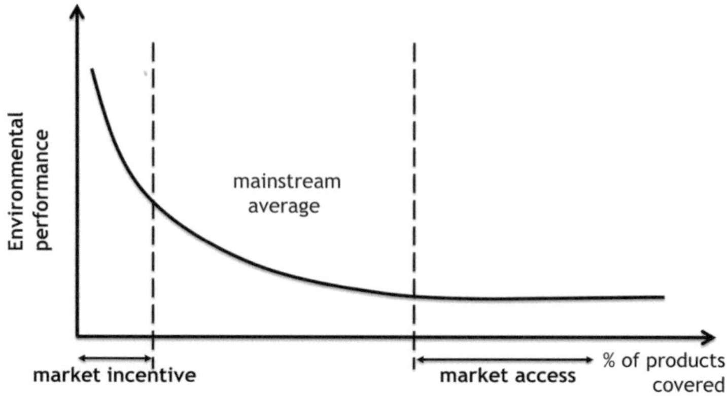

Fig. 15.1 Effects of "market access" and "market incentive" policy options on the share of environmentally preferred products on the market

and to the levers), rigor of LCA (referring to the extent of LCA use) and stakeholder acceptance (referring to an assumed acceptance or willingness of stakeholders for LC implementation). The prioritized policy options were subject of further detailed analyses, described in the following section.

2.2 *Identification and Description of Characteristics of the Policy Options*

Characteristics here refer to technical requirements for policy implementation. These requirements also determine (though not solely) further characteristics of the policy options like applicability (easiness) or acceptance and thus are relevant to identify for example potential strengths or threats of the particular options.

For this work, the authors specified six technical requirements, namely methodology, models, data, tools, quality assurance and communication and described them for the prioritized policy options. Based on this, they further analyzed the options using a SWOT[2] analysis to reveal their potential strengths, weaknesses, opportunities and threats. This analysis was complemented by the RACER[3] analysis, which originates from the European Commission's Impact Assessment Guidelines (EC 2009) to assess the value of scientific tools for use in policy making based on the criteria relevance, acceptance, credibility, easiness and robustness. For both the SWOT and the RACER analysis different stakeholder perspectives were considered obtained

[2] SWOT: strengths, weaknesses, opportunities, threats.

[3] RACER: relevance, acceptance, credibility, easiness, robustness.

from stakeholder meetings with policy makers, original equipment manufacturers (OEMs), material industries and academia in Europe and beyond (in the USA, Japan and China).

3 Results and Discussion

The following two sections present selected results of the ongoing research process. The results focus on a detailed theoretical description of the identified policy options for LCA deployment in legislation including existing examples from practice. In addition – but on a higher level – key findings from the analysis of the characteristics of the policy options are provided.

3.1 Developed Policy Options

Based on the theoretical approach, by combining different structural elements describing policy features, eleven policy options for LCA deployment were developed. The options range from voluntary to mandatory policies, describe legislations with and without direct effects on the product design, legislations which are either fully LCA based or which consider LCA or LCA data during policy development (without requiring full LCAs) and legislations addressing two different market roles i.e., "market incentive" and "market access".

An overview of the 11 policy options identified is presented in Fig. 15.2. For voluntary policy options, the feature "market access" is not foreseen because of the nature of the policy options, they are not legally binding but intend to have indirect effects on the market.

The following sections provide further details on the policy options, including examples of how these options could theoretically look like. Moreover, examples from practice (focusing on Europe) are provided, allocated to the theoretical options and showing that some of the options are in fact already implemented in real world legislation. The sections are differentiated between *mandatory-performance* based options, *mandatory-process* based options and *voluntary* policy options.

3.1.1 Mandatory-Performance Based Policy Options

These options cover mandatory policies, thus "hard" legislations, which are based on product performance. Consequently, they have a direct effect on the product design and may require a product re-design. The *mandatory-performance-direct* policy option (Table 15.2) is the most stringent solution for using LCA in policy, while the *mandatory-performance-indirect* option (Table 15.3) is less stringent as it does not require full LCAs. The latter can be seen as an intermediate step for full

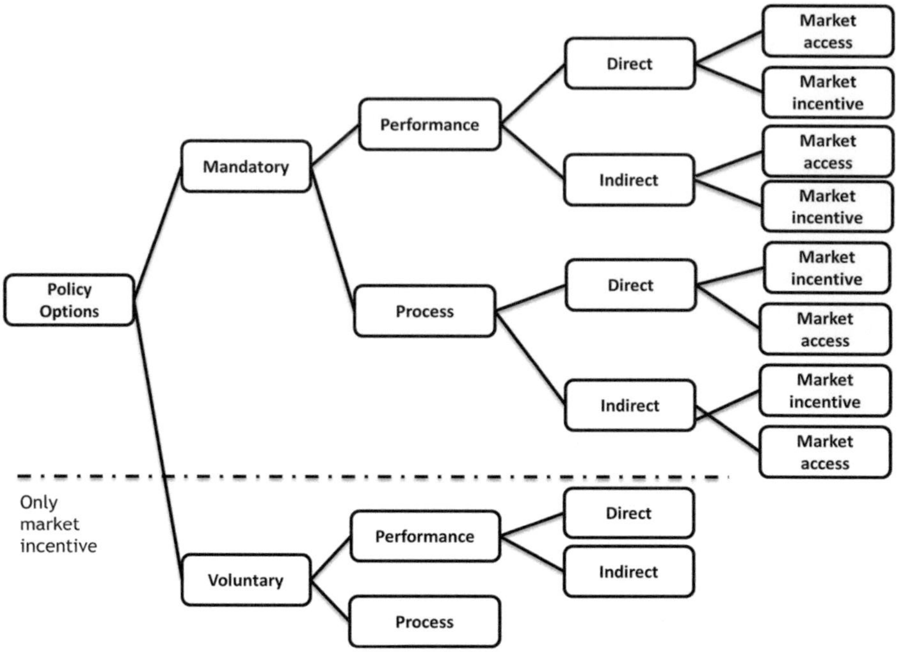

Fig. 15.2 Policy options for LCA deployment in legislation

Table 15.2 Mandatory-performance-direct policy options – theoretical and practical examples differentiated according to the market role

Mandatory-performance-direct policy options		
Market role	**Theoretical example (LC CO_2 legislation of cars)**	**Examples from practice (EU)**
Access	Company has to show a proof that LC CO_2 emissions do not exceed defined limit values; e.g. an EU fleet limit of 100 g LC CO_2/km or LC limit values per vehicle class	Renewable Energy Directive (RED): LC based limit values are provided for renewable energies; e.g. if the company shows that their biofuels meet these targets, they can enter the market as biofuels
Incentive	(a) Company has to provide LC CO_2 values; the performance is used for criteria in GPP	No policy in the EU
	(b) Company has to show LC CO_2 emissions, e.g. by using an "LC CO_2-label"; possible market advantage if consumers consider environmental performance for purchase decisions	

Table 15.3 Mandatory-performance-indirect policy options – theoretical and practical examples differentiated according to the market role

Mandatory-performance-indirect policy options		
Market role	**Theoretical example (LC CO_2 legislation of cars)**	**Examples from practice (EU)**
Access	Company has to show a proof that CO_2 emissions in relevant phases, e.g. in the use phase of a car do not exceed defined values, e.g. Xg CO_2/km in use phase or Yg CO_2/kg in the production phase of car	EU Ecodesign Directive: Uses LCA studies in the preparatory study to identify implementing measures which are not full LC based
Incentive	Company has to show/publish CO_2 emissions, which occur in a relevant phase, e.g. the use phase of a car (e.g. by using a non-LC CO_2 label)	(a) CO_2 label for vehicle: Information on tailpipe CO_2 emissions per km has to be available at the point of sale in EU
		(b) Clean Vehicles Directives: Requires that energy/environmental impacts linked to the operation of vehicles over their whole lifetime are considered in all purchases of road transport vehicles, as covered by the public procurement (PP) directives and the public service regulation

LCA based legislation. It also reflects existing policies, which are especially relevant for the automotive industry (e.g. CO_2-label for tailpipe emissions).

3.1.2 Mandatory-Process Based Policy Options

These options capture mandatory policies, thus "hard" legislations, which are not directly related to product performance but refer to technical aspects of process based policies, e.g. requirements on a company level for continuous improvement. Thus, they are less stringent than the performance based options. The *mandatory-process-direct* policy option (Table 15.4) can be seen as intermediate steps to use LCA as basis for performance based legislation. The *mandatory-process-indirect* policy options (Table 15.5) as the weakest form of mandatory legislation can be considered as an introductory step towards full LC based (direct) and/or performance based legislation.

3.1.3 Voluntary Policy Options

These options cover voluntary policies, thus "soft" legislations. The *voluntary-performance-direct* and *-indirect* options are based on product performance, and can have a direct effect on the product and may require a product re-design.

Table 15.4 Mandatory-process-direct policy options – theoretical and practical examples differentiated according to the market role

Mandatory-process-direct policy options		
Market role	**Theoretical example (LC CO_2 legislation of cars)**	**Examples from practice (EU)**
Access	Company uses LCA to determine LC CO_2-emissions, e.g. an environmental product declaration (EPD) must be available, but the product performance documented is not decisive for type approval	No policy in the EU
Incentive	LC CO_2 values must be available for GPP, e.g. for having an EPD a bonus in PP can be received, but the product performance documented is not relevant	No policy in the EU

Table 15.5 Mandatory-process-indirect policy options – theoretical and practical examples differentiated according to the market role

<table>
<tr><th colspan="3">Mandatory-process-indirect policy options</th></tr>
<tr><th>Market role</th><th>Theoretical example (LC CO_2 legislation of cars)</th><th>Examples from practice (EU)</th></tr>
<tr><td rowspan="2">Access</td><td>(a) Company has to show a proof that CO_2 emission values in relevant phases (e.g. use phase) are determined (e.g. average values/fleet)</td><td rowspan="2">End-of-Life Vehicles Directive (ELV):
Requires to provide recycling/dismantling information as part of the type approval</td></tr>
<tr><td>(b) Company has to demonstrate a “certified” management system/strategy to decrease CO_2 for cars which is based on LCT, but does not require full LCA implementation</td></tr>
<tr><td rowspan="2">Incentive</td><td>(a) Company gets more points if it has an Environmental Management Auditing system (EMAS) for GPP; the information must be available</td><td rowspan="2">No policy in the EU</td></tr>
<tr><td>(b) Company has to show that CO_2 emissions, in relevant phases (e.g. use phase) are determined and documented, e.g. using a non-quantitative label like “we have used LCA”; providing this could bring bonuses</td></tr>
</table>

The *voluntary-performance-direct* option is the strongest voluntary policy option and can be seen as an introductory step to reduce resistance against mandatory legislations. The same applies to the *voluntary-performance-indirect* option, which additionally can be considered as basis for full LC based legislations. The *voluntary-process* based option is the weakest policy option and represents a very first step towards performance based and mandatory legislations. Further details on the three voluntary policy options as well as theoretical and practical examples are provided in Table 15.6.

Table 15.6 Voluntary-performance-direct and -indirect and voluntary-process based options – theoretical and practical examples (market role: only incentive)

Policy options	Theoretical example (LC CO_2 legislation of cars)	Examples from practice (EU)
Voluntary-performance-direct	(a) Recommendation to show LC CO_2 emissions, e.g. using a label (LC CO_2 label)	GPP: Allows including full LC performance criteria in GPP
	(b) Recommendation to use performance criteria documented in EPDs for GPP	
Voluntary-performance-indirect	Recommendation to show CO_2 emissions which occur in relevant phases, e.g. the use phase of a car, e.g. by using a label	EU flower: LCA is used to develop criteria
Voluntary-process	Recommendation to present CO_2-emissions as part of e.g. EMAS/inclusion of information on CO_2 emissions in environmental program	EMAS: Recommends using LCA in the process of continuously improving the company's environmental performance

As described in Sect. 2.1, in a second step the identified policy options were prioritized. The following four policy options were selected:

- *Mandatory-performance-direct*
- *Mandatory-performance-indirect*
- *Mandatory-process-direct*
- *Voluntary-performance-direct*

These policy options cover the whole range of policy option features defined in Table 15.1: mandatory and voluntary options, options with and without effects on the product design as well as options which use full LCA or LCA as "backoffice" for legislation. For these four options, further detailed analyses were conducted. The high level results are summarized in the next section.

3.2 *Characteristics of the Prioritized Policy Options*

The different policy options require different solutions for implementation, which depend on their particular characteristics. The characteristics of the four prioritized policy options were described taking into account the technical requirements, methodology, models, tools, data, quality assurance and communication.

It was found that the differences between the policy options are partly not very significant and that some technical requirements are the same for both voluntary and mandatory policies. For example, all *direct* policy options, which consider the full LC, have some technical requirements, which are similar for all of them, e.g. the need of LCA data and LCA tool(s) and the need to specify the product system model and LCIA model as well as characterization factors. On the other hand, *indirect* policy options – with its "backoffice" use of LCA – generally may have less strict technical requirements than *direct* options. Moreover, it was shown, that sometimes, *voluntary* policy options have stricter requirements than *mandatory* options, for example with regard to communication as they require a specific communication format to reach the consumer.

The technical requirements of the particular policy options also determine their applicability (efforts), comparability, robustness, relevance or stakeholder acceptance, and thus are relevant for their feasibility and probability of implementation. The SWOT and RACER analysis from different stakeholder perspectives revealed that generally highest relevance regarding CO_2 reduction, but also highest efforts for implementation seem to be related to the *mandatory-performance-direct* option. Moreover, it was shown that robustness and credibility can principally be guaranteed by all policy options and that acceptance strongly depends on the perspective of the stakeholders. Generally, it can be assumed that if the policy options are implemented properly, the acceptance is high, but if there is a risk of poor implementation the acceptance would be low.

4 Conclusions and Outlook

Within the first phase of the research process, promising policy options were identified and their characteristics were described. It was found that theoretically a broad range of options for implementing LCA into policy exists and that practically (with focus on the EU) some of them are already implemented in real world legislation. Moreover, it was shown that there is no clear analytical, scientific overall preference for one single policy option: for example, possible trade-offs were revealed (e.g. showing that some technical requirements may be lower for a particular policy option, others may be higher) and naturally no "black-and-white" results from the SWOT and RACER analysis were obtained. Also, it was found that technical implementation strongly depends on the implementation level and that solutions for most technical requirements are already available,[4] but that a consensus on the proper setting of these requirements is missing.

The selected portfolio of policy options and the knowledge on their potential benefits and shortcomings allow the development of a concept of how an LCA methodology can be used in future automotive emission regulations. Preliminary implementation scenarios were already described (but are not addressed in this

[4]For example, different available LCA methodologies (e.g., ISO 14044, Product Environmental Footprint (PEF)), databases for background data (Sonnemann et al. 2011), e.g., European Reference Life Cycle Data System (ELCD) from the European Commission (EC n.d.), ecoinvent, PE International) or data collection formats (Finkbeiner et al. 2003), models (e.g., UCSB model by WorldAutoSteel or the model from the European Aluminum Association (EAA)) or communication formats (e.g., EPDs, labels).

chapter), outlining how current regulations could potentially be amended to include a LC perspective and/or how new co-existing LC based CO_2 regulations could look.

The research process continues in 2015 and includes a broader stakeholder dialogue in the EU and beyond to communicate and refine the policy options as well as to specify possible implementation pathways for policy options aiming at LCA deployment in legislation. Though the focus in this work is/was laid on CO_2 legislation in the automotive industry, the policy options developed can be used for other industries and other environmental impacts as well.

References

Baitz M et al (2012) LCA's theory and practice: like ebony and ivory living in perfect harmony? Int J Life Cycle Assess 18(1):5–13, http://link.springer.com/10.1007/s11367-012-0476-x. Accessed 25 Oct 2013

Daimler AG (2014) Mercedes-Benz cars: 2014. Environmental brochure. smart fortwo electric drive. https://www.daimler.com/Projects/c2c/channel/documents/2243139_Environmental_brochure_smart_fortwo_electric_drive.pdf. Accessed 24 Nov 2014

EC (2009) Impact assessment guidelines. SEC (2009) 92. http://ec.europa.eu/smart-regulation/impact/commission_guidelines/docs/iag_2009_en.pdf

EC (2015) CE marking. http://ec.europa.eu/enterprise/policies/single-market-goods/cemarking/. Accessed 2 Mar 2015

European Platform on LCA (EPLCA). http://eplca.jrc.ec.europa.eu/. Accessed 28 Feb 2015

European Union (2013) Commission Recommendation of 9 April 2013 on the use of common methods to measure and communicate the life cycle environmental performance of products and organisations. Off J Eur Union L124

Finkbeiner M (2012) From the 40s to the 70s – the future of LCA in the ISO 14000. Int J Life Cycle Assess 18(1):1–4

Finkbeiner M (2014) Product environmental footprint – breakthrough or breakdown for policy implementation of life cycle assessment? Int J Life Cycle Assess 19(2):266–271

Finkbeiner M et al (2000) Life cycle engineering as a tool for design for environment. In: Total life cycle conference of the Society of Automotive Engineers in Detroit, USA, vol 26, p 189, 28 Apr 2000

Finkbeiner M et al (2003) Data collection format for life cycle assessment of the German association of the automotive industry. Int J Life Cycle Assess 8(6):379–381

Galatola M, Pant R (2014) Reply to the editorial "Product environmental footprint – breakthrough or breakdown for policy implementation of life cycle assessment?" written by Prof. Finkbeiner (Int J Life Cycle Assess 19(2):266–271). Int J Life Cycle Assess, 19(6):1356–1360

Inaba A et al (2003) Practical tools and thoughtful principles for sustainability. Int J Life Cycle Assess 8(1):1–5

ISO 14024 (1999) Environmental labels and declarations – type I environmental labelling – principles and procedures

ISO 14044 (2006) International standard – environmental management – life cycle assessment – requirements and guidelines

Kendall A, Price L (2012) Emissions, incorporating time-corrected life cycle greenhouse gas emissions in vehicle regulations. Environ Sci Technol 46:2557–2563

Krinke S (2009) Environmentally friendly vehicles and environmental strategy of the brand Volkswagen, 19 Nov 2009

Lehmann A, Bach V, Finkbeiner M (2015) Product environmental footprint in policy and market decisions – applicability and impact assessment. Integr Environ Assess Manag, doi:10.1002/ieam

PE International (2013) Life cycle assessment of low carbon cars 2020–2030, London. http://www.lowcvp.org.uk/resource-library/reports-and-studies.htm

Reimann K, Finkbeiner M, Horvath A, Matsuno Y (2010) Evaluation of environmental life cycle approaches for policy and decision making support in micro and macro level applications. JRC-IES

Sonnemann G et al (2011) Process on "global guidance for LCA databases". Int J Life Cycle Assess 16:95–97

World Steel Association (2013) Steel industry calls on member states for life cycle approach to EU vehicle emissions from 2020. http://www.worldsteel.org/media-centre/press-releases/2013/2020-Vehicles-Emissions-Life-Cycle.html. Accessed 28 Feb 2015

Part IV
Mainstreaming and Capacity Building on Life Cycle Management

This part includes a series of chapters addressing the challenges of mainstreaming life cycle management. It discusses opportunities to build operational capability and the potential for mainstreaming LCM in emerging economies through capacity building, concluding on the need to enhance communication and collaboration within the global LCA community.

Chapter 16
Taking Life Cycle Management Mainstream: Integration in Corporate Finance and Accounting

Cornelis Theunis Van Der Lugt

Abstract What does it take for life cycle management (LCM) to capture the attention of the financial community? LCM experts face a window of opportunity as technological progress and economic developments lead to greater interest in mainstreaming the sustainability agenda and integrating it in business decision-making. This includes new approaches to the assessment of impacts on Natural Capital, environmental management as well as annual corporate reporting. Having highlighted these, this chapter employs a Green Business Case Model to define ways in which life cycle applications can be employed to link with core financial indicators of special interest to investors. It describes three hypotheses to illustrate where LCM tools can best make a difference, positively affecting core financial value drivers. The author suggests three hypothetical pathways to capture the attention of investors, linking environmental life cycle costing (LCC) and financial, activity-based costing. These are complemented with company case examples. It draws lessons from past work on the business case as well research on environmental versus financial life cycle costing (LCC).

Keywords Accounting • Corporate finance • Life cycle assessment • Life cycle costing • Life cycle management • Life cycle thinking • Sustainability

1 Evolution in Assessment, Management and Reporting Standards

Technological progress and improved understanding of the impact of a growing world population on resource use globally has set the scene for a gradual transformation of environmental assessment, management and reporting tools over

---

C.T. Van Der Lugt (✉)
Center for Corporate Governance in Africa, Stellenbosch University Business School, Cape Town, South Africa
e-mail: cornis.lugt@usb.ac.za

G. Sonnemann, M. Margni (eds.), *Life Cycle Management*,
LCA Compendium – The Complete World of Life Cycle Assessment,
DOI 10.1007/978-94-017-7221-1_16

the last decade. While globalization and trade liberalization has led to greater integration in the world economy, it has also been accompanied by increasing fragmentation in production (see OECD 2012; Elms and Low 2013). The life cycle assessment (LCA) community has seen a lively debate on, for example, the value of deepening research to the level of product subsystems versus widening research through systems expansion (cf Curran 2013), as well as combining process-LCA and Input–output (IO) LCA to deal with the complexities of long supply chains and product chain organization (see Gereffi et al. 2005; Finnveden et al. 2009; Koh et al. 2013; Lake et al. 2014; Eriksson and Olsson 2011; Baumann 2012).

Furthermore, global financial crises as well as dramatic cases of corporate failure has led to renewed questioning of the role of corporate reporting. Frustration about information overload in annual reports and apparent lack of what is really material or strategic information has led to the birth of an integrated reporting movement. The International Integrated Reporting Council (IIRC) seeks to promote "integrated thinking", akin to "life cycle thinking". In defense of the GRI Guidelines, all its environmental indicators reflect a life cycle approach. This includes impacts at the end of the useful life of the product, especially important for life cycle management (LCM) as environmental life cycle costing (LCC) takes into account use- and end-of-life phases and hidden costs (Klöpffer 2008). The LCA response to the integration challenge has been to define the emergence of life cycle sustainability assessment (LCSA) (see Finkbeiner et al. 2010; Guinee et al. 2011; Hellweg and Canals 2014)

Against this background of value chain complexity and initiatives in favor of integration, this chapter seeks to define pathways along which life cycle assessment (LCA) applications and life cycle thinking can be integrated with core business planning and strategic financial performance (Eun et al. 2009). Importantly, this chapter is not about moving from environmental and/or social LCA to economic LCA. Rather, it is about moving from environmental LCA to business finance and accounting. Its reflections will also be relevant for what integration in the form of sustainability LCA (a new integrated LCA or compilation of separate assessments) implies, and how life cycle sustainability assessment (LCSA) could be engrained in corporate financial planning.

2 Linking Life Cycle Assessment with Financial Value Drivers

The weighting of different impact categories and resource use in the impact assessment (LCIA) steps of LCA takes us to the heart of making the business case and linking it with corporate finance. This is where relative importance of impacts and dependencies are assigned, where the level of significance is determined, where ultimately the question of financial materiality is asked from a business perspective.

ISO recognizes that this involves subjective judgment, and is dependent on the overall goal of an LCA study (goal and scope definition). In the accounting and reporting domain, international standards from International Accounting Standard Board (IASB), IIRC (IIRC 2012, 2013), GRI and AccountAbility (AA1000) include recognized definitions of "materiality" and recommend procedures for determining "levels of significance" that involve various stakeholders to a greater or lesser degree.

Monetized costing provides an important way of weighing or prioritizing among various impacts and dependencies. A valuable field for making the link between LCA and business finance is life cycle costing (LCC), both financial LCC and environmental LCC (see Hunkeler et al. 2008; Reich 2005; Hunkeler and Rebitzer 2003). The former refers to financial economic analysis of a product or a function, in other words, conventional business financial analysis that would be done for investment decision-making considering the economic life cycle of the product or function. This contrasts with environmental LCC, which involves weighting the environmental impacts of an LCA system in monetary terms. Links between the two become evident when the environmental impacts have an economic impact on the system being analyzed, for example when environmental externalities are being taxed by local authorities. The environmental LCC may more scientifically reflect resource scarcities than the financial LCC, which when using mainstream economic system prices or market values for resources signal costs that may not adequately reflect real, absolute resource scarcities. This is where risk definition needs to be more science-based and reflect appropriate context. Furthermore, in as far as an investment decision for a specified number of years to come needs to be made in the face of uncertainty (such as future pricing or taxing of resource use or pollution), business managers can benefit from the application of real option (RO) theory in combination with LCA and LCC (cf Cucchiellaa et al. 2014).

Any attempt to integrate monetized values of significant life cycle impacts or dependencies in business decision-making has to address the indicators that are of greatest interest to chief financial officers (CFOs) and those who provide financial capital to enterprises. This is essential in making the business case, mapping out cause and effect relations between environmental or sustainability actions and financial results for the business. It can be illustrated by using a "Green Business Case Model" (Van der Lugt and Bertoneche 2013) that includes the core financial value drivers of special interest to financial managers.

The listed action areas and connectors included in the model (see Table 16.1) have been identified based on the review of over 60 research articles and business reports on the business case that have been published from 2002 to 2012 (see, for example, Margolis et al. 2007; Ambec and Lanoie 2007; Berger et al. 2007; EABIS 2009; Molina-Azorín et al. 2009; Business in the Community 2011). Considering the evolution of business case research over the last decade, it is evident that the indicators most commonly referred to can best be grouped in a three step model of (i) action areas, which lead to change in the area of what can be described as (ii)

Table 16.1 The Green Business Case Model (columns 1–3), with additional references to LCM linkages (column 4)

Environmental action areas	Connectors: lead indicators	Financial value drivers	*LCM linkages*
Eco-design (DfE)	Customer attraction	Growth of sales	*Product LCA*
Goods and services	Brand value, reputation	Duration of sales	*Ongoing LCM*
Standards (incl. life cycle) and technologies	Innovation	Operating margin	*LCC and CBA*
Supply chain management	Operational efficiency	Investment in fixed capital	*Maintenance, Remanufacturing*
Education, training	Human capital, productivity	Investment in working capital	*Product service systems*
Risk management	Risk profile	Cost of capital	*Impact risks*
Communications, reporting (incl. stakeholder engagement)	License to operate	Tax rate	*LC-based eco-taxes*

precursor or lead indicators, change that eventually affect (iii) the financial value drivers. For each of the rows of indicators a cause and effect relation can be hypothesized. For illustrative purposes three of these are discussed below, with examples including special reference to LCA applications. It shows pathways of making the connection between greening actions and business financial performance.

2.1 *Growth of Sales*

Hypothesis for Sales Growth As market and regulatory demand for sustainability grows, the business that (i) makes effective use of design for sustainability and delivers greener products and services will be in a position to (ii) boost its innovation ability and attract more customers, which (iii) will show positive results in its growth of sales.

This argument aligns with the proposition by management experts that sustainability has become the key driver of innovation, in new products and services as well as business models (Nidumolu et al. 2009). Take-up of such products and services is boosted by, among others, more reliable information (e.g., labeling). Of special significance for LCA-based product innovation, and LCM more broadly, is survey research by McKinsey with Boston College (2009) which cited innovation, new products, new customers, and new markets as specific areas where sustainability factors have demonstrable impact on overall organizational growth. The challenge for the LCM community is to illustrate how LCA can be employed in a convincing manner to deliver products with reliable certifications and product information. It may, for example, be asked whether the price of the product includes costs associated with the use and the end-of-life phases of the product. In as far as the greener product may be more expensive, consumers need education and guidance to realize that over the full life cycle the product is in fact cheaper. To start with, the LCA opportunity starts at the design phase. Eco-design has often been cited as one of the obvious applications of life cycle thinking, integrating business economics and the environmental sciences. Consider, for example, that most of the resource requirements and environmental impacts of products are predetermined during their design stage.

The need for reliable labels and certifications applies not only to manufactured products. Consider agrifood products and increasing consumer concern about where (local versus imported) and under what conditions (organic or sustainable) they have been produced. The retailer sector provides growing evidence of the use of green labels and the ability of major corporations to report impressive growth in the sales of such products. Whole Foods Market (2011), a Fortune 500 company and largest retailer of natural and organic foods in the USA, has seen its sales grow from $92.5 million in 1991 to $10.11 billion in 2011, at a compounded annual growth rate of 26 %.

Topical examples of product innovation and LCC can be found from companies such as Procter & Gamble (P&G), General Electric (GE), Puma, and Baxter. Mindful that consumers' product use causes the highest level of water consumption related to a detergent product's life cycle, P & G in 2007 switched all of its liquid detergents to a compact formula. By fall 2008, its Fabric Care and Home Care segment had seen a 10 % increase in net sales growth (CERES and Pacific Institute 2009). Launched in 2005 as an explicitly cleantech labeled product range, the Ecomagination products of GE reached sales of US$18 billion in 2009 – the size of a Fortune 150 company. GE has since predicted that revenues of Ecomagination products will grow at twice the rate of total company revenues over the coming 5 years (Porter and Kramer 2011). By 2015 the company reported sales revenue of Ecomagination products since 2005 was US$160 billion.

In 2010, the sporting goods company PUMA published an environmental profit and loss (EP&L) account, providing a monetary value on life cycle environmental impacts along its entire supply chain. The methodology in doing this involved a hybrid I-O process LCA. It enabled the company assess its overall environmental impacts valued at EUR 145 million in 2010 and define its most significant environmental impacts, where these impacts mainly occur (tiers 3 and 4 of the supply chain), in which regions most of the impact occur (Asia Pacific) and how they are spread in its product portfolio (e.g., mainly related to footwear). Today its annual report includes a Sustainability Scorecard that gives performance data by product, factories and company. From 2010 to 2012 its global brand sales grew from EUR 2.86 billion to EUR 3.45 billion (PUMA 2012).

The EP&L published by Puma can be compared with the Environmental Financial Statement published by healthcare company Baxter in its annual environmental reports since the mid-1990s. The statement by Baxter lists annual environmental costs as well as environmental income, savings or cost avoidance. These reflect actual costs associated with activities such as pollution control, environmental fees for packaging, energy and water consumption. The Baxter statement rather reflects an approach akin to financial LCC, as opposed to the Puma statement that reflects the application of an environmental LCC. In 2012, Baxter ranked number 86 in the Corporate Knights list of "Global 100 Most Sustainable Corporations in the World" and for the 11th year was recognized as the Medical Products Industry Leader of the Dow Jones Sustainability World Index (Dhanda 2012). In its annual sustainability report Baxter prominently discusses the application of LCA and LCM including the conducting of product sustainability reviews (PSRs) that are described as providing the foundation for its sustainable product design. From 2008 to 2012, its annual sales revenues grew from US$12.3 billion to US$14.2 billion.

2.2 Duration of Sales

Hypothesis on Duration of Sales Growth The business that (i) introduces greener goods and services to the market, backed up consistently by recognized standards and labels, will (ii) reap the benefit of greater brand value and reputation, enabling it (iii) to sustain a good growth of sales with longer duration.

Any assessment of the financial health of a company needs to consider not only its "growth of sales" over the last quarter or year. It also has to consider the "duration of sales," for example, trends over a 5-year period. The ability to not only reach new customers but also maintain their loyalty and trust over the longer term is determined by a range of factors, all of which serve to build company or product brand value. This is where credible use of LCM tools and consistent communication of product performance based on LCA applications can be critical. It may also require applying LCM in developing or assessing not just individual products or product lines but a broader product portfolio with longer-term customer relations in mind.

From surveys of senior managers and investment professionals in global firms it is evident that brand and corporate reputation tends to be a key area where they see a business case. The concern with reputation implies not only business to consumer (B2C) but also business to business (B2B) relations. If brand is to provide a guarantee of product safety and quality, the ongoing performance of all tiers of suppliers in the value chain becomes critical. This presents fertile ground for applying life cycle methods in business value chains. Accompanying the attributes of greener products and services with greener standards in operations will serve to further boost the reputation of both product and company (cf Iraldo et al. 2014).

A survey in 2012 of 1,375 consumers and 575 senior executives of companies with revenues of over US$500 million in China, Brazil, the USA and the UK found that 78 % of respondents indicated they do not buy a product if they do not like the parent company (Weber Shandwick and KRC Research 2012). In addition, 67 %

indicated they examine product labels to find the parent company, and 56 % would think twice if they could not find information about the company behind it. This illustrates possible limitations of only applying LCA to products, and the advantages of broadening the scope and applying life cycle thinking transparently to the broader enterprise.

Supporting duration of sales and continual improvement in quality is of course the training of employees and incentives for management in the use of LCM tools. Consider the example of life sciences and material sciences company Royal DSM (2010/2012). It links almost one-quarter of management compensation to the company's performance in eco-product development, energy efficiency and employee engagement. The company's 23,000 employees deliver annual net sales of more than €9 billion. ECO+ products constituted 40 % of running business sales in 2010 and 43 % in 2012.

2.3 Operating Margin and Capital Expenditure

Hypothesis on Profit and Capital Through (i) the use of recognized standards and cleaner technologies in its own operations to use resources more sustainably, as well as advancing those through its supply chain, a business can (ii) improve its operational efficiency – its ability to turn inputs into productive outputs in a cost-effective manner – as a result of which (iii) it will improve its net profit margin and optimize its capital expenditure.

The key role of resource efficiency in operations and its impact on earnings before interest and taxes (EBIT) is recognized today, especially in industries in highly competitive markets such as information and communications technology, car manufacturing and consumer goods. Traditional analysis on the business case has tended to start off by highlighting cost savings, in particular savings related to energy use. In as far as resources are wasted and polluting emissions not avoided, business earnings will additionally be taxed in more economies world-wide. An analysis of the carbon exposure of an emerging markets investment portfolio benchmarked against the S&P/IFCI LargeMidCap Index is illustrative. For 16 firms from emerging markets, analysis by Trucost (2010) has found that at US$108 per metric ton of carbon dioxide equivalent (CO_2e) by 2030, carbon costs could equate to more than 100 % of their EBIT (Carbon Disclosure Project 2010).

The costs of penalties for inaction contrasts with the benefits of preventative action. 3M has been running its Pollution Prevention Pays (3P) program for 40 years by 2015. In 2013 it estimated that since its inception the program has served to avoid 1.9 million metric tons of pollutants (waste, air and water pollution) and saved the company nearly US$1.8 billion based on aggregated data from the first year of each 3P project (3M Sustainability Report 2014). LCM is applied to all its products. Furthermore, LCM evaluation as a required component of its New Product Introduction process. It is also building on its LCM experience to develop new sustainability solutions. Clearly this is no longer just about operational efficiency in the name of cost savings. More companies have also started to focus on increased revenues and competitive market position.

McKinsey (2011) has found that 70 % of productivity opportunities today have an internal rate of return (IRR) of more than 10 % at current prices. As a result, some argue that business finds itself in the era of the Resource Revolution (Heck and Rogers 2014). Alongside opportunities are growing risks related to resource use. The cost of raw material inputs is impacted by growing natural resource constraints, which puts at risk the profit margins and EBIT of a range of sectors. The past decade alone has reversed a 100-year decline in resource prices. Analysis of fast-moving consumer goods companies by WRI and ATKearney in the late 2000s considered the impact of commodity price rises (Callieri et al. 2008). They calculated an ecoflation scenario in which natural resource constraints cause a reduction of 13–31 % in EBIT by 2013, and 19–47 % by 2018 for companies that do not develop strategies to mitigate the risks posed by environmental pressures. Examining data from six firms with a global presence in producing food, beverages, personal care and household care items, they found that, on average, raw materials and packaging costs each equaled 15 % of revenues (Von Falkenstein et al. 2010).

What then is the connection between cleaner production standards, operational efficiency and *capital expenditure*? Improved efficiency in the use of resources will drive more optimal use of *fixed assets* (e.g., land, buildings, equipment, machinery, vehicles). A challenge for the LCM community is to define how life cycle management of fixed assets can bring efficiency improvements through the use of approaches such as remanufacturing. There also exists an LCM opportunity with respect to *working capital*, a financial value driver with respect to which limited research on the green business case exists. It is related to the use of product service systems (PSSs) in the form of leasing rather than buying equipment, which can bring significant savings alongside its environmental benefits. This includes efficiencies due to services provided at scale, onsite or offsite, by an external business partner (see Willard 2012). In how far is LCA able to capture such benefits with different system boundaries involved? If LCA and LCC can quantify the benefits of PSSs in physical and monetary terms, the findings will also be of special relevance to working capital expenditure. PSS-related efficiency improvements can serve as a driver for innovation in the way inventory and customer or supplier relations (receivables or payables) are managed.

3 Conclusion

Amidst greater interest in how enterprises use different types of capital and generate value in a sustainable manner, LCM has to prove its value by illustrating linkages with corporate finance and strategic performance. Considering the core financial value drivers highlighted in this chapter, the necessary contribution of LCM is summarized below:

- *Sales growth and its duration*: LCM has to be used effectively in the design of products or product portfolios, services and business models that are convincingly sustainable. LCC experts have recognized that new business models may be required to develop more integrated (not fragmented) value chain systems (Swarr et al. 2011). Mindful of the lead indicators of customer attraction and brand reputation, LCA applications also have to be used credibly and consistently in a manner that enables the communication of reliable information via labels and the like to customers.
- *Operating margin*: The use of LCM standards, internally and through supply or value chains, need to effectively promote innovation and operational efficiency in order to boost operating margins. This is not simply about short-term profit. It is about defining avoided and opportunity costs in making business approaches that secure the longer-term sustainability of profit and cash flows.
- *Investment in fixed and working capital*: LCA experts need to consider in how far their methodologies can be used to define the value of closed loop manufacturing (CLM), in particular remanufacturing, as well as product service systems (PSSs) in enabling more optimal and sustainable capital expenditure. CLM and PSSs are highly under-estimated.
- *Cost of capital*: LCM needs to be effectively integrated with risk management, helping broadly to define hot spots in value chains and, specifically, through the collection of bottom-up data, risks of various kinds (including operational and regulatory) that may be associated with specific products, operations and organizational entities. In addition, the providers of financial capital need to be educated about the meaning of LCA findings.
- *Tax rate*: The LCA community needs to illustrate how its research can be used by regulators to define convincing eco-tax regimes that succeed to reward early adopters and penalize enterprises that persist with damaging products, services and business models.

It is theoretically convenient to state that environmental LCC is different from financial LCC and activity-based costing (ABC) in management accounting (cf Rebitzer and Nakamura 2008). One is focused on the costs of environmental damage and the other on business costs. It is, however, imperative today to define the link between these two and not leaving this to regulators. It means that LCA experts will need to support cost benefit analysis in which the incentives and cost structures for individual actors involved in whole life cycle systems or value chains are assessed. Furthermore, analysis will also need to show an ability to assess *future* costs and benefits likely to occur in the short, medium or long term. While seeking to meet these expectations, the LCM community will need to take cognisance of the preference among the mainstream investment community for "using a handful of the most important indicators and proxies to capture risk (that) can minimize complexity" (IIRC 2012). The three hypotheses examined in this chapter suggest pathways to capture the attention of investors in tackling this difficult task.

References

3M (2014) Sustainability report. http://solutions.3m.com/wps/portal/3M/en_US/3M-Sustainability/Global/Environment/3P/. Accessed 12 Feb 2015

Ambec S, Lanoie P (2007) Does it pay to be green? A systematic overview. Acad Manag Perspect 22(4):45–62

Baumann H (2012) Using the life cycle approach for structuring organizational studies of product chains. In: 18th greening of industry network conference, Linköping, 22–24 Oct 2012

Berger I et al (2007) Mainstreaming corporate social responsibility: developing markets for virtue. Calif Manage Rev 49(4):132–157

Boston College Centre for Corporate Citizenship (2009) How virtue creates value for business and society: investigating the value of environmental, social and governance activities. Carroll School of Management, Boston

Business in the Community (2011) The business case for being a responsible business. BITC, London

Callieri C et al (2008) Rattling supply chains: the effect of environmental trends on input costs for the fast-moving consumer goods industry. World Resources Institute (WRI)/A.T. Kearney, Washington, DC

Carbon Disclosure Project (2010) Carbon disclosure project 2010 – global 500 report (written by PWC). CDP, London

Coalition for Environmentally Responsible Economies (CERES) and the Pacific Institute (2009) Water scarcity and climate change – growing risks for business and investors. Ceres and Pacific Institute, Boston

Cucchiellaa F, D'Adamoa I, Gastaldia M, Koh L (2014) Implementation of a real option in a sustainable supply chain: an empirical study of alkaline battery recycling. Int J Syst Sci 45(6):1268–1282

Curran MA (2013) Life cycle assessment: a review of the methodology and its application to sustainability. Curr Opin Chem Eng 2:273–277

Dhanda K (2012) Case study in the evolution of sustainability: Baxter International Inc. J Bus Ethics. doi: 10.1007/s10551-012-1565-2

Elms DK, Low P (eds) (2013) Global value chains in a changing world. WTO with Fung Global Institute (FGI) and Nanyang Technological University (NTU), Geneva

Eriksson T, Olsson D (2011) The product chains of rare earth elements used in permanent magnets and NiMH batteries for electric vehicles. ESA report 2011:8. Chalmers University of Technology, Göteborg

Eun JH, Son JH, Moon JM, Chung JS (2009) Integration of life cycle assessment in the environmental information system. Int J Life Cycle Assess 14:364–373

European Academy for Business in Society (EABIS) (2009) Sustainable value. EABIS with Cranfield University School of Management, SDA Bocconi School of Management, Vlerick Leuven Gent Management School, Brussels

Finkbeiner M, Schau EM, Lehmann A, Traverso M (2010) Towards life cycle sustainability assessment. Sustainability 2:3309–3322

Finnveden G et al (2009) Recent developments in life cycle assessment. J Environ Manage 91(1):1–21

Gereffi G, Humphrey J, Sturgeon T (2005) The governance of global value chains. Rev Int Polit Econ 12(1):78–104

Guinee JB et al (2011) Life cycle assessment: past, present and future. Environ Sci Technol 45(1):90–96

Heck S, Rogers M (2014) Resource revolution. Amazon/McKinsey, Seattle

Hellweg S, Canals LM (2014) Emerging approaches, challenges and opportunities in life cycle assessment. Science 344:1109–1113

Hunkeler D, Rebitzer G (2003) Life cycle costing – paving the road to sustainable development? (Editorial). Int J Life Cycle Assess 8(2):109–110

Hunkeler D, Lichtenvort K, Rebitzer G (eds) (2008) Environmental life cycle costing. CRC Press, Boca Raton

International Integrated Reporting Council (IIRC) (2012) Capturing the experiences of global business and investors. The pilot programme 2012. IIRC, London

International Integrated Reporting Council (IIRC), Association of Chartered Certified Accountants (ACCA) and Netherlands Institute of Chartered Accountants (NBA) (2013) Capitals background paper. IIRC, London

Iraldo F, Testaa F, Bartolozzia I (2014) An application of life cycle assessment (LCA) as a green marketing tool for agricultural products. J Environ Plann Manag 57(1):78–103

Klöpffer W (2008) Life cycle sustainability assessment of products. Int J Life Cycle Assess 13(2):89–95

Koh L, Genovesea A et al (2013) Decarbonising product supply chains: design and development of an integrated evidence-based decision support system – the supply chain environmental analysis tool (SCEnAT). Int J Prod Res 51(7):2092–2109

Lake A, Acquayeb A, Genovesea A, Kumara N, Koh L (2014) An application of hybrid life cycle assessment as a decision support framework for green supply chains. Int J Prod Res, doi:10.1080/00207543.2014.951092

Margolis J, Elfenbein HA, Walsh JP (2007) Does it pay to be good? A meta-analysis and redirection of research on corporate social and financial performance. Harvard University working paper, July 2007

McKinsey & Co (2011) Resource revolution: meeting the world's energy, materials, food, and water needs. http://www.mckinsey.com/insights/sustainability/resource_revolution

Molina-Azorín JF et al (2009) Green management and financial performance: a literature review. Manag Decis 47(7):1080–1100

Nidumolu R, Prahalad CK, Rangaswami MR (2009) Why sustainability is now the key driver of innovation. Harv Bus Rev. https://hbr.org/2009/09/why-sustainability-is-now-the-key-driver-of-innovation

Organization for Economic Cooperation and Development (OECD) (2012) Mapping global value chains. Policy dialogue on aid for trade. Document TAD/TC/WP/RD(2012)9. OECD, Paris

Porter M, Kramer M (2011) Creating shared value. Harv Bus Rev. https://hbr.org/2011/01/the-big-idea-creating-shared-value

Puma (2012) Annual business and sustainability report. Puma, Herzogenaurach

Rebitzer G, Nakamura S (2008) Environmental LCC. In: Hunkeler D, Lichtenvort K, Rebitzer G (eds) Environmental life cycle costing. CRC Press, Boca Raton, pp 35–57

Reich MC (2005) Economic assessment of municipal waste management systems – case studies using a combination of life cycle assessment (LCA) and life cycle costing (LCC). J Clean Prod 13(3):253–263

Royal DSM (2010/2012) Integrated annual report 2010/2012. Royal DSM N.V, JH Heerlen

Swarr TE, Hunkeler D, Klöpffer W, Pesonen H-L, Ciroth A, Brent AC, Pagan R (2011) Environmental life-cycle costing: a code of practice. Int J Life Cycle Assess 16:389–391

Trucost (2010) Carbon risks and opportunities in emerging markets – Trucost study on the exposure of different regional equity strategies to carbon costs. Trucost Plc, London

Van der Lugt CT, Bertoneche M (2013) Finding the God particle of the sustainability business case: greener pastures for shareholder value. Harvard Business School finance working paper no 13–072 (HBS 2013)

Von Falkenstein E, Wellenreuther F, Detzel A (2010) LCA studies comparing beverage cartons and alternative packaging: can overall conclusions be drawn? Int J Life Cycle Assess 15:938–945

Weber Shandwick, KRC Research (2012) The company behind the brand: in reputation we trust. Weber Shandwick, New York

Whole Foods Market Inc (2011) Form 10-K, filed. www.wholefoodsmarket.com/company-info/investor-relations. Accessed 1 Mar 2015

Willard B (2012) The new sustainability advantage: seven business case benefits of a triple bottom line. New Society, Gabriola Island

Chapter 17
Building Organizational Capability for Life Cycle Management

Thomas E. Swarr, Anne-Claire Asselin, Llorenç Milà i Canals, Archana Datta, Angela Fisher, William Flanagan, Kinga Grenda, David Hunkeler, Stephane Morel, Oscar Alberto Vargas Moreno, and M. Graça Rasteiro

Abstract Corporations are being pressured to integrate life cycle thinking and practices across global supply chains. The UNEP/SETAC Life Cycle Initiative has been developing a life cycle management capability maturity model (LCM CMM) to help mainstream life cycle assessment (LCA) and life cycle management (LCM). Pilot projects in small-to-medium-sized enterprises (SMEs) to apply the model showed the companies were able to identify and implement projects that delivered both near-term business value and developed the organizational capability for LCM. A key benefit of the life cycle approach was enhanced cross-functional integration and collaboration with suppliers and customers. The projects did identify a need for more guidance on how to interpret the business impact of environmental

T.E. Swarr (✉)
Yale University, New Haven, CT, USA
e-mail: thomas.swarr@yale.edu

A.-C. Asselin • L. Milà i Canals
Division of Technology, Industry and Economics, UNEP, Paris, France

A. Datta
Federation of Indian Chambers of Commerce and Industry (FICCI), New Delhi, India

A. Fisher • W. Flanagan
General Electric, Ecoassessment Center of Excellence, Niskayuna, NY, USA

K. Grenda • M.G. Rasteiro
University of Coimbra, Coimbra, Portugal

D. Hunkeler
aquaTECH, Geneva, Switzerland

S. Morel
Renault, Alliance Technology Development, Guyancourt, France

O.A.V. Moreno
Secretaría Distrial de Ambiente (SDA), Bogotá, Colombia

G. Sonnemann, M. Margni (eds.), *Life Cycle Management*,
LCA Compendium – The Complete World of Life Cycle Assessment,
DOI 10.1007/978-94-017-7221-1_17

concerns and to align LCM efforts with company business strategy. Collaborative networks where more advanced companies can share their knowledge are a key enabler, particularly in developing economies.

Keywords Business • LCM capability maturity model • Life cycle assessment • Life cycle management • Life cycle thinking • Sustainability • UNEP/SETAC life cycle initiative • Value chains

1 Introduction

Advances in information and communication technology have enabled companies to rationalize their production systems across the globe for peak efficiency, providing a continuous stream of innovative and low-cost consumer goods. The dramatic expansion of the market economy has been extremely successful in raising the material standards of living. However, there is growing awareness that the material and energy intensive modes of current production systems are unsustainable as the size of the global middle class consumer market is expected to grow from 1.8 billion in 2009 to 4.9 billion by 2030 (Pezzini 2012). Business has recognized the challenge and is moving from a narrow focus on short-term economic value to a broader concept of shared value, defined as policies and operating practices that enhance the competitiveness of a company while simultaneously improving the economic and social conditions in its host community (Porter and Kramer 2011). Leading companies, as well as some new firms, are working to develop value chain indices that measure and price all externalities, from raw materials to final product disposal enabling direct comparison of products at the point-of-sale (Chouinard et al. 2011; UNEP/SETAC 2014, 2015). Life cycle assessment (LCA) is an internationally standardized method for quantifying environmental impacts of product systems and has been recognized as the primary methodology for helping decision-makers select effective improvement strategies while avoiding burden shifting between impact categories or life stages (Guinée et al. 2011; Hellweg and Milà i Canals 2014). However, there is concern that small-to-medium-sized enterprises (SMEs) lack the capacity to provide quality data for the various supply chain metrics, let alone conduct comprehensive LCAs.

The UNEP/SETAC Life Cycle Initiative has been working to promote the implementation of life cycle management, or the effective integration of life cycle thinking into the day-to-day routines of business (UNEP/SETAC 2013a) for more than a decade. During Phase I (2002–2007), researchers identified a need for training materials and technical resources to build capacity for life cycle management (LCM), particularly in developing economies and small-to-medium-sized enterprises (SMEs) (Saur et al. 2003). During Phase II (2007–2012), one of the work areas was life cycle approaches for capability development (including institutional empowerment, training, curricular development, etc.), with a particular focus on addressing challenges in implementing LCM in SME suppliers as part of sustainable value chain initiatives (Swarr et al. 2011). That effort was continued in Phase III (2012–2016) with the funding of a series of pilot projects to apply a LCM Capability Maturity Model (LCM CMM) to help guide the implementation of a life cycle based improvement project.

The LCM CMM builds on well-established methods from the process quality improvement discipline and provides a structured framework to help companies incrementally build the organizational capacity for more comprehensive and powerful life cycle (LC) methods (Swarr 2011). Previous research on organizational change has shown that ~70 % of failed change initiatives were attributed to inadequate attention to 'softer' issues of organizational health (Keller and Price 2011). Sustained high performance requires a balance between top-down initiatives for results-driven action and bottom-up organizational development efforts to instill a culture of learning and continuous improvement (Beer 2001). Thus, the LCM CMM complements top-down efforts to drive sustainability measures across global supply chains with bottom-up organizational development efforts to build the necessary skills in SMEs to enable them to provide high quality data and to adapt LCM objectives to meet their specific competitive and stakeholder pressures.

1.1 Chapter Outline

The capability maturity model (CMM) concept originates from a quality management maturity grid proposed by Phil Crosby (1979), subsequently developed for numerous functions, such as software engineering, integrated product development, systems engineering and more. The CMMI Institute (cmmiinstitute.com) integrated these models, but they are extremely complex and inappropriate for SME suppliers (Sukhoo et al. 2007). Development of a simpler and more practical LCM CMM is described under Methods. A series of company case studies are then presented to compare the theory of CMMs with the real world practice of LCM. In some cases, the company projects explicitly applied the LCM CMM developed by UNEP/SETAC. In other cases, the company projects made no direct application of CMM, but were simply efforts to apply LCM principles to address specific objectives. The focus of the case studies is on the change management issues that can promote or inhibit success of company improvement initiatives. These company experiences are then reviewed and reconciled with the LCM CMM concepts to develop a framework that companies can use to tailor LCM to their specific competitive context, environmental constraints and opportunities, and strategic priorities. The basic premise is that a structured approach to help address these softer management issues can be a useful strategy to mainstream LCM in business. Finally, the chapter concludes with recommendations for future research to expand business capacity for LCM and to promote its effective integration into routine business decision-making processes.

2 Methods

2.1 Development of the Life Cycle Management Capability Maturity Model

The life cycle management capability maturity model (LCM CMM) was based on a model developed at the Massachusetts Institute of Technology to guide enterprises implementing lean production practices (Nightingale and Mize 2002). Researchers, academics, consultants and business managers had developed a broad vision of the values, behaviors and practices that constituted a lean enterprise (Womak and Jones 2003). However, practitioners were faced with a confusing array of principles, tools and practices, but no help on the order or precedence to implementing various best practices in a cohesive management system. These same considerations describe the current challenge of deploying sustainability across global supply chains.

The LCM CMM was structured into three broad categories. Leadership processes set the direction for the organization and determine if there is sufficient motivation and organizational support to successfully achieve the stated goals. Life cycle processes provide operational excellence to design, build, deliver and support product offerings in a safe, clean, equitable and profitable manner. Enabling infrastructure assures resources are in place over the long-term to successfully implement the defined strategy. The intent is to accelerate learning by defining a logical sequence of skill-building improvement projects that gradually build robust decision-making processes necessary for effective implementation of LCM. The model is summarized in Table 17.1.

The LCM CMM framework simply expands the concept of customer to include other interested stakeholders and the 'voice of the environment' and looks beyond the immediate financial and efficiency goals to continually assess how the value chain is meeting the broader needs of civil society and preserving its resiliency to adapt to external disruptions (Hart and Milstein 2003). Building on lean concepts

Table 17.1 Life cycle management capability maturity model (LCM CMM)

Maturity level	Span of concern	Metrics	Appropriate projects
Qualified	Project or facility	Binary yes-no compliance; wastes	Basic work procedures/skills, unit process improvements, waste minimization
Efficient	Enterprise	Process inputs/outputs; eco-efficiency	Interconnected processes, pollution prevention, process redesign, collaboration with key supplier or customer
Effective	Value chain	LCA; cradle-to-grave	Eco-design, enterprise-wide initiative, value chain collaborations
Adaptive	Society	Sustainability, resiliency indices	Public-private partnerships, community development outreach, public policy reform

was also intended to leverage any previous investments in lean methods and to ally with the quality function within the company. A questionnaire[1] was developed that provided diagnostic questions for key processes, along with example practices that described the various maturity levels.

In May 2013, the initiative announced a call for proposals from companies "interested in implementing a project which adopts life cycle thinking in their business management and operation". The proposal application required companies to conduct a maturity self-assessment and propose improvement projects to advance both life cycle and business objectives. Applicant companies identified a "LCM CMM coach" to help with the maturity assessment and the eventual implementation of identified projects. There was no requirement that coaches had previous training in the LCM CMM, and training materials that were developed in earlier phases of the LCI were made available to support their preparation of proposals (Swarr et al. 2011). In addition, UNEP/SETAC contracted a "mentor" to provide remote technical support to the coaches. Eight projects were selected from 22 applications and were awarded small grants to help implement the life cycle based improvement projects (UNEP/SETAC 2013b).

2.2 *Company Case Studies*

2.2.1 UNEP/SETAC LCM CMM Pilots

There were several constraints in the application process, which complicated evaluation of the efficacy of the LCM CMM. Coaches were required to conduct a maturity assessment and propose specific improvement projects as part of the application process. Few coaches had any prior exposure to the LCM CMM, and thus had to conduct the maturity assessment based on their review of the training materials, a PowerPoint presentation and a workbook. However, the maturity assessments submitted all appeared to be of reasonable quality and identified existing organizational strengths and weaknesses. In general, identified projects seemed to be reasonably aligned with the maturity assessments (i.e. existing LCM capability within the organization), with the possible exception being an over-emphasis of LCA in the projects. A second constraint was a short timeline, approx. 6–9 months, which meant that many proposed improvement projects would extend beyond the contract end date, precluding a final evaluation of outcomes. A list of the projects is provided in Table 17.2.

Several projects in companies led by executives with visions of strong LCM programs raised interesting questions of just how much life cycle knowledge is required and where should it reside in the organization. RUCID processes tropical fruits into juice and dried crisps. NEHSU Foods produces and sells soya-based milk

[1]A copy of the questionnaire and an accompanying workbook are available at http://blog.pucp.edu.pe/media/2360/20130211-4__cmm_workbook_ilcm2012_s__valdivia.pdf (Accessed 24 Feb 2015).

Table 17.2 LCM CMM pilot projects

Pilot	Outcomes	
	Training	**Environmental**
Freudenberg, SA	Conducted procurement audit, supplier workshop, site visits to two suppliers	Sustainability criteria used to create supplier scorecard, built into contracts
RUCID, UG	Stakeholder workshop-23 individuals; staff trained	Established energy monitoring procedures; project plan for anaerobic digester and higher efficiency stove
NEHSU Foods, CM	Executive team trained in LCM; pilot results shared with three sister organizations	Began implementation of Food safety management system; identified KPIs for supply chain and defined requirements for management information system
SDA, Bogota, CO	Workshop 1–18 co Workshop 2–13 co Workshop 3–16 co Workshop 4–10 co	Eleven companies developed improvement plans; five projects completed within time frame of pilot study
Ediouro Grafica, BR	Completed baseline assessment of wastes	Book printing waste reduced 41 %, stapled magazine waste 36 % and glued magazine waste 32 %
Finca Mountain, PE	Baseline study and options assessment reports	Implementation plan in place, started on motor management system
Polygenta Technologies, IN	Training workshop for core team for LCA/ten individuals	Improved data systems to measure environmental parameters
India Glycols Ltd., IN	2 day workshop – 30 individuals	Baseline LCA completed, developed LCM manual and merged with Integrated Management System (IMS)

and yoghurt products; Finca Mountain cultivates, produces and exports specialty coffees. All rely on relatively low-skilled labor that may be seen as having little need for LCM expertise. RUCID's project was to develop proposals to obtain funding for energy upgrades. As part of the baseline assessment, employees were trained to monitor energy use on a regular basis to raise awareness and promote higher process efficiency. Finca Mountain's project was similarly focused on equipment upgrades to improve efficiency, but the company also started with a motor management system to build the foundation for improved operational controls and a more comprehensive energy management system. NEHSU Foods focused on defining key performance indicators to manage its supply chain and realized a need for more formal and robust systems to manage internal operations. The company started by focusing on a Food Safety Management system that directly supported the company strategic value proposition.

The companies did not seem to have any difficulty identifying relevant projects that provided business value and addressed life cycle environmental concerns. The companies in the Bogota, CO network assessed themselves as having limited understanding of LCM principles. However, several developed useful screening tools to prioritize their improvements. Somos K S.A., a bus transit company, used a screening LCA to focus on improved maintenance procedures to reduce fuel consumption, providing operating savings and reduced greenhouse gas emissions. Azul K S.A., a manufacturer of cleaning products developed a qualitative screening tool that considered design, materials, manufacturing, marketing, waste generation, disposal and market trends. Based on this screening assessment, a liquid dishwashing detergent was selected for further analysis based on a high qualitative impact ranking and a growing market. Ediouro Grafica rated its LCM maturity at a relatively low level, and thus, decided to focus on a waste minimization project that could be used to develop basic data systems and lay the foundation for more complete LCAs. Freudenberg Nonwovens South Africa manufactures interlinings for the garment industry and develops and produces nonwovens for the energy, car interior, hygiene, medical, building interiors, as well as for special applications. Their LCM maturity was rated higher and the company had support from the corporate parent. Thus, it conducted a more sophisticated evaluation of its supply chain, defining specific sustainability criteria to be integrated into a supplier scorecard.

There was some tendency for companies to conflate LCA and LCM. Many viewed LCAs almost as a compliance requirement, and it was necessary to reassure companies that their projects did not necessarily require cradle-to-grave studies. For example, Cristacryl de Colombia S.A. is a manufacturer of high quality acrylic products using 100 % virgin materials. The company completed a streamlined study of its products that showed the production of basic chemicals dominated the impact assessment, but they saw little value, or ability, to gather more detailed data on upstream processes they had virtually no ability to affect. Instead, the company developed a communication campaign to educate customers on how to improve yields, improve recycling rates and avoid landfill disposal of the acrylic scrap. This directly supported their strategy of positioning themselves as a quality supplier of superior products. However, during the pilots, it was necessary to reassure the company that LCM did not require a complete LCA with site-specific data from their chemical suppliers.

In other cases, companies conducted LCAs, but struggled to identify improvements that provided a financial benefit. India Glycols conducted a study of polyethylene glycol made from sugar cane molasses, a byproduct from a sugar refinery. Although the study showed benefits relative to petroleum-based products, their customers have limited understanding and appreciation for the environmental benefits, and the product currently competes in a commodity market. It will require significant marketing to build awareness and identify niches that will attach a premium to bio-based chemicals. Similarly, Multidimensionales S.A., a packaging company, conducted a LCA of disposable polypropylene (PP) and polystyrene (PS) cups. Although there would be significant benefits from improved recycling rates, there is a lack of recycling infrastructure in Bogota and a need to change end user behaviors. The company has identified shopping mall operators as a high leverage point, with sufficient volume of material that can be practically collected and a good venue for end user outreach programs. Azul K identified landfill disposal of the empty bottle as a significant impact, but did not have the ability to influence end-of-life recycling rates. The company was able to redesign a light-weight version that had the same look and feel necessary for customer acceptance, which would at least reduce the weight of material being landfilled. High concentration cleaning products were also identified as a potential improvement, but again would require a significant outreach effort to change consumer behaviors.

The Secretaria Distrital de Ambiente (SDA) project in Bogota, CO, offered a model for future efforts to disseminate LCM practice. SDA is the environmental agency for the city and had been working with a network of companies to advance their environmental programs. The companies had previously participated in programs on compliance, clean production and environmental management systems. Although not explicitly based on a CMM approach, these earlier programs were well aligned with the structure of the LCM CMM. In addition, the experience helped establish the technical credibility of SDA for this program and helped build trust among the member companies necessary for voluntary projects. A good example is a project Colcafé S.A.S. conducted in collaboration with Andria Logística, a waste hauler, and Santa Reyes, an egg producer, to find a better solution for managing a waste stream of coffee bean 'silver skins'. The waste silver skins were used as an additive to improve the anaerobic digestion of poultry manure, boosting efficiency, reducing odors and improving the logistics of waste hauling for both Colcafé and Santa Reyes. This type of project is only feasible for local companies that have a necessary level of trust to overcome transaction costs and to arrive at equitable sharing of the costs and benefits.

The Federation of Indian Chambers of Commerce and Industry (FICCI) coordinated the two pilots in India. A specific advantage of FICCI was that their quality group managed the pilots. Quality professionals are more familiar with CMM concepts, and this was important in helping to communicate the concepts to the company. The project was designed to facilitate adoption of LCM practices in business operations through three key activities including the LCM CMM maturity assessment by the senior management team, training of departmental heads and officers and joint review of results with the coach. The initial maturity self-assessment exercise lasted 10 days, as senior management needed several back and forth discussions with the project coach to understand the model and relate it to the company's day-to-day activities. It was clear that dedicated effort was needed to rearticulate the model for wider socialization. The model was therefore converted into a Yes/No questionnaire on LCM CMM and its extensions. Participants could choose 'Yes' or 'No' (with reasons) based on their own work/knowledge (categorized as 'Primary') as well as based on reliable information through other sources (categorized as 'Secondary'). For questions where participants were not dealing with that particular area nor had any reliable information from other sources, they could choose 'Don't Know' or leave a blank. The model was socialized through a specially designed training workshop on LCM CMM for officers from different departments of the organization. As part of the training, all generic terms in the questionnaire were mapped to company specific language and examples. The significant reduction in Blank/Don't Know responses shown in Fig. 17.1 demonstrates how the re-articulation and mapping exercises helped staff from different departments understand the practical application of LCM in the organization and in their own work areas after completion of the training workshop.

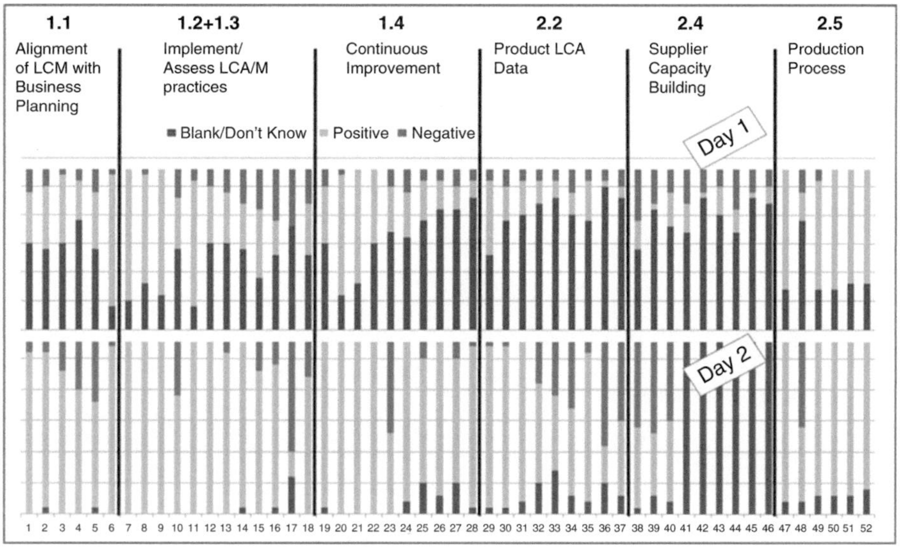

Fig. 17.1 Comparison of participant responses on day 1 and 2 of the training showing improvement in understanding of LCM concepts

2.2.2 Additional Case Studies of Mature Companies

Few companies have explicitly used a formal CMM to guide efforts to implement LCM practices. Thus, it is useful to survey experiences of companies with more mature programs to elicit lessons that might be applied to evaluating the efficacy of a capability approach. GE (General Electric) has been developing LCM and LCA since 2008. The GE Ecoassessment Center of Excellence (CoE) was founded to assess the environmental impact of products and technologies throughout the entire lifespan, from raw material extraction through reuse, recycling and disposal at end of life. The programs avoid a 'one-size fits all' approach by developing and applying a variety of tools and resources, from simple screening tools to detailed LCAs. The intent is to leverage the screening and streamlined approaches in early phases of product development, make strategic use of LCA and focus on value creation for customers, investors and society. The Ecoassessment CoE does not dictate the use of LCM, but rather works to build awareness, deliver tools and resources, provide guidance and support, and where necessary manage the detailed LCA studies. An environmental LCM portal is used to make tools and resources easily accessible to diverse business units. It is also necessary to tailor the tools to the diverse needs of different businesses. For example, a product LCM tool includes ten separate sets of customizable weightings that can be used to emphasize issues most relevant to a specific business context. When detailed LCA studies are warranted, the CoE provides trained professionals to perform the LCA or manage the work with external experts. It is necessary to demonstrate value to the business to promote adoption of LCM practices. An interactive tool was created to build awareness of how a LCM

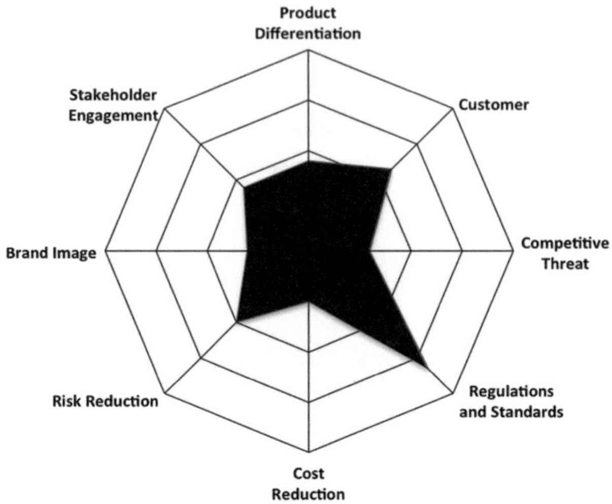

Fig. 17.2 Interactive tool showing perspectives on creating business value with LCM (example output: results shown are specific to a particular product)

perspective can create business value for a particular product or business activity, and the results are plotted in a spider chart easily interpreted by business managers (see Fig. 17.2.)

AQUA + TECH is a global leader in water purification systems and wastewater treatment technology for industrial and infrastructure markets. The company is collaborating with the University of Coimbra to explore the use of organic wastes to create coagulants for improved wastewater treatment. The majority of coagulants currently used are derived from iron and aluminum recovered from waste streams. In general, for each ton of carbon removed from the wastewater, one ton of sludge is generated. Even after thickening, the sludge is typically 70 % water. LCAs have shown us that the main environmental impact is, therefore, the transport of the sludge to its ultimate disposal (Rebitzer et al. 2004). If the coagulants were produced from organic sources, instead of the inorganic waste metals, wastewater clarification would be possible concomitant with biodegradability. If bacteria, already used in the wastewater plant, could metabolize the coagulant itself, then less sludge could be generated. AQUA + TECH is able to leverage an expanded technical base for a detailed LCA study through the collaboration, which is part of a SME-PhD program supported by EU FP7. An interesting caveat of making water treatment chemicals from waste streams is that the final product has high levels of water. Therefore, the extraction and transformation must take place near the waste stream and the final product can only be transported economically about 300 km. A key driver of the project is AQUA + TECH's vision of a zero-discharge facility. Therefore, any aqueous streams used in production have to be either re-used or recycled with minimum cost and effort, back into the process.

The Renault group conducted its first full vehicle LCA in 2005 (Cabal 2005). This first LCA project led to a standardized global assessment, and many later studies provided accurate and quantified information to document the improved environmental profile of new vehicles. In 2012, the company organized a cross-functional team from design, innovation, manufacturing, IT systems, marketing and more to benchmark Renault's life cycle thinking activities and propose recommendations to advance their efforts. Renault developed a CMM tailored to their specific needs. The model addressed five management areas:

1. Plan – set strategy and define targets
2. Enablement tools and competencies
3. Evaluate – measure product performance
4. Eco-design – create products with improved environmental performance
5. Value-clarify benefits and capital creation

Five stages of maturity were defined – Compliant (license to produce), Fragmented (localized environmentally efficient approaches), Integrated (company aligned behind common key indicators), Extended (company has integrated the key stakeholders) and Global (pioneer in innovation and sustainable value creation, i.e. human, natural, financial capitals). The LCM CMM was used as a resource to develop semantic descriptions for the maturity matrix. For example, an 'Integrated' maturity level for the management area 'Enablement' was described as 'Partial integration with traditional company IT system. Personal development in place for all employees, including sustainability training & development.'

The eco-transition matrix was used to survey the top 20 suppliers and proved to be a valuable aid for enhancing collaboration, which is not an intuitive action and need to be organized (Segrestin 2003). The transition from 'Integrated' level to 'Extended' is particularly ambitious, as the company needs to open its boundaries to engage stakeholders and integrate them into its activities. Specific projects, such as an LCA of electric vehicles (Renault 2011) were used to engage value chain partners in co-designing tools, making key decision or product specific rules in a Collaborative LCA scheme (Morel 2014). The matrix clearly documents company progress and helps identify additional opportunities to enhance sustainable development management. It has also proved to be a very good team-building tool within supplier training programs. The matrix also helps disseminate knowledge of company activities and reveal differing perceptions of various functional groups or business units.

3 Discussion

The case studies presented cannot be considered a rigorous evaluation of the LCM CMM. Case studies are always subject to concerns about the degree to which the observations can be generalized. Thus, the results should be viewed in terms of an ongoing action research project to further refine the approach (Kaplan 1998). To validate the efficacy of the LCM CMM it would be necessary to demonstrate the model incorporates the capabilities needed to effectively conduct LCM programs within a company; that the sequence of skill-building provided does speed learning from experiences of companies with more mature programs; and, that the balanced focus on both near-term performance objectives and long-term organizational development needs leads to sustained high performance in both financial and environmental dimensions.

There is a rich literature on capabilities required for sustainable business management (van Kleef and Roome 2007; Wiek et al. 2011; Silvius and Schipper 2014; Wesselink et al. 2014). While there is yet no broad consensus on a definitive list, there are common themes – system thinking, interpersonal skills, anticipatory or foresighted thinking, embracing diversity and strategic management. Additional insights can be gathered from research on organizational health and financial performance (De Smet et al. 2007). Concepts such as direction, leadership, accountability, innovation, coordination and external orientation were found to contribute to financial success. Although the UNEP/SETAC Life Cycle Initiative pilots were not intended to evaluate the structure of the LCM CMM, some useful insights can be elicited from the various projects. The basic structure of leadership, life cycle practices and enabling infrastructure appears adequate to capture the various elements identified in the literature. The process model is summarized in Table 17.3. Observations from the pilots, however, identified some gaps in the supporting training materials.

Leadership and vision are obvious success factors. The case studies presented, however, are a biased sample with all showing projects motivated by strong vision and leadership. Thus only one example will be presented to illustrate the significance of vision on results. The Colcafé project was viewed not as a waste minimization project, although it was selected to help meet a company zero waste goal. The 'waste' silver skins were re-imagined as a co-product valued at the price of the purchased coffee beans. Thus, the alternative of using the silver skins as an additive to improve the digestion of poultry manure was implemented. The silver skins, when initially viewed as a waste, were collected in bulky bags expensive to ship and difficult to handle by workers at the digester. The company evaluated options to compact the material for easier transport and redesigned the packaging for easier handling while feeding material into the digester. The redesign of the packaging leveraged their expertise in product logistics. Strong management support is a critical success factor for the implementation of these kinds of complex collaborations. The same can be said for AQUA+TECH where zero-discharge, as a policy, forced the firm to look at new products it could make from its "waste" streams.

One of the key training gaps identified in the pilot studies was a need for more guidance in tailoring the model to specific contexts, such as industry sector or geographic area. The model was developed primarily for an individual coach or change agent attempting to champion implementation of LCM at a manufacturing facility. Thus, application to service industries was particularly challenging. There was also a need for additional guidance on interpreting the business significance of environmental impacts. These gaps can be related to capabilities associated with strategic management. Companies achieve a competitive advantage by assembling a unique, valuable and difficult to imitate collection of resources and capabilities (Helfat and

Table 17.3 Summary of LCM CMM

Key business processes of LCM CMM
Leadership processes
Integrate LCM into business planning
Implement LCM practices
Assess performance and communicate to interested stakeholders
Focus on continual improvement
Life cycle processes
Business development and program management
Requirements definition
Design and develop products and processes
Manage the supply chain
Produce products
Distribute, support and retire products
Enabling infrastructure
Organizational enablers (structure and design)
Process enablers (systems, support groups and incentives)

Peteraf 2003; Sanchez 2004). The targeted capabilities need to be tailored to the specific competitive environment facing the company, linked directly to shareholder value and supported by organizational design (Hart and Milstein 2003; Heraty 2004; Sanchez 2004). The LCM CMM was designed to build the capability to measure and report a company's environmental performance, with an emphasis on LCA. It is possible that too much emphasis was placed on LCA to the detriment of mainstreaming LCM or LC thinking.

The training resources did cover the evaluation of a company's competitive context, but clearly were insufficient to meet the needs of the coaches. In part, this can be attributed to the dominance of environmental or LCA practitioners acting as coaches. However, it should be acknowledged that strong management leadership and cross-functional teams helped many companies identify appropriate projects that delivered business value as well as environmental benefits. The coach need not be the strategic expert, but does need sufficient understanding to tap the right people within the company. The guidance provided by the GE Ecoassessment CoE is a good model for supporting the LCM champions. In particular, the various weighting schemes available in the LCM tool and the interactive tool showing how LCM can add business value would help address this gap. These examples also demonstrate the importance of management vision – a clear expectation that LCM *will* add business value – that is supported with resources and enabling infrastructure (the environmental portal). The Renault initiative to deploy eco-design across its supply chain and the role played by SDA and FICCI in supporting the pilot studies are further evidence of the importance of strong leadership and focus on adding value.

Another common feedback from the coaches was that the maturity assessment tended to be somewhat abstract and difficult to translate into specific actions. The customized maturity assessments FICCI and Renault developed are noteworthy.

Renault adopted a model structured around management practices. The FICCI maturity assessment was tailored to the needs of its members, and the yes-no format was easier to connect to corrective actions. Another potential problem for the maturity assessment is that it can be viewed as just another environmental audit. This can drive a "check the box" strategy to achieve a "good" score across all elements. Research on organizational health has shown that the key to success is to focus on fixing broken processes and target a few key capabilities that were critical to the success of the company's selected strategy. Attempting to exceed benchmarks on all practices was ineffective and wasted resources (Keller and Price 2011). It is important to first identify the core tasks a company must perform to successfully execute its strategy, and then link competencies to these tasks to prioritize improvement efforts (Wesselink et al. 2014). The maturity assessment questionnaire did provide examples of key tasks and the LCM CMM focuses on decision-making processes to help target actions, but it was clear that additional guidance on how to effectively align LCM improvement projects with a company's business strategy would be helpful.

Another risk for any improvement project is that the softer organizational development objectives get lost due to a singular focus on performance outcomes. It is natural for any business to emphasize the immediate demands of customers, investors, regulators, etc. The LCM CMM is designed to promote a balanced approach that addresses both performance and organizational development goals in parallel. Change programs with well-defined goals for both were more than four times more likely to succeed (Keller and Price 2011). This can be challenging for the change agent. Typically, the organizational development efforts have a delayed impact. Support for the longer-term changes depends on delivering near-term performance gains. Sometimes resolving this apparent dilemma is easy, such as improved maintenance procedures at Somos K, which provided fuel savings, reduced emissions and better trained employees. Other cases can be more difficult and required a management commitment. The ability of Multidimensionales or India Glycols to benefit from improved data collection systems supporting their LCA studies will depend on the success of customer education programs. The ability to communicate the business value of the organizational systems and structures necessary to embed the new practices is vital to maintain ongoing management support.

The structure of the pilot studies, specifically requiring identification of a dual objective for each project to meet both performance and organizational development goals, was a strength that resulted in numerous positive outcomes, such as NEHSU developing a Food Safety Management System, Finca Mountain a motor management system, RUCID developing procedures for employees to monitor energy use and so on. Examples from the companies with more mature programs reinforce the importance of providing the enabling infrastructure to embed the improved procedures into normal routines. GE founded the Ecoassessment CoE to support the business units. AQUA+TECH tapped university support to obtain added capability for LCAs. Freudenberg Nonwoven also benefited from a corporate support function. For SME companies in particular, a local resource, such as FICCI or SDA, is a critical success factor for providing the enabling infrastructure to help institutionalize changes. This can be in the form of training workshops or simply nurturing informal networking and exchange of ideas among member companies.

These observations are particularly relevant to developing countries, where over 85 % of companies are small and medium-sized. Accordingly, most of these companies focus their efforts on compliance with environmental standards, and they have limited understanding of the added value of advanced environmental practices, such as EMS, LCA, green procurement or LCM. Interventions to strengthen the ability of companies to implement voluntary projects to advance environmental performance require two components: technical and collaboration. The technical component is related to creating knowledge, starting with basic issues as Cleaner Production, Eco efficiency, data management, environmental management systems before tackling LCM, which has been the focus of many programs to build capacity for LCM. This is a necessary element to ensure companies with low maturity can address issues such as LCM and develop them in a proper and technically valid way. But the technical knowledge must also be supported and nurtured through collaborative networks. Companies that have achieved leadership in environmental performance and have progressed in the implementation of LCM can share their experiences with other companies that are just beginning to develop these practices. This is a key point for SME companies, because they prefer real world examples over theory; and, on the other hand, companies that are sharing their experiences also deepen their knowledge through these workshops. Several examples also highlighted the important role of government policy to help drive the need for LCM.

4 Conclusions and Research Needs

The case studies represent a snapshot in time, so it is not possible to draw significant conclusions regarding the sequence of skill-building suggested in the LCM CMM. However, the programs SDA had conducted with its network of companies to incrementally develop compliance, clean production and environmental management systems were seen to be significant factors in the success of their pilots and were consistent with the suggested sequence of LCM CMM. It seems there is a logical order for acquiring LCM skills that is path dependent, but also a need to embed the lower level capabilities in more developed patterns of action (Hart 1995). Thus, Ediouro Grafica focused on wastes, but used the data in an LCA study. Freudenberg and NEHSU developed KPIs for supply chain management, but again within a LCM frame. The premise is that developing the capabilities with an appreciation of higher-level maturities (i.e. beyond the company's current level) will result in faster and more effective learning. Only time will tell if LCM CMM achieves that objective.

The pilot projects also revealed some key gaps in the training resources provided to coaches attempting to apply the model. Key needs are interpreting the business impact of environmental issues and aligning LCM initiatives with the company's business strategy. The pilots and observations from mature company programs highlighted the need to support efforts to develop the technical capacity for LCA and LCM with complementary resources to address softer management issues associated with individual and organizational development. The success of the pilots was correlated with strong management vision and direction and greater cross-functional representation on project teams. In fact, one of the greatest benefits of the LCA studies conducted might be the enhanced communication across functional groups, supply chain partners and customers. The Renault Collaborative LCA emphasized the importance of developing relationships that crossed organizational boundaries to improve value chain performance.

The collaboration among companies within a geographic region can be a powerful facilitator for disseminating LCM, particularly in developing economies. The networks facilitated by SDA and FICCI are models for future efforts to promote LCM. SMEs often lack the resources necessary to help champion and support change initiatives develop new procedures and systems due to the pressing competitive priorities, production schedules, etc. The external support of a credible organization can be a critical factor in overcoming the inertia to change. It is important that the human change management issues receive the same attention as concern for the technical methodological details of LCA and LCM. It is also important to recognize that companies face numerous market and regulatory barriers to implementing LCM. The social networks, and especially collaboration with government agencies, will be a necessary factor in negotiating policies that facilitate sustainable consumption and production.

A pressing need is more rigorous evaluation of the efficacy of LCM CMM in promoting faster learning in business and in producing outcomes that do move business toward more sustainable practices. The customized maturity assessments developed by FICCI and Renault raise an interesting dilemma. Is it better to tailor the maturity assessment to be more relevant to the local context and company priorities, or work towards a more standardized assessment that would facilitate correlating performance with maturity level? A fundamental challenge in evaluating the efficacy of LCM CMM is that there is no broad consensus on the desired end state. Views of sustainability are value-laden, and different groups will value the same outcome differently. It would be interesting to revisit these pilot studies in 6 months to a year for a retrospective evaluation of the benefits achieved.

Acknowledgements and Disclaimer The views expressed do not necessarily represent the decision or the stated policy of the United Nations Environment Program. Funding from the European Commission is kindly acknowledged for the case studies presented in Sect. 2.2.1.

References

Beer M (2001) How to develop an organization capable of sustained high performance: embrace the drive for results-capability development paradox. Organ Dyn 29(4):233–247

Cabal C (2005) Définition et implications du concept de voiture propre. Office parlementaire d'évaluation des choix scientifiques et technologiques:55

Chouinard Y et al (2011) The sustainable economy. Harv Bus Rev 89(10):53–62

Crosby P (1979) Quality is free. McGraw Hill, New York

De Smet A, Palmer R, Schaninger W (2007) The missing link: connecting organizational and financial performance, McKinsey & Company

Guinée JB et al (2011) Life cycle assessment: past, present, and future. Environ Sci Technol 48(1):90–96

Hart SL (1995) A natural-resource-based view of the firm. Acad Manage Rev 20(4):986–1014

Hart SL, Milstein MB (2003) Creating sustainable value. Acad Manag Exec 17(2):56–67

Helfat CE, Peteraf MA (2003) The dynamic resource-based view: capability lifecycles. Strateg Manag J 24(10):997–1010

Hellweg S, Milà i Canals L (2014) Emerging approaches, challenges and opportunities in life cycle assessment. Science 344(6188):1109–1113

Heraty N (2004) Towards an architecture of organization-led learning. Hum Resour Manag Rev 14:449–472

Kaplan RS (1998) Innovation action research: creating new management theory and practice. J Manag Account Res 10:89–118

Keller S, Price C (2011) Organizational health: the ultimate competitive advantage. McKinsey & Company

Morel S (2014) L'empreinte environnementale à l'ère de la société collaborative: de l'Analyse du Cycle de Vie comme outil expert à une instrumentation collaborative pour conduire une transition organisationnelle. Dissertation, Ecole Nationale Supérieure des Mines de Paris

Nightingale DJ, Mize JH (2002) Development of a lean enterprise transformation maturity model. Inf Knowl Sys Manage 3:15–30

Pezzini M (2012) An emerging middle class. OECD Observer

Porter ME, Kramer MR (2011) Creating shared value: how to reinvent capitalism -and unleash a wave of innovation and growth. Harv Bus Rev 89(1/2):62–77

Rebitzer G et al (2004) LCC-the economic pillar of sustainability: methodology and application to wastewater treatment. Environ Prog 22(4):241–249

RENAULT SAS (2011) Life cycle assessment of fluence Z.E. www.renault.com. Accessed 6 May 2013

Sanchez R (2004) Understanding competence-based management: identifying and managing five modes of competence. J Bus Res 57:518–532

Saur K et al (2003) Draft final report of the LCM definition study

Segrestin B (2003) La gestion des partenariats d'exploration: spécificités, crises et formes de rationalisation. École Nationale Supérieure des Mines de Paris

Silvius AJG, Schipper RPJ (2014) Sustainability in project management competencies: analyzing the competence gap of project managers. J Hum Resour Sustain Stud 2:40–58

Sukhoo A et al (2007) An evolutionary software project management maturity model for Mauritius. Interdiscip J Inf Knowl Manag 2:99–118

Swarr TE (2011) A capability framework for managing social and environmental concerns. Int J Life Cycle Assess 16(7):593–595

Swarr TE et al (2011) Life cycle management capability: an alternative approach to sustainability assessment. In: Finkbeiner M (ed) Towards life cycle sustainability management. Springer, Dordrecht, pp 35–41

UNEP/SETAC (2013a) Life cycle initiative. http://www.lifecycleinitiative.org

UNEP/SETAC (2013b) Life cycle management capability maturity model (LCM CMM) pilot projects. http://www.lifecycleinitiative.org/activities/lcm-pilot-projects/

UNEP/SETAC (2014) Life cycle thinking in Latin America: 16 p. http://www.lifecycleinitiative.org/wp-content/uploads/2015/01/LCA-LCM-company-case-studies-in-Latin-America.pdf
UNEP/SETAC (2015) Organizational life cycle assessment – guidance for organizations to conduct LCA considering their value chain. UNEP, Paris
van Kleef JAG, Roome NJ (2007) Developing capabilities and competence for sustainable business management as innovation: a research agenda. J Clean Prod 15:38–51
Wesselink R, Blok V, Van Leur S, Lans T, Denton D (2014) Individual competencies for managers engaged in corporate sustainable management practices. J Clean Prod. doi:10.1016/j.jclepro.2014.10.093
Wiek A et al (2011) Key competencies in sustainability: a reference framework for academic program development. Sustain Sci 6:203–218
Womak JP, Jones DT (2003) Lean thinking: banish waste and create wealth in your corporation. Simon & Schuster, London

Chapter 18
Promoting Life Cycle Thinking, Life Cycle Assessment and Life Cycle Management Within Business in Brazil

Marina Santa Rosa Rocha, Luiz Gustavo Ortega, Yuki Hamilton Onda Kabe, Maria da Graça C.B. Popi, Felipe Duarte, and Maria Luisa Nerys

Abstract The development of life cycle assessment in Brazil began in 1994. Business companies were involved in this process by the government, and in this way life cycle management developed. To support and disseminate LCA and LCM, a group of nine companies created the Brazilian Business Network for LCA in 2013. In 2014, the Brazilian Business Council for Sustainable Development became the coordinator of the Network. Despite the many challenges in promoting the theme in Brazil, the Network is advancing and benefits from the representativeness of large companies and their power to push governments, peers and their value chains.

Keywords Brazilian Business Council for Sustainable Development • Life cycle assessment • Life cycle management

1 Introduction

The development of life cycle assessment (LCA) in Brazil began in 1994 with a subgroup to discuss the subject within the Supporting Group for Environmental Standardization (GANA, acronym in Portuguese). This group was supported by the

---

M.S.R. Rocha (✉)
Brazilian Business Council for Sustainable Development, Rio de Janeiro, Brazil
e-mail: marinasrosa@cebds.org

L.G. Ortega • Y.H.O. Kabe
Brakem, São Paulo, Brazil

Maria da Graça C.B. Popi
Oxiteno, São Paulo, Brazil

F. Duarte • M.L. Nerys
Vale, Rio de Janeiro, Brazil

G. Sonnemann, M. Margni (eds.), *Life Cycle Management*,
LCA Compendium – The Complete World of Life Cycle Assessment,
DOI 10.1007/978-94-017-7221-1_18

Brazilian Association of Technical Standards (ABNT, acronym in Portuguese) (Ribeiro and Cherubini 2014). This represented the first step towards the dissemination of LCA in Brazil.

After this, a few important guidelines and studies were developed until the creation of the Brazilian Life Cycle Association (ABCV, acronym in Portuguese) in 2002. This association aims at the dissemination and consolidation of life cycle management in Brazil, working along with businesses, academic institutions, governments and organized civil society (ABCV 2015, see Ribeiro and Cherubini 2014).

Four years later, in 2006, a project for constructing the Brazilian Life Cycle Database started. The governmental institution responsible for this project was the Brazilian Institute of Information on Science and Technology (IBICT, acronym in Portuguese), which is a federal institute associated to the Science, Technology and Innovation Ministry. In 2010 this project led to the creation of the Brazilian Life Cycle Assessment Program (PBACV, acronym in Portuguese). This program was approved by the National Council for Metrology, Standardization and Industrial Quality (IBICT 2015).

The coordination of the program was delegated to INMETRO (National Institute for Metrology, Standardization and Industrial Quality) in cooperation with IBICT. The objective of PBACV is to "support the sustainable development and the environmental competitiveness of the Brazilian industrial production and promote the access to internal and external markets" (BRASIL 2010b, see Ribeiro and Cherubini).

Despite all these activities and the involvement of some industries in PBACV, the engagement of some key businesses was still lacking. Therefore, some key companies created the group "Brazilian Business Network for LCA".

2 Brazilian Business Network for Life Cycle Assessment

The Brazilian Business Network for LCA was launched in 2013 by nine large companies in Brazil, namely Braskem, Danone, Embraer, GE, Grupo Boticário, Natura, Odebrecht, Oxiteno and Tetra Pak, in partnership with the Brazilian Life Cycle Association and Akatu Institute.

The first activity of the Network was to conduct a survey among its members to understand the main gaps in LCA development in the Brazilian business. The conclusions of the survey showed that the lack of experts and reliable data prevented further use of LCA and LCM in business. Thus, the Network defined its main goals as follows:

- To create an environment of cooperation between companies interested in using LCA in Brazil, allowing the optimization of resources
- To educate and qualify companies regarding the concept, application, and the benefits of LCA

- To provide access and disseminate LCA information in Brazil through studies, good practices and specialists
- To influence and support governments to create a Brazilian LCA database

To be linked to an institution that have the representativeness in the sustainable development field and that could give support to the Network for achieving its goals turned out to be an interesting path. So, in 2014, the Brazilian Business Council for Sustainable Development (BCSD – Brazil) became the coordinator of the Network. BCSD-Brazil is a non-profit civil association that promotes sustainable development for companies that operate in Brazil by interacting with governments and civil society, being the representative in Brazil of the World Business Council for Sustainable Development (WBCSD) (CEBDS 2015).

Until May 2015, approximately 20 members joined the Network, amongst others, 3M do Brasil, Alcoa, Arcelor Mittal, BASF, Duratex, Klabin, Petrobras, Vale, Volkswagen and Votorantim Cimentos.

The company members have complex and extensive value chains which are key for the dissemination of life cycle thinking. So, they influence not only their value chains and peers, but also the governments and the civil society. In terms of actions to support and influence governments, the Network plays a key role in the development of the Brazilian Data Base which will enable access of trustful life cycle information regarding the Brazilian reality.

Through BCSD-Brazil, the Network has created a partnership with the United Nations Environmental Program (UNEP) and IBICT to build up the capacity of local specialists for the development and adaptation of datasets.

Not only specialists must be prepared to work with LCA but also corporate professionals should be aware of basic LCA and LCM concepts. In this sense, the Network has launched in October 2014 a 2-day course designed for a corporate audience which had Professor Cassia Ugaya, from Federal Technological University of Parana (UFTPR, acronym in Portuguese), as a trainer. The 2014 edition was well succeeded and had 36 participants from 23 companies from many different sectors like Adidas, KPMG and Caixa Econômica Federal (CEBDS 2014).

These two accomplishments were the result of the work developed by two of the Network's Working Groups (WG). In order to achieve its objectives the Network is divided into five WGs which are: Accreditation, Capacity Building and Professional Education, Communication, Data Base and Labelling. This division enables the development of specific projects on each key area for promoting LCM in Brazil.

3 Implementing Life Cycle Assessment and Life Cycle Management Within Companies

Incorporating life cycle thinking into company's management is a complex task, despite the benefits it can bring. The main challenges are: professional training, top leadership commitment, availability of sensitive data (regarding intellectual property) and mainly, for Brazil, to obtain realistic data. On the other hand, the major gain is the improvement of the decision-making process in order to achieve more sustainable practices, often related to an adequate use of resources and the risk prevention along the supply chain.

In this sense, Vale, a Brazilian mining company, has started to incorporate life cycle thinking into its processes. In 2013, an LCA study was completed for all pelletizing facilities in Brazil. With this study, Vale could identify internal best practices, which may become a reference for environmental issues assessed and be extended to other units, as well as adding value to pellet, raising its competitiveness in the medium- and long-term (Vale 2013).

Vale's goal is to work in an expanding environmental management strategy beyond the borders of Vale, positively influencing its supply chain. In 2015, the company is continuing to develop strategies to disseminate life cycle thinking across the whole company (Vale 2013).

Another example is Braskem, a petrochemical company, whose corporate strategy is based on the belief that chemistry and plastics make people's lifes better. This belief has led to the definition of a purpose: to develop sustainable solutions that make people's lives better. Both these statements are based on a life cycle perspective since it is in the use phase that most of the benefits of plastic products become apparent.

Life cycle management at Braskem began in 2004 with two eco-efficiency studies that led to improvements in their PVC production processes. Another eco-efficiency study led to the decision to invest in a bio-based polymer using sugar cane ethanol as feedstock: the I'm Green™ Polyethylene. Ten years later, more than 30 LCA studies have been conducted. These are used to aggregate sustainability as an added value to products and also to plastic solutions developed together with direct customers. Recently Braskem has used the company's Materiality Matrix to develop a single-score life cycle impact assessment method to help corporate decision making process by incorporating stakeholder opinion into a weighting scheme.

Braskem is also incorporating LCM. A qualitative assessment for the early stages of project development and for innovation processes has been developed. In addition, the company is currently working on extending it to incremental optimization projects of market development and application engineering.

4 Conclusions

Companies with a vision of the future are recognized as one of the main transformation drivers in our society. This has been proved by the development of LCA and LCM in the last years in Brazil, despite the many challenges it still faces.

The Brazilian experience is probably akin to several other countries, but it shows that transformation is greatly accelerated by the engagement of large companies which can move entire value chains. Societal transformation, however, requires academia, government, businesses and the general society working together.

References

CEBDS (2014) http://cebds.org/noticias/rede-de-acv-promove-curso-de-capacitacao/#.VOdwwPnF_OM

CEBDS (2015) http://cebds.org/en/about-cebds/about-us/

IBICT (2015) http://acv.ibict.br/portal/comunidade/historico-acv

Ribeiro P, Cherubini E (2014) Projeto Diálogos Setoriais Brasil – União Europeia sobre Análise do Ciclo de Vida. Brasília, 171 p

Vale (2013) 2013 Sustainability report. Available at http://www.vale.com/EN/aboutvale/sustainability/Pages/default.aspx

Chapter 19
Mainstreaming Life Cycle Sustainability Management in Rapidly Growing and Emerging Economies Through Capacity-Building

Sonia Valdivia, Sanjeevan Bajaj, Guido Sonnemann, Ana Quiros, and Cassia Maria Lie Ugaya

Abstract *Professionals in companies in industrialised countries as well as consultants and academics working for these companies have acquired capabilities in life cycle sustainability management (LCSM). Actors in emerging economies are beginning to develop those capacities as well, supported by training programmes.* This chapter highlights the work of the UNEP/SETAC Life Cycle Initiative which has worked toward strengthening LCSM/LCA capacities worldwide since its creation in 2002; presents the current conditions for mainstreaming LCSM following an analysis of available knowledge and infrastructure in 18 rapidly growing economies (including least developed countries, rapidly growing and emerging economies) based on four criteria; and discusses the results and proposes a framework to develop LCSM capacities and provides conclusions and an outlook with recommendations. As some capacities in a number of emerging economies are beginning to consolidate, the implementation and changes of current practices toward improved performance by implementing *LCSM* remain insufficient.

Keywords 10YFP Consumer Information Program (CIP) • Decision making • Life cycle assessment • Developing countries • Life cycle management • Life cycle

---

S. Valdivia (✉)
World Resources Forum, St. Gallen, Switzerland
e-mail: sonia.valdivia@wrforum.org

S. Bajaj
Federation of Indian Chambers of Commerce & Industry, New Delhi, India

G. Sonnemann
University of Bordeaux, Bordeaux, France

A. Quiros
Association for Life Cycle Assessment in Latin America (ALCALA), San Jose, Costa Rica

C.M.L. Ugaya
Universidade Tecnológica Federal do Paraná (UTFPR), Curitiba, Brazil

G. Sonnemann, M. Margni (eds.), *Life Cycle Management*,
LCA Compendium – The Complete World of Life Cycle Assessment,
DOI 10.1007/978-94-017-7221-1_19

sustainability assessment • Life cycle sustainability management • Sustainability • Sustainable Lifestyles and Education Program • UNEP/SETAC life cycle initiative

1 Introduction

Access to (elementary) education is a human right and technical and professional education shall be made generally available and equally accessible to all according to article 26 of the Universal Declaration of Human Rights (UN 1948). Education is also a fundamental determinant not only of health, demographic trends and individual income, but also of a country's aggregate level of economic growth (IIASA 2008) and thus provides the most promising route toward sustainable development.

Furthermore, the authors consider that education at all levels – from primary school via university to lifelong learning – is crucial to improving the understanding and minimization of environmental and socio-economic impacts. Thus, further promotion of life cycle sustainability management (LCSM) – which follows a holistic approach for better informed decisions – is paramount.

LCSM is a powerful approach that has been recognized by multinational companies for the last decade. It helps to structure decision-making processes to follow a life cycle perspective in business and industry across an ever-increasing number of organisations in industrialized countries.

In response to the need for improved understanding and technical skills on LCSM, a number of life cycle-focused initiatives worldwide have incorporated capacity-building in their missions.[1] To illustrate this development, it is worth highlighting the work of the UNEP/SETAC Life Cycle Initiative. It establishes the strengthening of LCSM/LCA capacities worldwide. Since its creation in 2002, more than 30 training events in about 18 developing countries (Sonnemann et al. 2015) have been organized, not to forget the support of the life cycle management (LCM) capability maturity model (CMM) implementation in developing countries between 2012 and 2013 (Swarr et al. 2015a, b).

In 2014, important developments at the international level include the launch of two global programmes under the 10-Year Framework of Programs (10YFP), which cover consumer information, education and awareness-raising on life cycle-based approaches[2] (10YFP 2015):

- The 10YFP Consumer Information Program (CIP) is defined as a range of tools and systems that seek to encourage consumers for more sustainable choices about goods and services, including their use and end-of-life phases (CIP 2015).
- The Sustainable Lifestyles and Education Program (SLE) seeks

[1] The term "capabilities" is used for individuals and "capacities" for organizations as it is more institutional.

[2] In the context of the 10YFP, life cycle approaches also include life cycle management.

– To develop sustainable ways of living common practice through overall systems change
– To promote demand for and enable use of sustainable infrastructures (e.g. related to transportation)
– To achieve well-being and improve quality of life for all through awareness-raising, education and capacity-building across all sectors of society (SLE 2015)

Capabilities on LCSM have been acquired by professionals in companies in industrialized countries as well as consultants and academics working for these companies. Actors in emerging economies are beginning to develop those capacities as well, supported by training programmes like the CMM implementation in developing countries between 2012 and 2013 (Swarr et al. 2015a, b). While local capacities are now available in the developing world, especially in rapidly growing and emerging economies, the implementation and changes of current practices toward improved performance is still insufficient (Valdivia et al. 2014; Sonnemann et al. 2015). This chapter seeks to present the current conditions for mainstreaming LCSM based on an analysis of available knowledge and infrastructure on life cycle approaches in 18 selected rapidly growing and emerging economies based on four criteria.[3] Furthermore, the chapter aims to discuss the results, propose a framework to develop LCSM capacities and provide conclusions and an outlook with recommendations.

2 Potential for Mainstreaming Life Cycle Sustainability Management in Developing Countries: A Global Status Assessment

The term "mainstream" can be understood as "ideas, attitudes, or activities that are shared by most people and regarded as normal or conventional" (Oxford Dictionaries 2015) or as "products and services readily available and appealing to the general public, as opposed to being of interest only to a very specific subset of the public" (Business Dictionary 2015). The authors acknowledge that the connotation of mainstreaming can be negative if the ideas, attitudes, products or activities ignore the individual and group values that are essential in free societies or, in a broader sense, that become obstacles in a path toward sustainable development. In order to reach substantial socio-economic changes in societies, mainstreaming is a key instrument, but cultural diversity and different backgrounds need to be taken into account before designing and implementing measures.

To illustrate the existence of different mainstreaming perceptions in a country and the ways to implement changes, the case of India is described as follows.

[3] (a) training activities in place, (b) local LCA studies available, (c) national LCA databases in place, (d.1) an active national network and aspects such as (d.2) size of the network and (d.3) gender balance.

In India, value systems that discourage non-conventional behavior are still "mainstream", but this is changing as being conventional and as weak and conformist. A need for mainstreaming social reform has been recognized in India and is being implemented through provisions in the Constitution and various laws. The social reform currently in place is questioning many traditional practices such as those perpetuating inequalities based on gender, caste or economic class. Although the social reform in India has legal support, substantial changes are still far behind.

While mainstreaming life cycle sustainability assessment (LCSM) in industrialized countries focuses on effective communication between opinion leaders and consumers on life cycle thinking, in developing countries it is capacity-building of companies and governments. Information-sharing platforms and demonstration projects are essential in developing countries.

2.1 *Methodology and Criteria for the Assessment of Mainstreaming Conditions*

The focus of the assessment of mainstreaming conditions concerns 18 rapidly growing and emerging economies that were selected according to geographical balance (Table 19.1).

Aside from gender, the criteria used to assess the mainstreaming conditions status of LCSM/LCA were inspired by the survey from AIST (2006) to analyze the situation of LCA implementation in selected countries, including Brazil, Malaysia and Australia. The following criteria were applied:

(a) LCM and LCA training activities in place. Results in Table 19.1 are interpreted as follows: 0 is equal to "not at all"; 1 to "only on LCA"; 2 to "multiple on LCM/LCA"; and 3 to "many good quality ones on LCM/LCA".
(b) LCA studies available. This implies that local data exist and that organisations start to implement LCSM. The scores have the following interpretations: 0: not at all; 1: few studies; 2: some, mainly by academics; and 3: many, by academics and big companies.
(c) National LCA database operating. This is the basic infrastructure needed to have the potential to mainstream LCSM. Three cases are considered: 0: there is no database; 1: a database is under development; and 2: local database is available for core sectors in the country.
(d) Active national life cycle network(s), including information on size of the network and gender balance aspects (Buckingham-Hatfield 2002). In general, these networks can function as a major multiplier for applications of LCSM by organisation. The first part (d.1) indicates the existence ("1") or not ("0") of a network in the country; 0/1 means that there is neither agreement on the existence of a network nor a website available. The second part (d.2) refers to the size of the network and the last part (d.3) to the percentage of women.

Table 19.1 Mainstreaming conditions criteria in selected rapidly growing and emerging countries (a–d.3) and status of the market for LCA professionals (e)

	(a)	(b)	(c)	(d.1)	(d.2)	(d.3)	(e)
Country	**Training activities**	**LCA studies**	**LCA database**	**LC network**	**Size**	**Gender**	**Market for LCA professionals**
Latin America							
Argentina	2	2	0	1	10–50	73 %	1
Brazil	3	2	1	1	50–100	31 %	1
Chile	2	1	1	1	10–50	38 %	1
Colombia	2	1	0	1	50–100	44 %	1
Peru	2	2	0	1	10–50	61 %	1
Mexico	2	2	2	1	100–250	37 %	1
South East Asia							
China	3	2	2	1	1000	42 %[a]	1
India	2	1	1	1	250–500	23 %	1
Indonesia	1	1	0	0/1	5–10	13 %	1
Malaysia	2	1	2	1	10–50	40 %	1
Philippines	0	1	0	0	0–3	nr	1
Thailand	3	3	2	1	100–250	56 %	1
Africa							
South Africa	2	2	1	1	10–50	35 %	1
Egypt	1	1	0	0	5–10	13 %	0
Morocco	1	1	0	0	0–3	nr	0
Eastern Europe							
Russian Federation[b]	0	1	0	0	0–3	nr	1
Turkey	2	2	1	0/1	5–10	24 %	1
West Asia							
Saudi Arabia[b]	0	0	0	0	0–3	nr	0
Total 18							

nr not representative. No more than three persons have been identified in that country; hence, the result cannot be considered representative

[a]Info estimated by Ping Hou, China

[b]Only one response was received from this country; hence, the results cannot be considered valid

(e) A market for LCA professionals. In this case, an answer equal to 0 means there is no market at all, 1 means that the market is still developing and 2 means that the market is fully developed. The authors conclude that the existence of training activities, LCA studies, LCA databases and life cycle networks are conditions for mainstreaming.

Between 2013 and 2014, a global and a Latin American survey took place (Sonnemann et al. 2015; Valdivia et al. 2014), which were the basis for the assessment (see results in Table 19.1). About 80 responses were received from the life cycle national networks. Gender was analyzed based on the 2000 followers of the UNEP/SETAC Life Cycle Initiative. A discussion of the main findings is presented in the next sections.

The analysis of progress in the use of LCA in policies (see Sect. 2.7) could not be completed, due to insufficient information. Nevertheless, Sect. 2.7.1 on "LCA in policies" illustrates the situation in Thailand, China, Brazil and Mexico.

2.2 *Capacity-Building for Life Cycle Sustainability Management*

Capacity-building for LCSM means a process of building (i) capabilities to use the range of LCSM topics (see Fig. 19.2), (ii) relationships among key stakeholders and users and (iii) awareness-raising on the importance of a holistic life cycle perspective covering social, economic and environmental aspects in decision-making. Training is just one element of the capacity-building process, which usually focuses on providing skills for a specific problem (e.g. by learning how to generate LC data or to use LCA software).

However, as can be seen from the example India, it does not make sense to provide technical training if the managers are not convinced that LCSM is the way to tackle sustainable development challenges in organizations. Therefore, a package of training courses is needed, including LCM CMM and technical assistance for its implementation in developing countries, as done between 2012 and 2013 (Swarr et al. 2015a, b).

The parameters analyzed in 18 countries show that multiple training activities on both LCA and LCM, so in Thailand and Brazil, found a critical mass of experts to contribute to mainstreaming LCSM. A critical mass is developing, due to emerging efforts, in South Africa, Turkey, the other Latin American countries analyzed and the Asian countries except Indonesia and the Philippines. In the two latter countries, the level of LCA training activities is at a very early stage.

In Egypt, Morocco, Indonesia, the Philippines, the Russian Federation and Saudi Arabia, the offer of LCA and LCM training activities does not exist or is very limited. Consequently, mainstreaming LCSM cannot be envisaged in the short term, and the international community is asked for support in form of basic training activities.

2.3 *LCA Studies and LCA Databases*

The existence of LCA studies indicates the availability of local resources in terms of life cycle experts and tools and minimum funding to cover the costs of performing these studies. Subsequently, it also indicates the availability of some local foreground data, which can feed a national LCA database, so in China, Mexico, Malaysia and Thailand.

Survey findings in the area of LCA studies and LCA databases (beyond the list of countries with a database) are presented below:

- In Brazil, Chile, India, South Africa and Turkey, a breakthrough is expected, as international organisations (e.g. UNEP and the European Commission jointly) or projects (the ecoinvent project on Internationalisation (www.sustainable-recycling.org)) as well as national organizations (e.g. Fundación Chile) are making progress with international and local resources.
- Colombia, Peru and Argentina are at an early stage of discussions on how to generate data and establish LCA databases (Quispe et al. 2014; see RPCV 2005).
- There is very little or no evidence of local LCA studies and no records at all concerning the development of databases in the remaining countries.

2.4 Networking

The survey results and additional literature sources confirm the existence of national life cycle networks in all countries covered by the survey, except Egypt, Morocco, Indonesia, the Philippines, the Russian Federation and Saudi Arabia. Where networks are operating, websites have been identified except in Turkey and Indonesia. A website helps to improve co-ordination and communication, and is a requirement for effective dissemination and communication.

The analysis of networking capacity is based not only on the amount of networks' members, but also on specific capacities measured as per the estimated number of life cycle practitioners per 10 million inhabitants in the country. 2014 population figures are used for these calculations (Worldometer 2014). The values presented in Table 19.2 range from 0.3 in Indonesia to 22 in Thailand.[4] This result suggests that a critical condition for mainstreaming LCSM is the number of experts available in the country.

2.5 Gender Aspects

Gender equity is a pre-requisite for sustainable development (UNCED 1992).

The survey results in Table 19.2 show differentiated developments in countries where valid information is available. The average indicates a value of 40 % female participation. Only in Thailand and Argentina did females comprise the majority. The figures were more balanced in Latin American countries, Malaysia and South Africa.

[4]The Russian Federation, Saudi Arabia and the Philippines were not considered in this part of the analysis, as the responses provided were not representative.

Table 19.2 Mainstreaming status in selected rapidly growing and emerging countries (a–d.1) showing the number of LCA practitioners per 10 million inhabitants and gender distribution in life cycle networks

Country	# of LCA practitioners per 10 Mio inhabitants	Proportion of women (%)	Mainstreaming conditions (Total = a + b + c + d.1) [0–9]
Thailand	22	56	9
China	7	42	8
Mexico	11	37	7
Brazil	4	31	7
Malaysia	10	40	6
South Africa	6	35	6
Argentina	10	73	5
Peru	7	39	5
Chile	18	38	5
Colombia	12	44	5
India	2	23	5
Turkey	4	24	4
Egypt	1	13	2
Indonesia	0.3	13	2
Philippines	nr	nr	1
The Russian Federation	nr	nr	0
Morocco	nr	nr	0
Saudi Arabia	nr	nr	0

nr not representative. No more than three persons have been identified in that country; hence, the result cannot be considered representative

2.6 *Mainstreaming Conditions in Selected Countries*

While Thailand ranks the highest score (with a total of 9) and has very favourable conditions for mainstreaming, the respondents argue that the market is still under development. With 22 LC practitioners per 10 million inhabitants, it is plausible that Thailand needs more capacity and experts to be able to cover the demand in more public policy areas and the private sector.

The conditions for mainstreaming LCSM in China, Mexico and Brazil, with scores of 7 or 8, are sufficient, but the fact that the market for LCA professionals is still relatively small suggests that the demand may need to be boosted by encouraging a strong uptake of LCSM within more industrial sectors and by enforcing current legislations or regulations. Expanding the number of LCA professionals per 10 million inhabitants in these countries, which rank below Thailand, is an urgent need in order to be able to grow with the market. This can be achieved, for instance, by having more universities provide education on LCA in natural science and engineering programs and on LCM in business schools and economic faculties.

The next set of countries – South Africa, Malaysia, Argentina, Peru, Chile, Colombia, India and Turkey – with 4–6 points, have conditions for mainstreaming LCSM in the right direction. However, they still lack two essential ingredients: the business case for companies and the political will (e.g. to establish a national LCA database or create a policy requiring the use of LCA) through which progress is acknowledged and financial resources are made available. Without this push, the impact of efforts made by the UNEP/SETAC Life Cycle Initiative, national life cycle networks or other groups at the global and regional level will go unnoticed (UNEP/SETAC 2009a).

There is a last group of countries that has not yet developed mainstreaming conditions (scores from 0 to 2). These countries require tailored strategies and outreach approaches by taking into consideration their major challenges, in co-operation with key players in that country. Funding a core group of potential leaders in the life cycle area and a nationally-recognized organization like the Chambers of Commerce – take the example of the Federation of Indian Chambers of Commerce & Industry (FICCI) in India – that can provide matching resources (e.g. experts' time, venues, etc.), is fundamental for any programme to be successful.

The survey results concerning the number of LCA practitioners per 10 million inhabitants, the proportion of women in life cycle networks and the overall rating regarding the conditions for mainstreaming LCSM are summarized in Fig. 19.1.

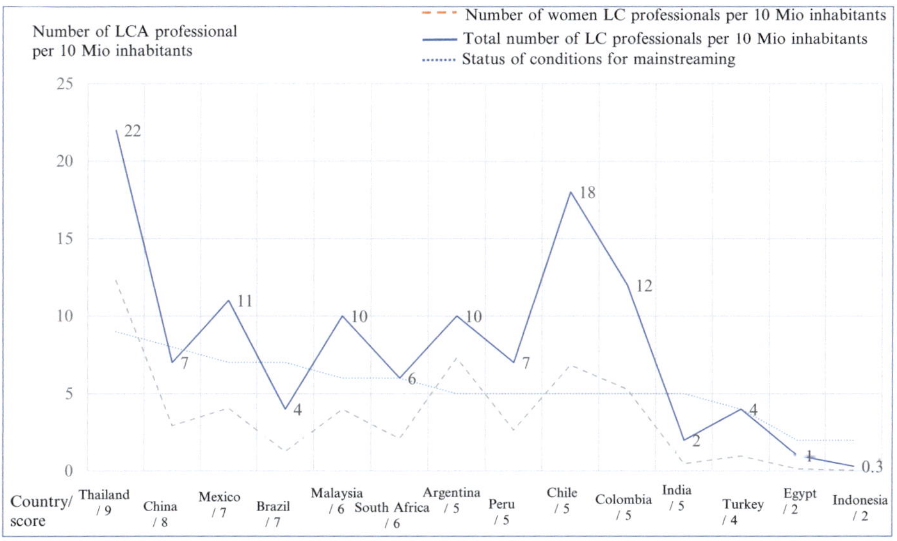

Fig. 19.1 LCA professionals per 10 million inhabitants vs. female participation and status of mainstreaming conditions

2.7 *Market for LCA Professionals*

The existence of a market for LCA professionals is a positive development in the countries and constitutes the proof of mainstreaming. If there are already experts in a country contributing to developing life cycle tools or generating LCA data for policy-making or assisting companies, then there is a good basis to move toward a better understanding and managing the countries' and companies' value chains.

In the sample analyzed, all respondents indicate that, at best, the markets are under development even in countries with LCA databases and progress on the use of LCA in policies (e.g. Thailand, China, Mexico and Brazil, as introduced in Sect. 2.7.1). This finding shows that, despite efforts made in those countries, additional mechanisms are needed to boost the market substantially. Options include more university education on life cycle matters, professional training, life cycle innovation by companies and more public policies.

2.7.1 Use of LCA in Policies

Policies based on LCA were identified in two Asian and two Latin American countries. The general developments started in 2010. Asia, China and Thailand evidenced a significantly *higher level of development and implementation* of policies using LCA. The most encouraging policy in China, the Eco-design of Industrial Products Guidance of 2013 (MIIT/MEP/NRDC 2013), is boosting mainstreaming of LCA by promoting its use in product design. Thailand is using life cycle inventory data to quantify the Green Gross Domestic Product (GDP) of its industrial sectors, as well as LCA and Life Cycle Costing to assess Phase 1 (2008–2011) of the Thai Green Public Procurement Plan (ORDER PRE/116/2008 2008) to decide whether and how to implement phase 2 (2014–2017) (Thumrongrut Mungcharoen 2013).

More recent examples of the use of LCA in regulations and programs have been identified in Brazil and Mexico. These examples have not covered all industrial sectors yet. In Mexico, a regulation for sustainable buildings (NMX 2013) on criteria and minimum environmental requirements needs the impact assessment of the whole life cycle of buildings (including the use phase). In case of the replacement of building materials, it is also demanded the use of third-party reviewed LCAs of alternative materials for comparative assertion purposes (Güereca et al. 2015). As for Brazil, there is no mandatory policy regarding the use of LCA, but there are recommendations to use it, for example, in the Solid Wastes National Policy (Federal Law No. 12.305, 2010, and Decree No. 7.404, 2010, MMA 2010), which calls for shared responsibilities among relevant stakeholders along the life cycle of wastes and the use of LCA to promote products with fewer environmental impacts.

The cases presented above suggest an emerging awareness by politicians in emerging economies. It means that an increasing number of policy-makers is demanding LCA and information for decision-making at the product and sectorial levels. Evidently, this will contribute to creating favourable conditions for mainstreaming LCSM.

3 Framework for Capacity-Building and Developing Capabilities to Mainstream Life Cycle Sustainability Management

Strengthening capacity-building and the development of capabilities to mainstream LCSM is a continual process and implies a step-by-step approach. A framework with the range of topics embraced by LCSM is presented in Fig. 19.2 and described below in this section.

The sequence is based on the flow of definitions and results that serve the next topic, but does not imply a fixed training sequence, except from LCM, which is recommended as the first training module because it introduces life cycle thinking and the various tools – not only LCA – and explains the way they could be articulated.

The tools covered have "continual improvement" as a central target message (see Fig. 19.2). If then there is an interest by an organization or an individual in environmental LCA (E-LCA according to ISO 14040/44 2006), further training on this tool can be provided. Other approaches, for which training could also be provided, include carbon, water and environmental footprinting, organizational LCA (O-LCA, UNEP/SETAC 2015b), hotspot analysis (Fava et al. 2014) as well as Social LCA (S-LCA, UNEP/SETAC 2009b), and Life Cycle Costing (LCC).

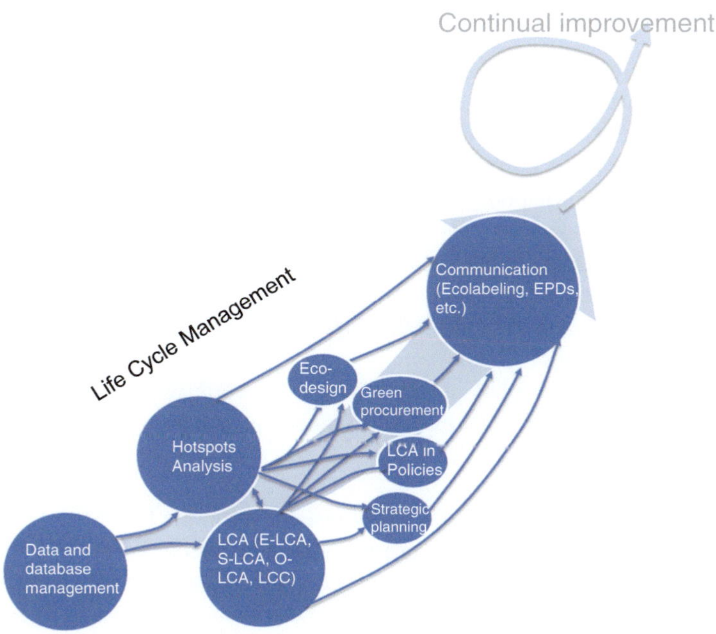

Fig. 19.2 Capacity-building and developing capabilities for mainstreaming LCSM

Since LCSM needs life cycle information, the availability of reliable data is also important. Due to the enormous amount of data needed to perform an LCA study, capacity-building regarding data development and database management can be also of interest. In this sense, a training kit has been developed, based on the Global Guidance for LCA Databases (UNEP/SETAC 2011). Data, however, are not enough for understanding data formation; impact assessment models and further training on the review of the LCA of products are highly recommended to be able to interpret the results.

The improved understanding of impact assessment models and frameworks facilitates the interpretation of results and their use in design choices (eco-design), sustainable consumption (sustainable procurement), strategic planning in business and LCA in policies. Further skills and abilities are necessary for communicating sustainability performance, aiming at disseminating best practices and improvements while avoiding greenwashing.

The choice of the training path and whether or not a topic is considered in a capacity-building program depends on the needs of the audience and on country and company priorities. No matter the choice, the continual improvement of capacities should remain a focus.

4 Conclusions, Recommendations and Outlook

4.1 Conclusions and Recommendations

1. General progress since 2002.
   Countries have advanced since 2002, when none of them counted on a life cycle network or a LCA database (UNEP/SETAC 2015b). The authors recommend reinforcing the trend in the countries analyzed with continual capacity-building programs, financial support and technical assistance endorsed by national and local governments and facilitated by agencies for international co-operation and inter-governmental organizations.
2. LC networks as a success in capacity-building.
   The authors' experiences confirm that national life cycle networks are key to successful capacity-building programs.
3. Insufficient number of LCA professionals in the country.
   The number of LCA professionals in all countries is low in relation to the size of the population.
4. Gender distribution.
   This indicator has not been sufficiently researched in the LCSM field. However, the respective findings of the current study are encouraging. In general, the higher the percentage of women, the better the score concerning the status of mainstreaming conditions. Only in two countries, Argentina and Thailand, did women represent the majority.

5. Presence of LCA in policies.
   The cases from Thailand, China, Mexico and Brazil reinforce the results of the status of mainstreaming condition assessment in which these countries rank in the top four. This shows that the use of LCA in policies is possible in emerging and rapidly growing economies.
6. Thailand with favourable LCSM/LCA mainstreaming conditions.
   Of the 18 countries researched, Thailand excels in mainstreaming LCA, as it uses LCA-based results in policies on public procurement. However, the authors cannot conclude that LCSM is mainstream in this country. Nevertheless, the case of Thailand demonstrates how far a country with limited resources can go when strong political acceptance is given.
7. Countries with no conditions for mainstreaming.
   Countries at the bottom of the ranking are in the position where the others were in 2002. The first step for them is to address awareness-raising events for decision-makers from business as well as municipal and national governments, and basic training for academia and consultants. Support from international organizations is key to assisting these countries in paving the way for the conditions in which mainstreaming LCSM can be envisaged.

4.2 Outlook

Capacity-building programs on LCA/LCM/LCSM are needed to intensify and deepen their activities on a greater pace and to connect them to global programs such as CIP and SLE.

Ongoing international co-operation as well as new forms of inter-regional co-operation are key to improving the conditions for mainstreaming LCSM (UNEP/SETAC 2015b).

According to current experiences regarding capacity-building and capability development trends, LCM, followed by LCA and footprinting approaches as well as data and database management, appear to be the most sought-after areas of skills development. While LCA appeals more to academia and consultants, LCSM is at the core of what companies are interested in.

Echoing the reflections of Paul Hohnen (2014), more awareness from politicians is essential to boost implementation of LCA in developing countries. Additionally, streamlined communication is needed at all levels. The implementation of LCSM and the tools and data required can no longer be presented as a complicated or expensive approach, but rather as a valuable and powerful management way toward innovation as well smart and responsible decision-making for more sustainable consumption and production (Hohnen 2014).

Acknowledgements The authors would like to thank the following for valuable feedback provided during the development of the study and writing of the report: Sohini Gupta (FICCI), David Cozac and Sandra Mendez (EMPA).

Disclaimer The views and opinions expressed in this chapter are those of the authors and do not necessarily reflect the views of their organizations.

References

10YFP (2015) 10-year framework of programmes on sustainable consumption and production. www.unep.org/10yfp/. Accessed 28 Feb 2015

AIST (2006) 5th AIST workshop on LCA for Asia Pacific Region: summary of country reports in Asia Pacific Region (incl. Australia, Brazil & Chile). www.aist-riss.jp/old/lca/asianetwork/symposium/report/2006/powerpoint/02sagisaka.pdf. Accessed 28 Feb 2015. ISO 14040/44 (2006) Environmental management – life cycle assessment – principles and framework. ISO. Geneva

Buckingham-Hatfield S (2002) Gender equality – a pre-requisite for sustainable development. Geography 87(3):227–233. www.jstor.org/discover/10.2307/40573738?sid=21105989898753&uid=2490615183&uid=2490615193&uid=70&uid=3&uid=2134&uid=60&uid=2&uid=3737760. Accessed 28 Feb 2015

CIP (2015) The 10YFP consumer information program (CIP). www.unep.org/10yfp/Programmes/ProgrammeConsultationandCurrentStatus/Consumerinformation/tabid/106265/Default.aspx. Accessed 28 Feb 2015

Fava et al (2014) Hotspots analysis: mapping of existing methodologies, tools and guidance and initial recommendations for the development of global guidance. www.lifecycleinitiative.org/wp-content/uploads/2014/12/UNEP-Hotspots-Mapping-Project-Final-Report-Phase-1.pdf. Accessed 28 Feb 2015

Güereca PL, Ochoa R, Gilbert H, Suppen N (2015) Life cycle assessment in Mexico: overview of development and implementation. Int J Life Cycle Assess 20(3):311–317

Hohnen P (2014) Life cycle thinking and the next industrial revolution. The smarter business blog (in Theory). http://tobiaswebb.blogspot.in/2014/12/life-cycle-thinking-and-next-industrial.html. Accessed 28 Feb 2015

IIASA (2008) Economic growth in developing countries: education proves key. Policy Brief # 3, International Institute of Applied Systems Analysis (IIASA). http://iussp2009.princeton.edu/papers/92600. Accessed 28 Feb 2015

ISO 14040/44 (2006) Environmental management — life cycle assessment. Requirements and guidelines. ISO, Geneva

MIIT/MEP/NRDC (2013) Eco-design of industrial products guidance. Published in 2013 by the Ministry of Industry and Information Technology and the Ministry of Environmental Protection of the People's Republic of China, and the National Development of Reform Commission (NDRC). http://www.indigodev.com/sepa_eip_guideline.html. Accessed 28 Feb 2015

MMA (2010) Politica Nacional de Resíduos Sólidos – Ministério do Meio Ambiente. Lei no 12.305/2010, Decreto No. 7.404/2010. www.abdi.com.br/. Accessed 28 Feb 2015

NMX (2013) Norma Mexicana NMX-AA-164-SCFI-2013: Edificacion Sustentable – Criterios y Requerimientos Ambientales Minimos (Sustainable building – minimum environmental criteria and requirements). http://biblioteca.semarnat.gob.mx/janium/Documentos/Ciga/agenda/DOFsr/DO3156.pdf. Accessed 28 Feb 2015

ORDER PRE/116/2008 (2008) [Thai] Green public procurement plan of the state general administration and its public bodies, and the managing bodies of the social security. Published on 21 January. www.measures-odyssee-mure.eu/public/mure_pdf/transport/SPA44.PDF. Accessed 28 Feb 2015

Oxford Dictionaries (2015) Mainstream. www.oxforddictionaries.com/definition/english/mainstream. Accessed 28 Feb 2015

Quispe I, Santillan J, Valdivia S (2014) VI International conference on life cycle assessment in Latin America: CILCA 2015—Lima, Peru, July 13 to 16, 201. Int J Life Cycle Assess 19(3):718

RPCV (2005) Red Peruana de Ciclo de Vida. www.red.pucp.edu.pe/ciclodevida/. Accessed 28 Feb 2015

SLE (2015) The 10YFP sustainable lifestyles and education program. www.unep.org/10yfp/Programmes/ProgrammeConsultationandCurrentStatus/Sustainablelifestylesandeducation/tabid/106266/Default.aspx. Accessed 28 Feb 2015

Sonnemann G, Ciroth A, Valdivia S (2015) Opportunities for national life cycle networks creation and expansion around the world. Life-Cycle Initiative, United Nations Environment Programme and Society for Environmental Toxicology and Chemistry, Paris (in press)

Swarr et al (2015a) Building organizational capability for life cycle management. In: Sonnemann G, Margni M (eds) LCA compendium, book life cycle management. Springer, Dordrecht, pp 239–256

Swarr T et al (2015b) Building organizational capability for life cycle management. Int J Life Cycle Assess (in press)

Thumrongrut Mungcharoen (2013) LCA applications in Thailand. In: Proceedings of the 2014 Chinese conference on life cycle management (CLCM), Beijing (www.clcm.org.cn/). 2014

UN (1948) Universal declaration of human rights. http://www.un.org/en/documents/udhr/. Accessed 28 Feb 2015

UNCED (1992) United Nations conference on environment and development, 1992. 'Rio Declaration on Environment and Development' (Rio Declaration). Rio de Janeiro, 2012

UNEP/SETAC (2009a) LCM: how business uses it to decrease footprint, create opportunities and make value chains more sustainable. Paris

UNEP/SETAC (2009b) Guidelines for a social life cycle assessment. Paris

UNEP/SETAC (2011) Global guidance principles for LCA databases. Paris

UNEP/SETAC (2015a) UNEP/SETAC life cycle initiative. www.lifecycleinitiative.org. Accessed 28 Feb 2015

UNEP/SETAC (2015b) Organizational life cycle assessment – guidance for organizations to conduct LCA considering their value chain. Life-Cycle Initiative, United Nations Environment Programme and Society for Environmental Toxicology and Chemistry, Paris (in press)

Valdivia S, Mila I, Canals L (2014) A evolução do pensamento sobre Ciclos de Vida na América Latina. www.brasilpnuma.org.br/informativo/index_137.html. Accessed 28 Feb 2015

Worldometer (2014) Countries in the World. http://www.worldometers.info/world-population/population-by-country/. Accessed 28 Feb 2015

Chapter 20
Communication and Collaboration as Essential Elements for Mainstreaming Life Cycle Management

Philip Strothmann, Jodie Bricout, Guido Sonnemann, and Jim Fava

Abstract This chapter addresses two major challenges for mainstreaming life cycle management that are intrinsically linked: collaboration and communication. To this end it is argued that in order to radically increase the take up of life cycle based approaches in business and government, life cycle professionals need to enhance global collaboration among themselves, as well as with others and communicate to a wider set of stakeholders. The chapter makes the case that the life cycle community does not have a home, and thus currently does not exist as one coherent and clearly identifiable stakeholder. It concludes that successful communication on behalf of and with the community can only be achieved when the community is formally organized. To this end the newly established Forum for Sustainability through Life Cycle Innovation is presented as a possible way to overcome the outlined gaps and challenges.

Keywords Forum for sustainability through life cycle • Life cycle assessment • Life-cycle based approaches • Life cycle community • Life cycle innovation • Life cycle management • Life cycle sustainability management • Sustainability

P. Strothmann (✉)
FSLCI, Berlin, Germany
e-mail: philip.strothmann@fslci.org

J. Bricout
[avniR] Platform – cd2e, Rue de Bourgogne, Base du 11/19, 62750 Loos en Gohelle, France

G. Sonnemann
University of Bordeaux, Bordeaux, France

J. Fava
thinkstep, West Chester, PA, USA

G. Sonnemann, M. Margni (eds.), *Life Cycle Management*,
LCA Compendium – The Complete World of Life Cycle Assessment,
DOI 10.1007/978-94-017-7221-1_20

1 Introduction

The life cycle community has come far since SETAC and ISO first started working on standardizing methodologies in the 1990s. Tools have been developed, studies have been undertaken and some forward thinking companies and governments have integrated life cycle thinking into their policies and strategies (United Nations Environment Programme 2012).

Despite this positive development, the widespread take up of life cycle based methodologies and use of information has yet to happen. Decision makers for example still tend to simplify, zooming in on one environmental problem to find a quick fix that may have broader implications over time.

Some LCAs are also done just to tick the box for e.g. a building standard, or because a client needed it, without any further consideration. Above all, since the global economic crisis in 2008, a number of companies and governments have simply taken anything "environmental" off the agenda or reduced their efforts (Geels 2013).

2 Missing Links

As sustainability challenges continue to increase along with a rise in attention and global recognition of the associated issues, some life cycle based approaches, such as for example the Green House Gas Protocol or the European Commission's Product Environmental Footprint (PEF), are starting to be tested or are already applied (European Commission 2015). However, despite this positive development, life cycle based approaches are yet to be applied on a large scale to achieve tangible results. Currently, only a limited number of companies apply life cycle methodologies in their businesses to enhance the sustainability performance of their products. And due to the lack of a broad understanding of the public about life cycle approaches and methodologies, these companies are struggling to communicate their efforts in a way that resonates with the average consumer. In return, companies' efforts are currently not properly recognized by customers which reduces their efforts' value for marketing purposes. A coordinated and strong push to promote the uptake of life cycle based methodologies and concepts in companies is thus needed more than ever, coupled with an equally strong communication effort.

While the life cycle community has matured enough to address remaining methodological challenges, it is, however, still not good at explaining what it does to people who do not have a thesis in environmental chemistry or chemical engineering. Which is, however, exactly what is needed: explaining life cycle based approaches in clear terms and how they help provide solutions to many challenges facing business and government today. And it needs to be done with a global voice that takes into account modern, collaborative ways of working together, as neither individual companies nor select circles of researchers and scientist or life cycle assessment (LCA) networks are currently able to successfully drive the communication effort needed.

According to Bjørn et al. (2013), over 100 LCA networks had been developed by 2012, with at least seven new networks per year since 2008. However, 40 % of these networks have less than 20 members, indicating that collaboration is still happening in rather small circles. Furthermore, 68 % of these networks operate at a local level and over a third are in Europe. So these small circles are mostly local, and concentrated on one continent.

Still, the positive growth of LCA networks, particularly in emerging economies over the past few years is a success of efforts undertaken, for example, by the UNEP/SETAC Life Cycle Initiative. It was launched in 2002 and has been key in boosting collaboration around methodologies and tools since then. Similarly, SETAC has provided the LCA community with some space, which has also been used primarily to work on methodological issues. SETAC's main focus, however, remains first and foremost on issues around environmental toxicology and chemistry (SETAC 2015). And while these global players have invaluable strengths and played a great role in advancing life cycle thinking around the world (Quiros 2014), they have so far not focused on working closely with industry and businesses on establishing a global push for consumer information on life cycle aspects.

In order to be successful, such a push should also be backed and driven by the entire life cycle community, instead of a number of different stakeholders with different priorities and interests. However, as outlined before, the current state of the life cycle community is characterized by a flurry of small circles, or nodes, at a local level, and a handful of (historical) global players. The community as one clear identifiable and organized stakeholder thus currently does not exist, mostly because it has no home.

The lack of such a more organized and structured community that is visible and has its own voice has, however, implications not only from a communications perspective, but also from a collaborative point of view. As outlined above, collaboration happens mostly in small circles. Due to the discussed lack of national, regional or even global coordination efforts, synergies get lost and opportunities for collaboration are missed.

While this is mostly an issue for the community itself, the consequences are relevant beyond the community. Communication is absolutely essential to help spread the word, to inform people about the benefits of applying a life cycle perspective and enable companies to put life cycle management into practice by more easily communicating their efforts.

As shown over the past two decades, life cycle based thinking and methodologies can help to operationalize sustainability efforts and are applied by companies to identify cost and resource effective sustainability improvements and highlight them to their potential clients (Box 20.1). However, in order to mainstream LCM and thus have a tangible impact on the world, the life cycle community has to get out of the small niche in which it is currently operating. Only then it can become a powerful partner to companies and industry stakeholders and work towards effectively informing consumers on the concept and advantages of taking a life cycle perspective.

Box 20.1: A Practical Example: Renault's Efforts to Inform Their Consumers About Their Life Cycle Management Work

To operationalize its commitment to the environment, Renault is using the life cycle management approach, focussing especially on incorporating sustainability concerns into the design and manufacturing process as well as the use phase of the car and its end-of-life treatment. In order to assess its cars' environmental performance, Renault conducts LCAs, such as, for example, for its Fluence Z.E. The LCA as well as its critical review is publicly available from Renault's website, as are brief introductions into the LCM and LCA concepts.

To help its consumers identify easily which of its cars have the best environmental credentials and are also the most economical to drive, Renault introduced in 2007 the Eco2 label. The label is based on Renault's environmental policy and on the analysis of the impact of their cars in the three key stages of their cars' life cycle:

- Manufacturing: Eco2 vehicles must be built in ISO 14001-certified plants
- Utilization: Eco2 vehicles must emit less than 120 g of CO_2/km
- Recycling: 7 % of the plastic in Eco2 vehicles must come from recycling and Eco2 vehicles must be 85 % recyclable

Renault's application of life cycle management and life cycle assessment concepts thus helps it to provide interested customers with an internationally accepted and scientifically sound decision-making basis when making purchasing decisions (Renault 2015).

As outlined in other chapters of this book, to be successful, life cycle information needs to be applied on a significantly larger scale to have an impact in management decisions. Only then the goal to enhance sustainability efforts and eventually achieve regeneration will be successful. Essentially following the vision of "A sustainable global society where life cycle approaches are established and decision makers have fully integrated them into their regular decision making processes." To this end it is understood that life cycle approaches contribute to the overall goal of a sustainable society through providing data and knowledge for:

- Reducing a product's resource use and emissions to the environment, and improving its social and socio-economic performance throughout its life cycle
- Enabling consumers to choose more sustainable products
- Information policy to optimize trade-offs
- Influencing and validating choices for disruptive improvements, such as new economic models or radical technology, to move towards possibilities to regenerate the system and not only be "less bad", but towards "more good"

However, for life cycle approaches to make a significant contribution to sustainability, they need to be used in a massive and systematic way, especially by companies and businesses within a framework and infrastructure provided by governments. Success will be when there is a broad understanding of the concepts around life cycle thinking by the general public and when these concepts are integrated into decisions by organizations in a systematic way through life cycle management (for example, integrated into standard design software or purchasing criteria).

3 Moving Towards Better Collaboration and Communication to Mainstream Life Cycle Management

As outlined before, it is now on the life cycle community to help making this vision come true. Key elements on this route will be achieved through:

- Better collaboration with each other across the globe as well as with other stakeholders
- Better communication to a wider set of stakeholders

3.1 Improve Collaboration Among the Life Cycle Community Across the Globe

"The community" as it is understood here is a group of people having a particular interest in common: the use of life cycle information. The community includes:

- Professionals who provide life cycle data and information to be used in internal processes or commercial or public databases
- Users of life cycle data and information as part of regular and existing decision making processes in business and governments
- Scientists and researchers who are actively working on developing new methods and approaches, methodological standards or advancements

- Life cycle "practitioners" who apply the tools developed above in specialist consultancies, in industry (often in EHS, sustainability, R&D departments), or in government (technical agencies like environmental agencies or policy makers)
- Life cycle advocates, who may not be able to undertake an LCA themselves, but who understand their utilization and importance
- Students in a variety of disciplines learning about how to apply life cycle thinking and approaches

Traditionally, global and local LCA networks have been dominated by academic stakeholders, but data providers, practitioners and users of life cycle data and information in industry and businesses are crucial. They are the ones who use the tools and methods to actually assess and interpret products' sustainability performance and thereby help establishing a scientifically more accurate decision-making basis for companies. As shown in Box 20.2, the LCM conference series is currently the only prominent place where companies and businesses get to exchange on their challenges and advancements. However, the LCM conference series is focused on Europe alone and organized only every 2 years, thus not well suited to facilitate a regular exchange. Other initiatives driving collaboration among the life cycle Community include:

- International life cycle networks such as the UNEP/SETAC Life Cycle Initiative and the International Society for Industrial Ecology
- Local or regional life cycle networks for practitioners, such as ACLCA in the US, ALCALA in Latin America, or ALCAS in Australia
- Life cycle conferences, such as the before-mentioned LCM, SETAC case study symposium, Ecobalance, [avniR], LCA Food, PEF World Summit
- Online collaborative platforms such as the PRe LCA discussion list and various LinkedIn groups (e.g. on GaBi)
- Specific workgroups targeting technical subareas of the discipline such as Social LCA, Food LCA or Toxicity and Water issues

These many "circles of collaboration" tend to be extremely dynamic and productive, and in most cases very complementary. Many work on similar issues in different geographical spaces or with different types of stakeholders. They all have their place and are an important part of nurturing this constantly evolving field. However, they do not represent the community as a coherent and clearly identifiable stakeholder, but limited possibilities for exchange.

Consequently a global effort will be needed to bring these small groups together and thereby not only facilitate knowledge sharing and the establishment of links, but more importantly create a shared sense of belonging. In the end, a community only functions if its members can identify themselves with it.

Uniting the community behind a commonly shared set of ideas, principles or perspectives is thus a key step towards creating a community as such and thus lay the ground for collaboration and co-creation.

Box 20.2: A Practical Example: Making the Bridge Between Life Cycle Management Conferences

Since the first LCM was held in Copenhagen in 2001, this conference series has established itself as one of the leading conference series worldwide in the field of environmental, economical and social sustainability. It is now held every second year in Europe, with the seventh LCM held in Bordeaux, France.

The unique feature of the LCM conference series is developing practical solutions for the implementation of life cycle approaches into strategic and operational decision-making in business, industry and beyond. The LCM Conferences bring together international decision-makers from science, industry, NGOs and public bodies. Almost half of the participants came from the private sector in recent gatherings. In this way it is clearly differentiated and complementary to more academic events such as conferences organized by SETAC or the International Society for Industrial Ecology.

The LCM conferences provide an excellent start to transfer knowledge from the scientific world to real world applications by the private sector or governments. Whilst some uptake by the application side for the use of life cycle information is happening (e.g. building and construction, food and beverages, cars and other transports means, materials and chemicals, packaging, consumers goods, etc.), there is no real global mechanism to accelerate this awareness and uptake by the private sector and government to apply life cycle approaches as part of their normal practices. Also, given that the LCM conference series has no formal structure and is organized only every 2 years and by different host organizations, it is currently rather limited in its ability to function as the central space to facilitate ongoing and continuous exchange of ideas and challenges.

Establishing a link between a better-organized community and the LCM conference series would thus be a natural fit that could help to improve the transfer of knowledge, experience and resources. One way to achieve this could be to establish a small secretariat that could build links between LCM in Europe and other LCM related conferences taking place around the globe, such as those in China, Brazil and India, so that the conversations continue rather than being disconnected (LCM 2015).

The life cycle community will also need to actively engage and collaborate with colleagues working in related fields, such as climate change, toxicity and biodiversity, so that cooperation and collaboration result in joint efforts instead of "competing" for the same air time. In the end, life cycle concepts should be understood as a key element of the circular economy and the energy transition, and hence they should be seen as the backbone of any effort to address sustainability issues, be it climate change, resource efficiency or other issues.

The community will also need to collaborate more closely with decision makers in industry and government to co-develop life cycle management solutions that respond to their needs, as will be demonstrated in Sect. 3.2.

3.2 *Effective Communication with a Wider Set of Stakeholders*

Widespread take up of life cycle management is impossible without a widespread understanding of what life cycle approaches are and why they are important. The audiences for this communication can be grouped into two broad categories.

3.2.1 Anyone Looking for Information on How to Operationalize Sustainability

Decision-makers in industry and government are key to applying life cycle thinking to operationalize sustainability in the real world. While the community develops and works directly with life cycle based tools, decision-makers in industry and government interact with life cycle based information on a different, yet from a societal perspective arguably more important level.

This audience first needs to make the step-change towards systemic, life cycle thinking in how they understand sustainability issues (including notions of arbitrage between different impact areas, and over different stages of the life cycle). They then need to have enough knowledge and understanding of what life cycle management is and how life cycle based information can be used to facilitate decision-making processes, to integrate it into their core activities.

3.2.2 General Public

As clearly demonstrated throughout the past decades, change is often only achieved if a strong push from the demand side has been established (Baron et al. 2009). To achieve this, the general public needs to understand that sustainability decisions need to be based on a systemic framework that takes into account impacts from a life cycle perspective. In return, the demand for and availability of life cycle based information will increase, once companies can communicate on their efforts without being burdened by the need to explain the concepts first.

While a growing number of consumers already consider sustainability aspects during their purchase decisions, they are increasingly confused by the vast amount of different standards, labels and other information tools (Atkinson 2014). Here it will be important to collaborate with existing programmes and initiatives which work on these issues, such as the Consumer Information Programme under the UN's 10 Years Framework on Sustainable Consumption and Production and others (CIP 2015).

An effort to simplifying and streamline communication on product's sustainability performance based on life cycle information is thus paramount to enable consumers to better interpret different kinds of information and help them understand why taking a life cycle perspective is a key step towards operationalizing sustainability. As previously outlined, educating consumers about the advantages and necessity of taking a life cycle perspective to avoid trade-offs is also highly relevant to companies and businesses that seek to explain their efforts. Only if a general understanding of the concept is present, efforts of companies applying life cycle methodologies will be understood and welcomed. This understanding will help businesses and industry to respond to green washing claims and demonstrate real sustainability progress by implementing life cycle management (in line with Box 20.1).

In this context opinion leaders will also need to be targeted to help spread the word and generate sufficient attention. Al Gore's "An inconvenient truth", for example, helped to put the issue of global climate change on the world's agenda, at least from a communications perspective. Thanks to the film and following public appearances, the understanding of the effects of climate change has increased (Jacobsen 2011). A messaging that is neutral, scientifically sound and not driven by company-interests thus needs to be developed that could be picked up by opinion leaders to help spreading the word. As with the IPCC process, the messaging, however, will have to be backed by the life cycle community to have both credibility and relevancy on a global level.

3.3 Audiences and Lessons Learned

In order to effectively communicate with the different target audiences identified in Table 20.1, the life cycle community needs to agree on key messages and deliver them with a global voice, albeit adapted to local conditions. It should be noted in this context that a communication effort should focus first and foremost on decision makers in business and government to drive change. In a second step, together with successful stakeholders in the field of consumer information, the focus will need to be extended to target also the general public.

In this context, bringing together the life cycle community as a clear stakeholder is critical to establishing a global voice, but the collaboration with other stakeholders is equally important in developing communication tools and capability that resonate with people outside of the LC community.

It is also important to consider that communication is a two way street. Whilst the life cycle community needs to educate decision makers and the general public about the advantages and benefits of taking a life cycle perspective, communication also needs to work in the other direction. In this context it is crucial that the community is visible, approachable and receptive for feedback from different voices in society.

In the end it will be important to acknowledge that if societal change is the objective, it is of utmost importance to interact closely with society. Here it will be important to proactively identify opportunities to discuss sustainability challenges and support other relevant communities and actors.

Table 20.1 Target groups and desired outcomes

Target group	Who are they?	Desired outcomes
Decision makers in business and public sector can be reached through • Cross cutting business federations (e.g. chambers of commerce, WGBC) • Sector based business federations • NGOs • Education centers (universities and schools) (future decision makers) • Specialist media • Funding agencies • Standard and certification bodies	Influential professionals (middle management) and final decision makers in various departments: (communication, sustainability, finance, product development, procurement, economic development, etc.) Business • Large businesses • SMEs Public sector • Federal/national level government • Local level government • Public departments	• Develop and implement LC approaches in their business or policy making (e.g. policy making, green procurement, product development) • Support national LCI database development and management following global guidance • Provide case studies of LC implementation • Collaborate with the LC community to ensure their developments are well aligned with needs • Encourage other businesses and governments to engage
General public Can be reached through • NGOs • Governmental programs • Education centers (universities and schools) • General media (including social)	The general public as individuals to influence their behavior as consumers, business stakeholders and active members of civil society	• Understand the concept of taking a life cycle perspective and its advantages • Create a market for sustainable consumer products • Build demand for reliable and transparent information about life cycle impacts

4 Making It Happen

For all of the reasons outlined above, it is of utmost importance for the life cycle community to become a stakeholder of its own, in order to be seen and have a global voice that is heard. This will only happen, if the community starts to structure itself in the form of a new organization. This organization should provide the actions in the following sections.

4.1 Space to Coordinate Efforts Around the World

Currently a number of small circles and a few international organizations exist in the life cycle space. A life cycle community organization would thus need to build close relationships with each of these existing organizations to become a central information and network hub that connects these circles and organizations again and facilitates coordination.

This does not only apply to stakeholders as such, but also and importantly to conferences. Over the past years, a number of conference series have been organized in addition to the LCM conferences in Europe and those in key emerging economies as mentioned before ranging from EcoBalance in Japan, the Australian LCA Conference to CILCA in Latin America, InLCA in the USA and [avnir] in France. Coordination among those conferences would help not only with regards to dates, but also on content focus and international visibility, for instance by ensuring that key messages and white paper with topics relevant for LCM are spread quickly around the world.

4.2 Space to Identify New Contacts, Apart from Conferences

While conferences are a great place to learn about new projects, latest research and to network, they do not facilitate regular exchange. In addition, while conferences technically offer a great way to get to know new people and identify possible collaboration opportunities, the reality is different. Most participants have a pretty full schedule and have little to no time to identify and meet potentially interesting colleagues and stakeholders, outside of their already existing networks and small circles. As a consequence, a life cycle community organization would need to establish a mechanism that facilitates the process of finding interesting connections outside of conferences.

4.3 Space to Collaborate with Colleagues

Another important element to foster the identification of stakeholders with the community would be to establish a way that fosters and enhances collaboration. To this end a platform where new ideas could be pitched to the community or research shared would need to be put in place. Enabling interaction among community members needs to be a crucial element, as it has been established before that better ways to collaborate and co-create are needed.

4.4 Space to Develop a Coherent Messaging

Finally, the platform identified before should also be used to work jointly on a clear and coherent messaging, for instance on recommendations on how to put life cycle management in business practice. In this context, in order to establish a shared sense of community and belonging, a specific set of principles should be established that community members cannot only subscribe to, but have co-developed.

Above all, the community organization should seek to be as inclusive as possible to provide members with a sense of empowerment. The idea should be to enable members to speak for and on behalf of the community and thus instead of being represented, represent themselves.

5 Conclusion

As outlined throughout the previous sections, two major challenges for mainstreaming life cycle management exist that are intrinsically linked: collaboration and communication. In order to radically increase the take up of life cycle based approaches in business and government, life cycle professionals need to enhance global collaboration among themselves, users of life cycle information, as well as with others and communicate to a wider set of stakeholders.

To facilitate this process, a home for the community is needed that enables it to become one coherent and clearly identifiable stakeholder but also acts as a central information and networking hub within the community. Such a home would need to respond to a variety of issues and gaps, as identified before. It would also need to make an effort to provide not only a space for collaboration and communication, but more importantly be rooted in a shared set of ideas and principles. Community members will need to be able to identify easily with the new organisation and be interested to engage with the organization and its members to jointly shape its future.

The newly established Forum for Sustainability through Life Cycle (FSLCI) is built around these considerations and will need to demonstrate in going forward that it can live up to and address the issues outlined throughout this chapter. Its success will be largely depended on its acceptance by the community, in particular those life cycle practitioners that make changes in real life by putting LCM theory outlined in the first part of this book into practice. At the moment, it remains to be seen, whether the life cycle community is ready to become a stakeholder of its own. Whether it is ready to get out of the niche, and aim for actual global change towards creating a global sustainable society, where decision makers have fully integrated life cycle management into their regular decision making processes.

References

Atkinson L (2014) Wild west' of eco-labels: sustainability claims are confusing consumers. The Guardian. http://www.theguardian.com/sustainable-business/eco-labels-sustainability-trust-corporate-government. Accessed 13 Mar 2015

Baron DP, Harjoto M, Jo H (2009) The economics and politics of corporate social performance. Graduate School of Stanford Business. Working paper no. 1993

Bjørn A, Owsianiak MI, Laurent A, Molin C, Westh TB, Hauschild MZ (2013) Mapping and characterization of LCA networks. Int J Life Cycle Assess 18:812–827

CIP (2015) The 10YFP consumer information program (CIP). www.unep.org/10yfp/Programmes/ProgrammeConsultationandCurrentStatus/Consumerinformation/tabid/106265/Default.aspx. Accessed 28 Feb 2015

European Commission (2015) The product environmental footprint pilots. http://ec.europa.eu/environment/eussd/smgp/pef_pilots.htm. Accessed 16 Mar 2015

Geels FW (2013) The impact of the financial-economic crisis on sustainability transitions: financial investment, governance and public discourse. Environ Innov Soc Trans 6:67–95

Jacobsen GD (2011) The Al Gore effect: an inconvenient truth and voluntary carbon offsets. J Environ Econ Manag 61:67–78

LCM (2015) About the conference. http://lcm2015.ism.u-bordeaux1.fr/?page_id=7. Accessed 16 Mar 2015

Quiros A (2014) Communicating life cycle thinking: a worldwide application. In: Proceedings of the Indian conference on life cycle management (ILCM 2014, New Delhi, 29-30 Sept.)

Renault (2015) Environmental policy. http://group.renault.com/en/commitments/environment/environmental-policy/. Accessed 16 Mar 2015

SETAC (2015) SETAC's mission. http://www.setac.org/?page=Mission. Accessed 16 Mar 2015

United Nations Environment Programme (2012) Greening the economy through life cycle thinking ten tears of the UNEP/SETAC life cycle initiative

Part V
Implementation and Case Studies of Life Cycle Management in Different Business and Industry Sectors

This part provides a few practice examples of life cycle management in different economic sectors, namely electricity sector, applied urban fabric planning, automotive industry and food and beverage industry.

Chapter 21
Exploring Challenges and Opportunities of Life Cycle Management in the Electricity Sector

Miguel Fernandez Astudillo, Karin Treyer, Christian Bauer, and Mourad Ben Amor

Abstract Electricity supply is often cited as a significant hot spot in life cycle assessment results, and consequently in life cycle management results. Despite its importance, however, practitioners continue to overuse generic LCI data and different simplified methodologies regarding electricity supply modeling. Such simplifications and inconsistencies can result in difficulties, e.g. to compare the findings of various studies. This chapter is intended to highlight issues on electricity supply modeling, methodological choices and data set selections. Attributional and consequential perspectives as well as systemic aspects of the electricity sector are also reflected. Finally, key challenges and opportunities are summarized and suggestions on how to deal with such problems are provided when possible.

Keywords Energy policy • Electricity supply • Modeling choices • Life cycle inventory data • Life cycle assessment • Life cycle management

1 Introduction

Life cycle management (LCM) aims to minimize the environmental and socio-economic burdens associated with a product or product portfolio throughout its entire life cycle and value chain (Remmen et al. 2007). From an environmental perspective, electricity is a major consideration in many life cycle assessment (LCA) studies. Indeed, electricity supply is often highlighted as a significant hot

---

M.F. Astudillo • M.B. Amor (✉)
Department of Civil Engineering, Université de Sherbrooke,
J1K 2R1 Sherbrooke, QC, Canada
e-mail: ben.amor@usherbrooke.ca

K. Treyer • C. Bauer
Laboratory for Energy Systems Analysis, Paul Scherrer Institut (PSI),
5232 Villigen, Switzerland

G. Sonnemann, M. Margni (eds.), *Life Cycle Management*,
LCA Compendium – The Complete World of Life Cycle Assessment,
DOI 10.1007/978-94-017-7221-1_21

spot in LCA results for a majority of product life cycles (Curran et al. 2005; Treyer and Bauer 2013). It has also been shown, from the LCA perspective, that electricity sector as such is an important originator of environmental burdens worldwide and that energy policies can result in burden-shifting (Laurent and Espinosa 2015). As LCAs are being conducted frequently, it is key that suitable life cycle inventory (LCI) data become in general more readily available (Soimakallio et al. 2011).

This book chapter aims to highlight issues on electricity supply modeling, methodological choices and data set selections. Attributional and consequential perspectives as well as systemic aspects of the electricity sector are also reflected. Finally, this chapter summarizes the key challenges and opportunities and provides suggestions on how to deal with such problems.

2 Identifying the Issues: Major Methodological Challenges

2.1 Data Issues: Generating Electricity Life Cycle Inventory Datasets

According to ISO 14044, electricity inventories shall take into account electricity mixes, fuel efficiencies, as well as transmission and distribution losses. Given the heterogeneity of electricity LCI data, representativeness is an important aspect when conducting an LCA. Referring to ISO 14044 (ISO 14044: 2006), data representativeness is the qualitative assessment of the degree to which the data set reflects the true population of interest for a specific application: geographical coverage, time-related coverage, and technology coverage. Other quality indicators are such as completeness, consistency and uncertainty are also addressed. Aspects covering special challenges in the electricity sector are highlighted in the following sections.

2.1.1 Geographic Coverage

The geographical coverage is the geographical area from which data for unit processes should be collected to satisfy the goal of the study. In the context of electricity process, LCA practitioner can face two main challenges: the first one refers to the situation where no regionalized electricity data is available (Sect. 2.1.1.1) and the second one refers to the grid delimitation (Sect. 2.1.1.2).

2.1.1.1 Extending the Geographical Coverage: Improving Production Data Accuracy

Regionalization of inventory data is recognized as an important need to increase the accuracy of LCA results, even if it is disputed down to which level the regionalization should go. Recent efforts have been undertaken to increase the geographical scope of inventory data and using country-specific statistics. The life cycle inventory database ecoinvent version 3 covers nearly 85 % of global electricity production in 2008 (Treyer and Bauer 2013) with country (or even region-specific) LCI data showing substantial differences in LCI data between specific countries and regions.

Despite the geographical coverage increase, gaps in LCI data keep existing and are in general more pronounced in non-OECD countries, where often extrapolations are unavoidable, increasing uncertainty (Treyer and Bauer 2013; Schmidt et al. 2011; Laurent and Espinosa 2015).

2.1.1.2 Grid Mix Boundaries: From Production Mixes to Supply Mixes

Once available in the grid, it is not possible to know where the electricity is coming from (Dones et al. 1998; Itten et al. 2014; Weber et al. 2010). This tracking issue becomes even more challenging as electricity grids are increasingly getting interconnected, and hence makes selecting a grid mix boundary a complicated task for the practitioner.

The common approach is to use national electricity mixes and accounting for imports from the neighboring jurisdictions. The underlying justification is that neighboring countries have either physical connections or administrative contracts to trade (Treyer and Bauer 2014). However, the boundaries selection is to some extent arbitrary and raises equity issues. As an example, if we take the North-American electricity grid, different resolutions are available: national, interconnect, Jurisdiction-average production and consumption mixes (US countries, Canadian provinces, etc.), ISO/RTO, EPA's eGrid subregions, and EIA region (Weber et al. 2010).

On top of that, congestion can effectively limit electricity transmission within a national boundary (an administrative barrier), which even makes the common approach selection (i.e. using the national energy mixes) unrealistic. A recent study developed an approach creating clusters of data according to the congestion status and its location within the Ontario (Canada) grid-mix. As an example, the avoided greenhouse gas emissions varied, for uncongested (i.e. using as a common approach selection the production energy mix: Ontario mix) and congested hours, between 280 and 390 kg/MWh. Even if these empirical estimates cannot be generalized to other contexts, the study underscored the importance of congestion in defining the grid mix boundary (Amor et al. 2014a).

2.1.2 Temporal Aspects of Electricity

Our capacity to store electricity is very limited, and in practice demand is dynamically (hour by hour) matched with a range of production technologies. Obtaining past yearly-averaged country supply mixes to be used in attributional LCA (ALCA) is relatively straightforward by using national statics. Typically organizations such as International Energy Agency (IEA) provide these data, even if the often-rough categorization of fuel and power plant categories in the statistics calls for assumptions and extrapolations increasing uncertainties. However, predicting and capturing changes in time of the electricity sector – being relevant in consequential LCA (CLCA) – is a challenging task, for both temporal scopes: short-term and long-term horizon.

2.1.2.1 Short-Term

In countries with de-regulated energy sectors, an independent system operator coordinates most of the markets by using price based dispatch systems. Price bids from generators, defining the supply curve would be ideal for analyzing the short-term variation of power plant production following different resolution: from hourly to annually and then be able to consider intermittency of renewable energy, intertemporal arbitrage, spinning and non-spinning reserves or ramp-rate limitations of producers.

However, price bids are not always publicly available. In the absence of such data, a procedure for integrating the short-term time variations of technologies is missing. Such a procedure could play an important role in increasing the robustness of LCA studies and refining their environmental impact estimates. Additionally, not all electricity markets have the same extent of de-regulation. As an example, leading players like China, the world largest consumer of electricity, still relies on a considerably more complex multi-level dispatch hierarchy partially based on generator output planning (Kahrl and Wang 2014).

2.1.2.2 Long-Term

In the long-term, additional capacity would need to be installed to cover increases in demand. Changes in the electricity sector depend on political, environmental and economic considerations that are substantially uncertain and country specific.

Different techniques are available to estimate prospective electricity mix. These techniques are useful from the average ALCA perspective and also from the CLCA perspective.

Future supply mixes can be estimated from national forecast, such as IEA annual energy outlook (e.g. Hertwich et al. 2015). In the absence of such available data, specific models (e.g. partial equilibrium models) can be used to estimate future average supply mixes: LEAP, TIMES, (see Pfenninger et al. 2014 for a review). CLCA often follows the step-wise procedure presented by Ekvall and Weidema (2004) and updated in Weidema et al. (2009) to identify marginal technologies but its application to the electricity sector is not yet satisfactory (Treyer and Bauer

2014). Marginal changes in the electricity sector are likely to affect a range of technologies (Pehnt et al. 2008; Mathiesen et al. 2009) and it is not straightforward task to consistently identify them with a heuristic approach (Zamagni et al. 2012; Earles and Halog 2011; Menten et al. 2015). Energy system models such as TIMES or LEAP can help to overcome such difficulties.

2.1.3 Technology Coverage

The main challenges in technology data coverage concerns currently used technologies and those, which will be installed in the future and are not yet commercially available (e.g. on a pilot plant level).

2.1.3.1 Actual Technologies

There is a wide variation among generation stations in terms of emissions and inputs per unit generation across and even within fuel types. Such variation becomes even more challenging with the differences among statistics sources (e.g. Eurostat and EIA) treating the same technology using a given fuel type at a given geographic location (e.g. jurisdiction). Specification of the time frame of LCI data can also be challenging: statistic sources often refer to different years and the availability of up-to-date data is not always given, depending on the type of environmental exchanges. In addition to that, there is considerable uncertainty over certain emission factors, even for mature technologies, such as hydropower or coal (Hertwich 2013; Henriksson et al. 2014). Moreover, not only LCI data for power plants as such can substantially vary, also specific fuel supply as well as infrastructure manufacturing chains can have important effects on LCA results (Bouman et al. 2015; Yue et al. 2014).

2.1.3.2 Prospective Technologies

Prospective LCA studies often rely on LCI data of current electricity generation, even if technology performance of current power generation chains is likely to improve in the future and new technologies will emerge (IEA 2014). Modeling how technology performance will change over time is particularly difficult for nascent technologies (Curran et al. 2005) such as organic photovoltaic panels or carbon capture and storage (Volkart et al. 2013). Moreover, disruptive technologies can bring improvements in efficiency, but also have implied changes in infrastructure and user behavior, which are more difficult to predict (Miller and Keoleian 2015).

2.2 User Perspective: Using Electricity Life Cycle Inventory Data

LCA practitioners, especially in industry, frequently rely on generic LCI data for electricity generation and supply from commercial LCA background databases such as ecoinvent,[1] GaBi,[2] or ELCD[3] in their daily LCA business. Therefore, the quality of these data is often crucial for the quality of LCA applications (e.g. certification, Eco-design, decision making) and users are confronted with the data issues discussed above.

Usually, use of generic electricity from the grid, i.e. electricity supply mixes, and from the generic electricity market, respectively, is sufficient in the industrial LCM context, if companies do not have own power plants or specific contracts for power supply.

However, today's industrial value chains are for many global products, i.e. industrial production, manufacturing, sales and research activities are distributed around the globe. In such cases, LCI data for electricity supply in different countries need to be used for LCM applications.

Background electricity LCI data need to represent actual local or regional/national power supply as close as possible in order to be useful for LCA practitioners. "As close as possible" means that the background data need to represent specific supply regions and markets, respectively, and need to be up-to-date in terms of technology market shares, technology performance as well as fuel and infrastructure supply chains. They also need to be available in a modifiable way on the unit process level in order to allow users to adapt the data to their specific needs in case they have more specific information concerning their electricity supply. This requirement calls for a comprehensive and transparent documentation on the unit process level as well.

Today, most LCAs in the LCM context are attributional (Zamagni et al. 2012; Masanet et al. 2013). However, taking into account goal and scope and change-oriented questions to be answered by LCA carried out in industry, consequential LCA should be performed more frequently. Ideally, such CLCA will be based on consequential background data as well, also for electricity supply. Unfortunately, the availability of consequential background data is currently limited to the ecoinvent database and even these data need to be used cautiously (Treyer and Bauer 2014).

Change-oriented and prospective LCA also calls for background data taking into account technology development, i.e. representing future technologies and economies. However, such "future background data" are almost unavailable, at least in a consistently implemented way. The only known attempt to create consistent LCI data for future electricity production was carried out in the European research project "NEEDS" and LCI data are available online.[4]

---

[1] www.ecoinvent.org

[2] http://www.gabi-software.com/international/databases/gabi-databases/

[3] http://eplca.jrc.ec.europa.eu/ELCD3/

[4] http://www.needs-project.org/needswebdb/search.php

3 Key Opportunities

3.1 *Opportunities: The Research Perspective*

3.1.1 Integration with Other Modeling Methods

3.1.1.1 Modeling Electricity Mixes

The environmental assessment of different strategies and policies in the energy and electricity sector, respectively, calls for integration of energy and electricity scenario modeling and LCA. Such integration needs to take into account systemic aspects, e.g. continuously matching power demand and supply, and should be performed with a relatively high temporal resolution (Messagie et al. 2014). Existing attempts to model electricity sector dynamics and their effect on electricity mixes have relied on standard economic models, both in the long-term (Pehnt et al. 2008; Lund et al. 2010) and short-term (Amor et al. 2014b). Electricity spot markets have been modeled assuming partial equilibrium conditions (i.e. ignoring feedback loops with markets outside the evaluated system). Electricity mixes were calculated minimizing cost, reproducing the bid-based clearance of electricity markets.

These approaches allow the integration of short-term dynamics characteristic of the electricity sector in long-term projections. The effect of public policies, such as taxes or subsidies can be included in the analysis with the help of life cycle costing. The scenarios can be, however, highly sensitive to uncertainties (Menten et al. 2015), and a thoughtful scenario analysis and a clear account of the limitations of studies is therefore needed. The integration of economic models increases substantially the number of variables, especially if general equilibrium models are used. In this case uncertainty assessments start to be computationally challenging and new approaches would need to be explored (Dandres et al. 2014). Initiatives such as the ENTSO-E[5] transparency platform of the pan-European electricity market can be very helpful to reduce uncertainty and effectively integrate short-term dynamics. Finally, Frischknecht and Stucki (2010) have proposed the so-called decisional approach and demonstrated it for electricity supply in industry, but this approach has not received wide attention and deserves more studies.

3.1.1.2 Filling Data Gaps

In the absence of specific data from facilities, statistic techniques can be used to estimate emission factors and associated uncertainties. Using the regression approach, emissions factors can be estimated from available data such as plant age, fuel type and country gross domestic product (see e.g. Steinmann et al.

[5] https://transparency.entsoe.eu/

2014b). Statistical interpolation methods such as kriging have also been tested and appear to be superior to regression analysis (Moreau et al. 2012) but their application in LCI need to be extended. The same observation remains regarding the effects of technological change to be accounted for in prospective LCA. Experience curves have been used to predict improvements in efficiency and effect of investment (Sandén and Karlström 2007) but more work is needed on the treatment of related uncertainties (Yeh and Rubin 2012; Miller and Keoleian 2015).

3.1.2 Uncertainty Management

Recent advances in the uncertainty management can be used to improve the models of the electricity sector. At the inventory level, data inaccuracy and lack of representative data are typically addressed using the semi-quantitative approach of the pedigree matrix, assuming log-normal distributions. New advances in LCI implementation now allow to model parameter uncertainty with different types of probability distributions, allowing the modeler to use the distribution that best fits the data (Muller et al. 2014).

In addition, it is recommended to distinguish between uncertainty and variability, as just uncertainty can be reduced with better data. Recent studies suggest that the spread in emissions of power plants can be dominated by uncertainty (Henriksson et al. 2014) or variability (Steinmann et al. 2014a) depending on the region under study. To support the modeling, novel methodological developments propose ways of tackling this issue (Steinmann et al. 2014a).

3.2 Opportunities: The User Perspective

3.2.1 Greater Use of Consequential Analysis

Most LCAs performed to date on electric power and systems can be classified as attributional. However, beyond the coming methodological improvement, ALCAs are not designed to assess the total environmental implications (i.e. consequences) of decisions. Alternatively, CLCAs are particularly useful for estimating the energy and emissions implications of policies such as a Renewable Portfolio Standard or those of policies that encourage carbon taxes (Amor et al. 2010; Masanet et al. 2013). Thus, increasing CLCA use help in understanding and improving the effects of a given energy policy.

There is an urgent need to introduce large improvements in the electricity sector, which will entail major economic and environmental consequences. The analysis of future scenarios will especially suit CLCA, which should receive more attention to overcome the challenges and profit from the opportunities outlined in this chapter.

3.2.2 Best Practices

In cases where it is important to accurately estimate the environmental impacts associated with electricity use with high temporal resolution, using a methodology where the electricity dynamics is considered (as developed by Amor et al. 2014b) is recommended. Such approaches are particularly relevant in assessing the implication of choosing different electricity supply modeling approaches during decision-making, e.g. in estimating the avoided environmental impacts as a consequence of renewable distributed generation.

In other cases, one must keep in mind the often-limited resources of LCA practitioners; generating specific LCI data for electricity supply at a high level of detail can be arduous, and often relying on generic background data from databases can be sufficient. However, these generic data should be adapted through modification of key parameters in order to represent specific electricity supply more closely.

In any case, knowing the implications associated with electricity supply modeling choices, it is highly recommended that practitioners exercise caution and sensitivity analyses should be systematically conducted using different electricity supply scenarios in order to take into account the complexity of electricity systems.

Finally, interpretation of LCA results of electricity generation technologies deserves attention, especially if used for decision support. Often, results show substantial variations without transparently documented and easily traceable underlying reasons. In this context, the harmonization approach by NREL (Brandão et al. 2012) should be considered.

4 Conclusions

This chapter summarizes key challenges and opportunities of LCM in the electricity sector, with focus at inventory level. Despite the advances, the challenges are numerous and span from gaps in geographical and technological coverage, uncertainties over emission factors, to complexities on the identification of marginal technologies.

There are many opportunities to improve the inventories that are being currently explored and some are described here: Economic models are being used to identify marginal technologies and assess the effect of policies, statistical techniques such as regression analysis are useful to fill inventory data gaps and experience curves can be used to assess novel technologies. As any methodological development, one should however, be very clear about their limitations, provide transparent documentation and uncertainty estimates together with the results.

Some of these developments are already available for practitioners, including consequential inventories, greater capabilities in uncertainty modeling and parameterized datasets. Other developments are not yet available in databases and would require that data providers take methodological decisions to maintain consistent inventories. These decisions could include questions where interpretation of ISO standards are less clear such as how to foresee marginal technologies, which data sources are more reliable or how to deal with the traceability of electricity. Dialogue between users and providers is therefore suggested to overcome methodological differences and help to build a sustainable electric sector.

References

Amor MB, Lesage P, Pineau P-O, Samson R (2010) Can distributed generation offer substantial benefits in a Northeastern American context? A case study of small-scale renewable technologies using a life cycle methodology. Renew Sustain Energy Rev 14:2885–2895

Amor MB, Billette de Villemeur E, Pellat M, Pineau P-O (2014a) Influence of wind power on hourly electricity prices and GHG (greenhouse gas) emissions: evidence that congestion matters from Ontario zonal data. Energy 66:458–469

Amor MB, Gaudreault C, Pineau P-O, Samson R (2014b) Implications of integrating electricity supply dynamics into life cycle assessment: a case study of renewable distributed generation. Renew Energy 69:410–419

Bouman EA, Ramirez A, Hertwich E (2015) Multiregional environmental comparison of fossil fuel power generation – assessment of the contribution of fugitive emissions from conventional and unconventional fossil resources. Int J Greenhouse Gas Control 33:1–9

Brandão M, Heath GA, Cooper J (2012) What can meta-analyses tell us about the reliability of life cycle assessment for decision support? J Ind Ecol 16:S3–S7

Curran MA, Mann M, Norris G (2005) The international workshop on electricity data for life cycle inventories. J Clean Prod 13:853–862

Dandres T, Gaudreault C, Tirado-Seco P, Samson R (2014) Uncertainty management in a macro life cycle assessment of a 2005–2025 European bioenergy policy. Renew Sustain Energy Rev 36:52–61

Dones R, Ménard M, Gantner U (1998) Choice of electricity-mix for different LCA applications. In: 6th LCA case study symposium SETAC-Europe, 1998. SETAC-Europe, Brussels, pp 1–4

Earles JM, Halog A (2011) Consequential life cycle assessment: a review. Int J Life Cycle Assess 16:445–453

Ekvall T, Weidema BP (2004) System boundaries and input data in consequential life cycle inventory analysis. Int J Life Cycle Assess 9:161–171

Frischknecht R, Stucki M (2010) Scope-dependent modelling of electricity supply in life cycle assessments. Int J Life Cycle Assess 15:806–816

Henriksson PJG, Zhang W, Guinée JB (2014) Updated unit process data for coal based energy in China including parameters for overall dispersions. Int J Life Cycle Assess 10:185–195

Hertwich E (2013) Addressing biogenic greenhouse gas emissions from hydropower in LCA. Environ Sci Technol 47:9604–9611

Hertwich E, Gibon T, Bouman EA et al (2015) Integrated life-cycle assessment of electricity-supply scenarios confirms global environmental benefit of low-carbon technologies. Proc Natl Acad Sci 112(20):6277-82. doi: 10.1073/pnas.1312753111

IEA (2014) Electricity information 2014. International Energy Agency/OECD

ISO (2006) ISO 14044 Environmental management – life cycle assessment – requirements and guidelines. Geneva

Itten R, Frischknecht R, Stucki M (2014) Life cycle inventories of electricity mixes and grid version 1.3. http://www.esu-services.ch/fileadmin/download/publicLCI/itten-2012-electricity-mix.pdf. Accessed 24 Feb 2015

Kahrl F, Wang X (2014) Integrating renewables into power systems in China: a technical primer – power system operations. Regulatory Assistance Project, Beijing. www.raponline.org/document/download/id/7459. Accessed 24 Feb 2015

Laurent A, Espinosa N (2015) Environmental impacts of electricity generation at global, regional and national scales in 1980–2011: what can we learn for future energy planning? Energy Environ Sci 8:689-701

Lund H, Mathiesen BV, Christensen P, Schmidt JH (2010) Energy system analysis of marginal electricity supply in consequential LCA. Int J Life Cycle Assess 15:260–271

Masanet E, Chang Y, Gopal AR et al (2013) Life-cycle assessment of electric power systems. Annu Rev Environ Resour 38:107–136

Mathiesen BV, Münster M, Fruergaard T (2009) Uncertainties related to the identification of the marginal energy technology in consequential life cycle assessments. J Clean Prod 17:1331–1338

Menten F, Tchung-Ming S, Lorne D, Bouvart F (2015) Lessons from the use of a long-term energy model for consequential life cycle assessment: the BTL case. Renew Sustain Energy Rev 43:942–960

Messagie M, Mertens J, Oliveira L et al (2014) The hourly life cycle carbon footprint of electricity generation in Belgium, bringing a temporal resolution in life cycle assessment. Appl Energy 134:469–476

Miller SA, Keoleian GA (2015) Framework for analyzing transformative technologies in life cycle assessment. Environ Sci Technol 49(5):3067-75 doi:10.1021/es505217a

Moreau V, Bage G, Marcotte D, Samson R (2012) Estimating material and energy flows in life cycle inventory with statistical models. J Ind Ecol 16:399–406

Muller S, Lesage P, Ciroth A et al (2014) The application of the pedigree approach to the distributions foreseen in ecoinvent v3. Int J Life Cycle Assess. doi:10.1007/s11367-014-0759-5

Pehnt M, Oeser M, Swider D (2008) Consequential environmental system analysis of expected offshore wind electricity production in Germany. Energy 33:747–759

Pfenninger S, Hawkes A, Keirstead J (2014) Energy systems modeling for twenty-first century energy challenges. Renew Sustain Energy Rev 33:74–86

Remmen A, Astrup-Jensen A, Frydendal J (2007) Life cycle management. A business guide to sustainability. UNEP. http://www.unep.fr/shared/publications/pdf/DTIx0889xPA-LifeCycleManagement.pdf. Accessed 24 Feb 2015

Sandén BA, Karlström M (2007) Positive and negative feedback in consequential life-cycle assessment. J Clean Prod 15:1469–1481

Schmidt JH, Thrane M, Merciai S, Dalgaard R (2011) Inventory of country specific electricity in LCA – consequential and attributional scenarios. Methodology report v2. 2.-0 LCA consultants, Aalborg. http://lca-net.com/files/Inventory_of_country_specific_electricity_in_LCA_Methodology_report_20110909.pdf. Accessed 24 Feb 2015

Soimakallio S, Kiviluoma J, Saikku L (2011) The complexity and challenges of determining GHG (greenhouse gas) emissions from grid electricity consumption and conservation in LCA (life cycle assessment) – a methodological review. Energy 36:6705–6713

Steinmann ZJN, Hauck M, Karuppiah R et al (2014a) A methodology for separating uncertainty and variability in the life cycle greenhouse gas emissions of coal-fueled power generation in the USA. Int J Life Cycle Assess 19:1146–1155

Steinmann ZJN, Venkatesh A, Hauck M et al (2014b) How to address data gaps in life cycle inventories: a case study on estimating CO_2 emissions from coal-fired electricity plants on a global scale. Environ Sci Technol 48:5282–5289

Treyer K, Bauer C (2013) Life cycle inventories of electricity generation and power supply in version 3 of the ecoinvent database – part I: electricity generation. Int J Life Cycle Assess, doi: 10.1007/s11367-013-0665-2

Treyer K, Bauer C (2014) Life cycle inventories of electricity generation and power supply in version 3 of the ecoinvent database – part II: electricity markets. Int J Life Cycle Assess, doi: 10.1007/s11367-013-0694-x

Volkart K, Bauer C, Boulet C (2013) Life cycle assessment of carbon capture and storage in power generation and industry in Europe. Int J Greenh Gas Control 16:91–106

Weber CL, Jaramillo P, Marriott J, Samaras C (2010) Life cycle assessment and grid electricity: what do we know and what can we know? Environ Sci Technol 44:1895–1901

Weidema B, Ekvall T, Heijungs R (2009) Guidelines for application of deepened and broadened LCA. ENEA, The Italian National Agency on new Technologies, Energy and the Environment. http://www.leidenuniv.nl/cml/ssp/publications/calcas_report_d18.pdf. Accessed 24 Feb 2015

Yeh S, Rubin ES (2012) A review of uncertainties in technology experience curves. Energy Econ 34:762–771

Yue D, You F, Darling SB (2014) Domestic and overseas manufacturing scenarios of silicon-based photovoltaics: life cycle energy and environmental comparative analysis. Sol Energy 105:669–678

Zamagni A, Guinée J, Heijungs R et al (2012) Lights and shadows in consequential LCA. Int J Life Cycle Assess 17:904–918

Chapter 22
Life Cycle Management Applied to Urban Fabric Planning

Xavier Gabarrell, Joan Rieradevall, Alejandro Josa, Jordi Oliver-Solà, Joan Manuel F. Mendoza, David Sanjuan-Delmás, Anna Petit-Boix, and Esther Sanyé-Mengual

Abstract Due to the rapid urbanization and the large contribution of cities to the global environmental impact, urban policies integrate sustainability in the public space design. Current literature has accounted for the environmental impact of the main elements of the urban fabric, although studies have dealt with them individually. This chapter aims to optimize the environmental performance of the urban fabric for supporting planning processes, based on existing life cycle assessment (LCA) data of the main elements of urban fabric: sidewalks, pavements, and the gas, water and wastewater networks. Material selection and lifespan are key issues in the

---

X. Gabarrell (✉) • J. Rieradevall
Sostenipra (ICTA-IRTA-Inèdit; 2014 SGR 1412)-Institut de Ciència i Tecnologia Ambientals (ICTA), Universitat Autònoma de Barcelona (UAB), Bellaterra, Spain

Department of Chemical Engineering, Universitat Autònoma de Barcelona, Bellaterra, Spain
e-mail: Xavier.Gabarrell@uab.cat

A. Josa
Department of Geotechnical Engineering and Geosciences, School of Civil Engineering, Universitat Politècnica de Catalunya-Barcelona Tech (UPC), Barcelona, Spain

Institute of Sustainability, UPC, Barcelona, Spain

J. Oliver-Solà
Sostenipra (ICTA-IRTA-Inèdit; 2014 SGR 1412)-Institut de Ciència i Tecnologia Ambientals (ICTA), Universitat Autònoma de Barcelona (UAB), Bellaterra, Spain

Inèdit. Inèdit Innovació, S.L. UAB Research Park, Cabrils, Spain

J.M.F. Mendoza
Sostenipra (ICTA-IRTA-Inèdit; 2014 SGR 1412)-Institut de Ciència i Tecnologia Ambientals (ICTA), Universitat Autònoma de Barcelona (UAB), Bellaterra, Spain

Sustainable Industrial Systems (SIS) Group, School of Chemical Engineering and Analytical Science, University of Manchester, Manchester, UK

D. Sanjuan-Delmás • A. Petit-Boix • E. Sanyé-Mengual
Sostenipra (ICTA-IRTA-Inèdit; 2014 SGR 1412)-Institut de Ciència i Tecnologia Ambientals (ICTA), Universitat Autònoma de Barcelona (UAB), Bellaterra, Spain

G. Sonnemann, M. Margni (eds.), *Life Cycle Management*,
LCA Compendium – The Complete World of Life Cycle Assessment,
DOI 10.1007/978-94-017-7221-1_22

environmental profile of the paved skin, while the installation accounts for the greatest share of the burdens in subterranean networks. The best design consists of concrete sidewalks, asphalt pavements, HDPE (high density polyethylene) gas pipes, PVC (polyvinyl chloride) water pipes, and concrete sewer pipes. Pavements and sidewalks are the most contributing elements to the overall environmental burdens of streets.

Keywords City developments • Life cycle assessment • Life cycle management • Life cycle thinking • Pavements • Sidewalks • Sustainability

1 Introduction

More than 50 % of world's population is concentrated in cities, although they occupy less than 2 % of the Earth's surface. Because of the intense activity of cities, which play an essential role in the global socio-economic development, they consume over 75 % of the world's resources, between 60 % and 80 % of total energy, and are responsible for approximately 75 % of global greenhouse gas (GHG) emissions (Ash et al. 2008; Kamal-Chaoui and Robert 2009; Pacione 2009; UN Population Division 2010; European Union 2011; Lazaroiu and Roscia 2012). Environmental awareness of the urban metabolism has raised due to the rapid urbanization patterns. As a result, policy and planning highlight the importance of promoting environmental strategies that increase the sustainability of cities (UN 2013).

Within cities, public spaces play a key role in supporting daily urban life. According to UN-HABITAT (2013a, b, c, d), four main issues might be considered. First, streets may become a matrix which increases the urban connectivity between people and activities thereby making mobility more efficient. Second, the street pattern hosts urban basic services, such as water supply. Third, public space is a key element for the cultural and political dimensions of cities. Finally, street design might enable the pedestrian and road mobility in a safe manner. Thus, streets are composed of multiple elements from diverse natures that satisfy the different functionalities of the public space (UN-HABITAT 2014).

The urban fabric is in constant change because of new city developments, maintenance and partial renewals due to the intense urban activity. The World Wide Fund for Nature (WWF) accounted for $350 trillion of expenditure on urban infrastructures in the coming decades. Assuming business-as-usual (BAU) conditions, these projects would contribute to around 465 Gt of GHG emissions (WWF 2008). Then, if no environmental criteria are applied in the urban planning design (i.e., aesthetic or economic criteria are prioritized), the global environmental burdens of cities can dramatically increase. Furthermore, the consideration of all the life cycle stages to account for the environmental burdens of urban elements is basic for understanding their environmental performance, as demonstrated by Oliver-Solà et al. (2009a, b, c), Mendoza et al. (2012a, b), Petit-Boix et al. (2014), Sanjuan-Delmás et al. (2014).

The general aim of this chapter is to incorporate the life cycle thinking approach into the design of the urban fabric for supporting planning processes, thereby optimizing the environmental performance of cities. The objective is to identify the best environmental practices by comparing existing LCA data of the main elements of the urban fabric (sidewalks, pavements, and the gas, water and wastewater networks) and to integrate these practices into the environmental impact accounting of the entire street profile.

2 Methods

2.1 Environmental Benchmarking

Life cycle assessment (LCA) (ISO 2006) is a tool for the systematic compilation and evaluation of the environmental aspects and the potential environmental impacts of products systems throughout their life cycle, from raw material acquisition through use until final disposal (UNEP/SETAC 2012). In this sense, the environmental benchmarking of multiple alternatives for urban fabric planning was conducted by reviewing LCA studies on their environmental impacts. To do so, different constructive solutions for sidewalks, pavements, and gas, water and wastewater networks were compared in terms of global warming potential (GWP; kg of CO_2 eq.) (IPCC 2007).

2.2 The Urban Fabric

Multiple elements compose the urban fabric where daily urban life is supported. In this study, the authors considered the paved skin (sidewalks, road pavement) and the subterranean pipelines that supply basic services (gas, water, wastewater). Figure 22.1 displays a scheme of the urban fabric system. The functional unit of the assessment is a 1×8 m street section in a time frame of 50 years, representative of a European medium-sized city. In particular, the scenario considers a street profile located at the city centre where light-weight traffic and pedestrian areas are preferential. Superficial structures such as benches or streetlights were excluded.

3 Environmental Profile of Urban Fabric Elements

The following section describes the environmental profile of multiple design options for the urban fabric. Once compared, the best practices are assembled to show an optimized design of the paved skin (sidewalks, road pavement) and the subterranean pipelines that supply basic services (gas, water, wastewater) from an environmental point of view.

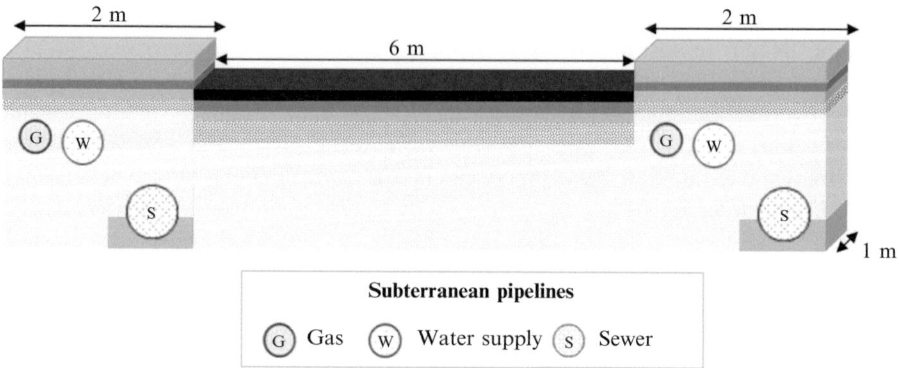

Fig. 22.1 The urban fabric: profile of the paved skin and subterranean pipelines under assessment

Table 22.1 Global warming potential (GWP) (kg of CO_2 eq.) of different pavement solutions for a functional unit of 1 m^2 in a time frame of 50 years and renovation scenarios

Pavement designs	Reference service life	High renovation rate (lifespan: 15 years)	Low renovation rate (lifespan: 40 years)
Concrete top-layer	79	129	82
Granite top-layer	109	221	117
Asphalt top-layer	84	84	–

Note: The reference service life of sidewalks designs are 15 years for asphalt and 45 years for concrete and granite

3.1 Paved Skin: Sidewalks and Light-Weight Traffic Road Pavement

Table 22.1 shows the environmental impact of conventional designs of concrete, asphalt and granite sidewalks to support pedestrian and light-weight traffic in cities. The design solutions have different service lives: 15 years for asphalt and 45 years for concrete and granite. However, the GWP of each sidewalk changes according to the variability of the service life of the constructive solutions. The shorter the service life, the higher the number of maintenance and removal operations (top-layer replacement) to restore the serviceability of the sidewalks during the period of analysis (50 years). Thus, two maintenance scenarios were considered, e.g., high and low maintenance rates. During this time frame, the most environmentally friendly pavement design is the concrete top-layer, which had a GWP 6 % and 28 % lower than asphalt and granite, respectively.

However, the environmental performance of pavements strongly depends on their service life and a sensitivity assessment was conducted for a range of lifespan from 5 to 45 years. When the same service life is considered for the three solutions (i.e., lifespan is equivalent under certain exposure conditions), asphalt becomes the

Table 22.2 Global warming potential (GWP) (kg of CO_2 eq.) of a natural gas network design for a time frame of 50 years for low and high density scenarios

Urban density	Neighborhood	Buildings	Dwellings	Waste treatment	Total
Low	6290	905	1160	866	9221
High	6290	30,000	75,100	4030	115,420

best practice: in urban areas with a high renovation rate of sidewalks and light-weight traffic pavements (≤15 years), asphalt is recommended for both sidewalks and light-weight traffic pavements: in contrast, concrete is the most suitable option in urban areas with a low renovation rate of sidewalks (>40 years), whereas asphalt has a maximum lifespan of 15 years and a high renovation rate is unfeasible. In general, granite sidewalks are less environmentally due to the high resource intensive manufacture of granite tiles (Mendoza et al. 2012a, b, 2014a).

3.2 Gas Network

Table 22.2 shows the environmental impact of natural gas distribution networks for low and high density neighborhood scenarios. Contrary to other elements, the gas network only considers a standard constructive design – i.e., HDPE (high density polyethylene) pipes for medium-pressure gas distribution – and instead of multiple materials, scenarios compared urban densities. The studied system is a standard local neighborhood. The scenarios considered recreate one low density detached house neighborhood (4 dwellings) and a high density Mediterranean neighborhood (24 dwellings). The assessment includes the network in the neighborhood (100 m), the elements of the gas distribution in the buildings and the dwellings, and the waste treatment of the materials. The results show that the distribution of the environmental load between subsystems changes radically according to urban density. This means that in low-density areas the neighborhood network is the subsystem that gives raise to most impact (68 %), while in high-density neighborhoods the building and dwelling subsystems are those that are responsible for more than 95 % of the GWP (Oliver-Solà et al. 2009a, b). For the purpose of our study, the impact of the neighborhood network was considered in the profile of the street (see Table 22.5).

3.3 Water Supply Network

Table 22.3 displays the environmental impacts of the drinking water network considering constructive solutions with different pipe materials and diameters. The environmental burdens were related to a functional unit of 1 linear meter of network. High and low density polyethylene (HDPE, LDPE) and polyvinyl chloride (PVC) were compared for a pipe with a diameter of 90 mm, and HDPE, PVC, glass fiber

Table 22.3 Global warming potential (GWP) (kg of CO_2 eq.) of different water supply constructive solutions for a functional unit of 1 m of pipe in a time frame of 50 years

	90 mm-diameter pipes			200 mm-diameter pipes			
Stage	**HDPE**	**LDPE**	**PVC**	**HDPE**	**PVC**	**GFRP**	**DI**
Production	4.03	6.93	3.56	13.6	12.3	49.9	131
Transport	8.73	8.94	8.70	9.83	9.71	9.73	16.2
Installation	12.5	12.5	12.5	13.3	13.3	13.3	13.3
Total	25.3	28.4	24.8	36.7	35.4	73.0	161

Table 22.4 Global warming potential (GWP) (kg of CO_2 eq.) of different sewer constructive solutions with a diameter of 300 mm for a functional unit of 1 m of pipe in a time frame of 50 years

	HDPE (lifespan: 50 years)		PVC (lifespan: 50 years)		Concrete (lifespan: 100 years)	
Life-cycle stage	**PP1**	**PP2**	**PP1**	**PP2**	**CP1**	**CP2**
Pipe production	23.7	23.7	7.2	7.2	4.8	4.8
Transport	35.1	21.1	34.8	20.8	10.2	11.0
Installation	33.2	125	33.2	125	42.3	65.0
Demolition	0.95	1.20	0.95	1.20	0.60	0.60
Total impact	93.0	171	76.2	154	57.9	81.4

reinforced polyester (GFRP) and ductile iron (DI) for 200 mm (Sanjuan-Delmás et al. 2014). The life cycle stages included were the production, transport, and installation. The operation was excluded because it depends on the specific network. The end of use phase was also omitted since the supply network is left underground after its use.

In water supply network solutions, the installation stage contributed the most to the GWP (around 50 % for 90 mm; 10–40 % for 200 mm). Furthermore, the relevance of the installation phase was higher for smaller diameters, since the pipe required a lower amount of material and, thus, production is less impacting (15–25 % for 90 mm, 35–80 % for 200 mm). Consequently, previous studies that only analyzed the pipe instead of the whole constructive solution were omitting a significant part of the impacts. No significant differences were found between constructive solutions with different plastic pipe materials. In contrast, GFRP and DI had a much larger impact than HDPE and PVC for 200 mm constructive solutions (around two times higher for GFRP and 5 for DI) due to the higher impacts of these materials.

3.4 Sewer Network

Table 22.4 shows the environmental impact of concrete, PVC, and HDPE sewer pipes with different trench designs. Sewer pipes with a diameter of 300 mm were considered because they are widely used in medium-sized cities. The environmental burdens of sewer pipes were related to a functional unit of 1 linear meter of network for a timeframe of 50 year, considering the different lifespan of the materials (i.e., 100 years for concrete and 50 for PVC and HDPE). The sewer operation was excluded from the assessment, given that the electricity required for pumping wastewater depends on the configuration of the city.

In terms of GWP, concrete pipes scored better than plastic pipes, mainly because of their longer durability. The composition of plastic pipes (i.e., oil-based materials) derived in a larger amount of CO_2 eq. emissions than in the case of concrete (i.e., mainly cement). There were also differences related to the trench designs that can be applied to each pipe material. In a first attempt to compare constructive solutions, two configurations were assigned to each pipe material: sand beddings (PP1), concrete beddings (PP2 and CP2), and mixed beddings (CP1). PP2 and CP2 resulted in greater environmental impacts because of the contribution of concrete to the GWP (>80 %). In general, the installation stage accounted for more than 70 % of the total life cycle impacts when PP2 and CP2 were applied. At the end, it was determined that concrete pipes with CP1 trenches were the most environmentally friendly alternative (Petit-Boix et al. 2014).

4 Discussion

4.1 *Optimizing the Urban Fabric Design*

Table 22.5 shows the GWP of two street profiles, which were composed considering the best and worst designs identified in Sect. 3. The most environmentally-friendly design option for the street profile resulted in a carbon footprint of 1.1 tones of CO_2 eq. The best design included concrete sidewalks, asphalt pavements, HDPE gas pipes, PVC, water pipes and concrete sewer pipe with CP1 trench. In general, the environmental impacts were reduced by 23 % with respect to the worst street profile. The use of granite in sidewalks, LDPE in water supply pipes and HDPE in the sewer system with PP2 trenches were the least recommended design options due to the increase in the GWP.

Table 22.5 Global warming potential (GWP) (kg of CO_2 eq.) of the best and worst designs of a 1 × 8 m street section in a time frame of 50 years

	Sidewalk	Pavement	Gas	Water	Wastewater	Total
Best design	Concrete	Asphalt	HDPE	PVC	Concrete CP1	–
GWP	316	504	125.8	49.6	115.8	1111.2
(%)	28.4	45.3	11.3	4.6	10.4	100
Worst design	Granite	Asphalt	HDPE	LDPE	HDPE PP2	–
GWP	436	504	125.8	56.8	342	1464.6
(%)	29.8	34.4	8.6	3.9	23.3	100

When focusing on the different sub-systems, the asphalt pavement had the greatest contribution to the total GWP because the light-weight traffic road had the largest surface in the street profile (6 m^2). Nevertheless, there are two elements that presented the greatest room for improvement. In both scenarios, the concrete sidewalk accounted for almost 30 % of the impacts. Generally, this material is widely applied to pedestrian sidewalks, although asphalt is also used given its low initial cost. Thus, the elements of the paved skins contributed the most to the urban fabric environmental profile, due to a more intensive maintenance of these exposed areas. However, this might depend on the traffic density, the type of mobility and the land use (e.g., pedestrian areas).

The second element that presented variations in the best and worst designs was the sewer network. In this case, the lifespan (i.e., required reposition) also played an important role: when concrete pipes (lifespan: 100 years) were included in the best scenario, they accounted for 10 % of the impacts; in contrast, the contribution of the sewer increased up to 23.3 % when plastic pipes were considered (lifespan: 50 years). Hence, the integration of life-cycle environmental data and service-life planning information is essential for urban planners for identifying long-term environmentally friendly constructive solutions (Mendoza et al. 2012b). Finally, it was detected that the installation of subterranean pipelines had a great contribution to the total GWP and this trend could be extrapolated to other subterranean systems such as telecommunication and electricity networks.

Another key issue is the variation in the street configuration. This study presented a standard street profile, which consisted of common pipe diameters and trench designs. Nonetheless, the population density and the configuration of the city might demand larger pipe diameters because of a more intensive gas and water consumption and wastewater production. Therefore, the contribution of these elements to the total impact of the urban skin might increase. As a result, urban planners must focus on the possible material and installation alternatives that best suit their case studies (Petit-Boix et al. 2014).

4.2 Towards Smart Grids and Self-sufficiency

Because networks present a relevant environmental contribution to the impacts of the street, the cities of the future should consider approaching smart grids and self-sufficiency. In this sense, decentralization of urban services is essential to improve the environmental performance of cities. This approach aims to reduce the required networks to supply services while increasing the independence of individual neighborhoods and buildings. From an environmental point of view, this would contribute to reducing the environmental burdens of the subterranean profile of cities. This strategy is particularly meaningful in low-density settlements where longer networks are installed. In the case of gas distribution, when an isolated house requires more than 69 m of neighborhood network, the installation of individual propane tanks becomes favorable in terms of GWP (Oliver-Solà et al. 2009a). Furthermore, the gas supply through district heating systems can be considered as an alternative at the neighborhood scale (Oliver-Solà et al. 2009b).

In the framework of decentralization, self-supply of endogenous resources such as water or energy contributes to the environmental improvement of urban areas. In this case, besides being independent of the central network, local and renewable resources substitute the consumption of nonrenewable ones. For instance, rainwater harvesting systems are increasingly implemented for nonpotable purposes and play a key role in countries dealing with water scarcity. In addition, wastewater recycling can also be integrated in the water metabolism of buildings.

5 Conclusions

LCA literature has dealt with the environmental impacts of multiple elements of the urban fabric. However, studies only showed the individual performance of certain elements. This study compiled LCA data in an attempt to quantify the environmental burdens of an entire street profile as a whole, thereby combining the impacts of the elements that constitute the urban fabric: the paved skins and the subterranean networks. The best and worst practices were identified.

The best practices in the design of streets accounted for a carbon footprint of 1.1 t of CO_2 eq. This design consisted of concrete sidewalks, asphalt pavements, HDPE gas pipes, PVC water pipes, and concrete sewer pipe with CP1 trench. When considering this configuration in urban planning, the global warming potential of streets can be reduced by 23 %. The most impacting scenario included the following practices: granite in sidewalks, LDPE in water supply pipes and HDPE in the sewer system with PP2 trenches. Pavements and sidewalks are the most contributing elements to the overall environmental burdens of streets, mainly because their exposition intensifies their maintenance.

Proper urban fabric design must consider three key aspects: the material selection, lifespan and, in the case of subterranean networks, the installation procedures. First, promoting cleaner production in the construction materials sector (e.g., granite production, Mendoza et al. 2014b) is essential to reduce the environmental burdens of constructive assets and achieve major environmental improvements at city level. Second, lifespan must be included in the design parameters of long-term urban planning. Third, the installation of subterranean networks has a relevant contribution to the total impacts and must be adapted to the technical and environmental requirements of the construction site.

Studies that quantify the environmental burdens of urban elements provide basic information for the decision-making process when environment is considered in urban planning. Thus, the application of life cycle thinking implies the inclusion of environmental criteria during the conception of cities. Towards a sustainable design of cities, urban planning studies might also integrate the social and economic dimensions in the decision-making process. From a life cycle perspective, Social Life Cycle Assessment (S-LCA) (UNEP/SETAC 2009) and Life Cycle Costing (LCC) (ISO 2008) methods may provide a standard quantitative way to assess urban elements.

References

Ash C, Jasny BR, Roberts L, Stone R, Sugden A (2008) Reimagining cities – introduction. Science 319(5864):739–739

European Union (2011) Cities of tomorrow: challenges, visions, ways forward. Report KN-31-11-289-EN-C. Unit C.2 – Urban Development, Territorial Cohesion, Directorate General for Regional Policy, Publications Office of the European Union, Luxembourg

IPCC (2007) Contribution of working groups I, II and III to the fourth assessmevnt report of the intergovernmental panel on climate change. Intergovernmental Panel on Climate Change, Geneva

ISO (2006) ISO 14040: environmental management – life cycle assessment – principles and framework. International Organization for Standardization, Geneva

ISO (2008) Buildings and constructed assets – service-life planning – part 5: life-cycle costing. ISO, Geneve

Kamal-Chaoui L, Robert A (2009) Competitive cities and climate change. OECD Regional development working papers 2009/2. Organisation for Economic Cooperation and Development (OECD), Directorate for Public Governance and Territorial Development

Lazaroiu GC, Roscia M (2012) Definition methodology for the smart cities model. Energy 47:326–332

Mendoza J-MF, Oliver-Solà J, Gabarrell X, Josa A, Rieradevall J (2012a) Life cycle assessment of granite application in sidewalks. Int J Life Cycle Assess 17:580–592

Mendoza J-MF, Oliver-Solà J, Gabarrell X, Rieradevall J, Josa A (2012b) Planning strategies for promoting environmentally suitable pedestrian pavements in cities. Transp Res Part D Transp Environ 17:442–450

Mendoza J-MF, Feced M, Feijoo G, Josa A, Gabarrell X, Rieradevall J (2014a) Life cycle inventory analysis of granite production from cradle to gate. Int J Life Cycle Assess 19(1):153–165

Mendoza J-MF, Capitano C, Peri G, Josa A, Rieradevall J, Gabarrell X (2014b) Environmental management of granite slab production from an industrial ecology standpoint. J Clean Prod 84:619–628

Oliver-Solà J, Gabarrell X, Rieradevall J (2009a) Environmental impacts of natural gas distribution networks within urban neighborhoods. Appl Energ 86:1915–1924

Oliver-Solà J, Gabarrell X, Rieradevall J (2009b) Environmental impacts of the infrastructure for district heating in urban neighbourhoods. Energ Policy 37:4711–4719

Oliver-Solà J, Josa A, Rieradevall J, Gabarrell X (2009c) Environmental optimization of concrete sidewalks in urban areas. Int J Life Cycle Assess 14:302–312

Pacione M (2009) Urban geography. a global perspective. Routledge, Oxon

Petit-Boix A, Sanjuan-Delmás D, Gasol CM, Villalba G, Suárez-Ojeda ME, Gabarrell X, Josa A, Rieradevall J (2014) Environmental assessment of sewer construction in small to medium sized cities using life cycle assessment. Water Resour Manag 28:979–997

Sanjuan-Delmás D, Petit-Boix A, Gasol CM, Villalba G, Suárez-Ojeda ME, Gabarrell X, Josa A, Rieradevall J (2014) Environmental assessment of different pipelines for drinking water transport and distribution network in small to medium cities: a case from Betanzos. Spain J Clean Prod 66:588–598

UN (2013) Towards sustainable cities. World Economic and Social Survey 2013. Sustainable Development Challenges. United Nations (UN), Department of Economic and Social Affairs, New York

UNEP/SETAC (2009) Guidelines for a social life cycle assessment. UNEP, Paris

UNEP/SETAC (2012) Greening the economy through life cycle thinking – ten years of the UNEP/SETAC Life Cycle Initiative. United Nations Environment Programme (UNEP) and Society of Environmental Toxicology and Chemistry (SETAC), Nairobi

UN-HABITAT (2013a) The relevance of street patterns and public space in urban areas. UN-Habitat working paper, United Nations Human Settlements Programme (UN-Habitat), Nairobi

UN-HABITAT (2013b) Streets as public spaces and drivers of urban prosperity. United Nations Human Settlements Programme (UN-Habitat), Nairobi

UN-HABITAT (2013c) Urban planning for city leaders. United Nations Human Settlements Programme (UN-Habitat), Nairobi

UN-HABITAT (2013d) Revised compilation for sustainable cities and human settlements in the sustainable development goals (SDGs) within the Post-2015 development agenda. United Nations Human Settlements Programme (UN-Habitat), Nairobi

UN-HABITAT (2014) Public Space in the global agenda for sustainable urban development: the "global public space toolkit". Urban Planning and Design Branch, United Nations Human Settlements Programme (UN-Habitat), Nairobi

UN Population Division (2010) World urbanisation prospects: the 2009 revision. United Nations (UN), Population division, Department of Economic and Social Affairs, New York

WWF (2008) Reinventing the city: three prerequisites for greening urban infrastructures. World Wildlife Fund (WWF) International, produced in cooperation with Booz and Company, Gland

Chapter 23
Implementing Life Cycle Engineering in Automotive Development as a Helpful Management Tool to Support Design for Environment

Florian Broch, Jens Warsen, and Stephan Krinke

Abstract This chapter describes the implementation of life cycle engineering, a life cycle management component that focuses on the environmental performance improvement, in the context of automotive design for environment. The purpose of life cycle engineering is to derive measurable technical targets from life cycle assessment (LCA). This approach is described using the example of lightweight design. The progress in this methodology is the ability to calculate measurable targets – such as weight reduction, fuel reduction on a vehicle level, or the amount of secondary material – on the basis of LCA results. It is important to note that LCA is not used here for comparing the environmental performance between competing materials or technologies. Instead, life cycle engineering, as a helpful management tool to support design for environment, shows the technical roadmap of measures that must be taken in order to assure environmental progress over the entire life cycle. In doing so, this tool supports putting life cycle assessment results into business practice.

Keywords Life cycle engineering • Automotive • Design for environment • Lightweight design • Life cycle management

1 Introduction

Life cycle management (LCM) is a management concept used by businesses to ensure a continuous sustainable performance improvement of their activities (UNEP/SETAC 2009). It aims at minimize environmental and socioeconomic burdens in the

F. Broch • J. Warsen • S. Krinke (✉)
Volkswagen AG, Environmental Affairs Product,
Letter box 011/1774, 38436 Wolfsburg, Germany
e-mail: stephan.krinke@volkswagen.de

G. Sonnemann, M. Margni (eds.), *Life Cycle Management*,
LCA Compendium – The Complete World of Life Cycle Assessment,
DOI 10.1007/978-94-017-7221-1_23

entire life cycle of a product (Remmen et al. 2007). The mobility sector is a key branch for the development of technologies that help to minimize environmental impacts. For the Volkswagen Group, the motivation for this is not only driven by external factors like environmental regulations, financial markets, customer requirements, competitors' behavior and the volatility of energy and resource prices, but also because, as a major automobile manufacturer, the Volkswagen Group takes seriously its responsibility for the sustainable development of the economy, the environment and the society. Based on this, the Volkswagen Group set itself the goal to become the world's most sustainable automobile manufacturer by 2018. In order to achieve this, a set of ambitious environmental targets has been elaborated to be pursued continuously in different business units like technical development, production, and sales division. The environmental strategy applies not only to all brands of the Volkswagen Group in all regions, but also to the whole value chain.

1.1 Life Cycle Management at Volkswagen

The aim of life cycle management initiatives in general and environmental management in particular at Volkswagen is to continuously improve the environmental performance of both the company itself and of the products. Consequently, the Environmental Strategy is based on four target areas: Top in intelligent mobility, leaders in eco-friendly products, top in lifecycle-based resource conservation, and consistent anchoring throughout the company (Fig. 23.1).

As leader in eco-friendly products Volkswagen is committed to reducing the CO_2 output of the European new car fleet to emissions below the threshold of 120 g CO_2/km by 2015 and furthermore to 95 g CO_2/km by 2020. Beyond and considering the lifecycle-based approach regarding resource conservation, a holistic view on the full lifecycle of the products is considered. This means that not only environmental issues relating to products are addressed, but the minimization of environmental

Fig. 23.1 Volkswagen Group environmental strategy (Source Volkswagen AG 2013)

impacts along the whole value chain is considered. The goal is to create products that have better environmental properties over their entire lifecycle in comparison to their predecessors. The tool chosen by Volkswagen to implement this approach is life cycle assessment (LCA) in line with ISO standards 14040 and 14044 (ISO 14040: 2006, ISO 14044: 2006).

1.2 Life Cycle Assessment as a Tool to Implement Life Cycle Management at Volkswagen

The LCA is one of the LCM tools used by organization to understand the consequences of their business operations. The LCA methodology was first developed 30 years ago, and since then the technique has been practiced in many different product fields. Early on, Volkswagen was engaged in this process and performed LCAs aimed at optimizing products and processes ever since the early 1990s. As early as 1996, the company was the first carmaker to prepare and publish a life cycle inventory (LCI) for the Golf III (Schweimer and Schuckert 1996). In the following years, LCIs were published for various vehicles of the Volkswagen Group (e.g., Schweimer and Levin 2000).

Conducting LCAs at Volkswagen means to collect all the important facts over the entire life cycle of a vehicle, component, or technology and back them up with relevant figures, e.g., the volume of raw materials, energy for production processes. The data collection process is based on vehicle parts lists, material and weight information stored in the company's own Material Information System (MISS), technical datasheets and drawings. For modeling the use phase, fuel consumption and the resultant emissions are worked out based on the legally prescribed New European Driving Cycle (NEDC). In addition, the amount of energy consumed during the dismantling and/or recycling of the vehicle parts is calculated.

Since 2007 the Volkswagen brand consequently publishes Environmental Commendations which inform customers and the general public about the ecological progress at life cycle level made by new models in comparison to their predecessor. These comparisons are based on detailed LCA studies which are certified by independent auditors in accordance to 14040 (ISO 2006a) and 14044 (ISO 2006b). By the end of 2013, Volkswagen Passenger Cars and Volkswagen Commercial Vehicles had published a total of 18 Environmental Commendations.

2 From Life Cycle Assessment to Life Cycle Engineering

Volkswagen uses the LCA methodology as an effective LCM tool to analyze the environmental profiles of products and processes and to identify ecological hotspots therein. Based on this knowledge it is determined which improvements will have the greatest effect and thus can develop targeted innovations. This is what is called life cycle engineering (LCE). It is one of the components of LCM. While LCM as a product management system aims to improve the sustainability performance of a company through minimizing environmental and socioeconomic burdens associated with an organization's value chain (Remmen et al. 2007), it has a narrower focus on the management and controlling of measures to improve the environmental profiles of a company's products. It is a helpful management tool to support design for environment.

2.1 Success Factors for Life Cycle Engineering Within the Company

The fundamental aim of life cycle engineering is the management and controlling of measures for the improvement of the environmental profiles of products. However, in order to implement and integrate LCE into corporate processes, there are some key success factors to be considered:

Organization: In order to help assure a consistent implementation of life cycle engineering into the companies' processes, the commitment of the top management is crucial. Thus, the life-cycle based improvement of the vehicle fleet in terms of environmental impact and resource conservation forms an integral part of Volkswagen's corporate policy and environmental strategy.

Analysis: The LCA methodology provides a comprehensive tool to both gain a detailed insight in the environmental profile of a product and analyze the potentials for improvements. However, to be applicable, the LCA must be based on a framework that ensures reliable and robust results derived within a reasonable time frame, despite of the degree of complexity of the product under study.

Optimization: In order to transfer conclusions from the LCA into realistic and convertible improvements, it is necessary to translate the results into technical goals. These have to be expressed in a form that is sufficiently specific to allow decisions and measures from an engineering point of view.

Communication: Environmental improvements in vehicles must be visible for customers and other stakeholders. It is therefore crucial to communicate about new features and to inform customers and other relevant stakeholders about the activities and successes in developing environmentally friendly technologies and products. For this purpose, the Volkswagen brand developed "Environmental Commendations" (Warsen and Krinke 2012).

2.2 Integration of Life Cycle Engineering into Company Processes

To successfully support the development of an ecologically advantageous new vehicle, it is crucial to obtain top management awareness and to implement life cycle engineering in the company's development processes.

Most decisions affecting the environmental performance usually have a financial impact on a vehicle project. Furthermore it is obvious that an environmental decision support should be implemented as early as possible in the development process. The earlier decisions can be supported, the more influence can be exercised.

Life cycle engineering is implemented in the environmental strategy of the Volkswagen Group and in the environmental objectives for technical development of the Volkswagen brand. These objectives are set and tracked by the environmental officer during the development of a vehicle. With the goal in mind to develop each model in such a way that, over its entire life-cycle, it presents better environmental properties than its predecessor, the environmental officer is present in decisive boards and supports decisions from the initiation of project on.

3 Automotive Life Cycle

Like many other products, the automotive life cycle consists of three main phases. In the following the automotive life cycle is described based on the greenhouse gas emission profile of a Golf VII, 1.6 TDI for an assumed running distance of 20,0000 km. Three phases are differentiated: production phase (20 % of GHG emissions), use phase (79 % of GHG emissions), and end-of-life phase (1 % of GHG emission). The production phase covers the raw material extraction to semi-finished products or components and finally the car's production and assembly. Within the production phase roughly 21 % of a car's production CO_2-eq. emissions are emitted at Volkswagen plants. The other 79 % are emitted over the entire supply chain back to the extraction of raw materials like iron ore for steel production or bauxite for aluminum production (Fig. 23.2).

The use phase covers the tailpipe emissions (tank-to-wheel) as well as the emissions for fuel extractions and production (well-to-tank). At the end-of-life phase, the vehicle is partly dismantled and then shredded for the reuse of the materials, which accounts for around 1 % of the total greenhouse gas-emissions.

In accordance with the drivers for environmentally compatible product design, the main effort is put into the reduction of emissions during the use.

This is achieved by developments like the electrification of the car, more efficient combustion engines and complex emission control systems. Furthermore, the lowering of running resistances, like mass and aerodynamic drag, are addressed.

But these measures can also increase the emissions in production. This can result in a shift of the hot spots within a car's lifecycle. The usage of energy-intensive technologies, like lithium ion accumulators or lightweight materials, can lead to a higher burden in the production and recovery phase, combined with a lower burden in the use phase.

Therefore the task for life cycle engineering is to assure that, in total, environmental impacts of cars over their entire life cycle is lower than that of their predecessor.

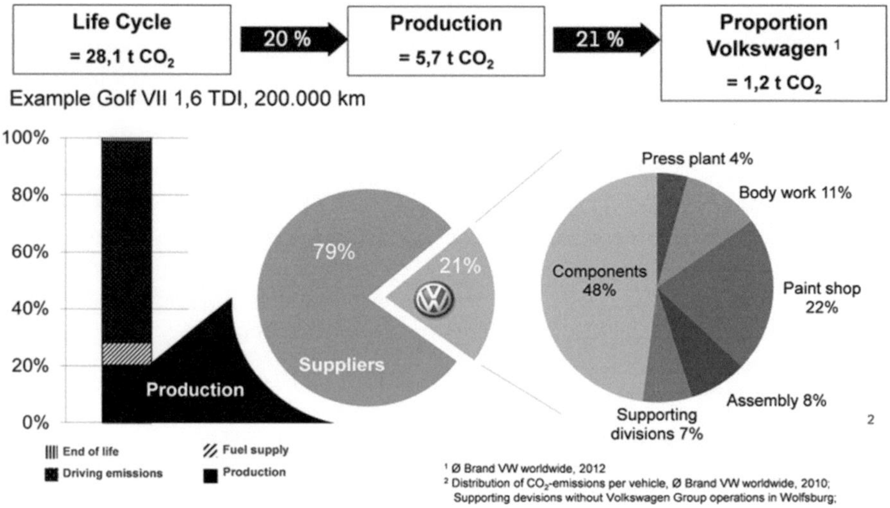

Fig. 23.2 Life cycle perspective: A car's CO_2 equivalents

3.1 *Lightweight Design*

Lightweight design is one relevant measure for lowering the car's fuel consumption and driving emissions, as the car's mass has the biggest single influence on the running resistances. However, from an environmental life cycle perspective, it is crucial to choose the "right" lightweight concepts and materials in order to avoid the shift of environmental burdens (Warsen and Krinke 2012).

From the environmental point of view, a ground-breaking success factor for lightweight design depends on the realization of secondary weight effects. Reversing the spiral of increasing weight can and should lead to an adaption of powertrain size. For example, the reduction of 100 kg in a car powered by a turbocharged petrol engine results in a reduction of tailpipe-emissions by 3.6 g CO_2/km, which is equivalent to a fuel reduction value (FRV) of 0.15 l/100 km. With an adapted powertrain (adapted engine displacement and gear ratio), the improvement more than doubles to 8.2 g CO_2/km (Rohde-Brandenburger 2014). At this point it is important to bear in mind that the choice of a powertrain is made on a vehicle perspective and depends on the available powertrain portfolio (Krinke et al. 2010, p. 38).

3.2 *Example: Hot Stamped Steel*

Usually the most common way to assess the environmental impact of lightweight design is the comparison of two materials in the context of a real application. On the one hand the specific constraints and assumptions are set, but on the other hand the assessment is not valid outside these constraints and assumptions.

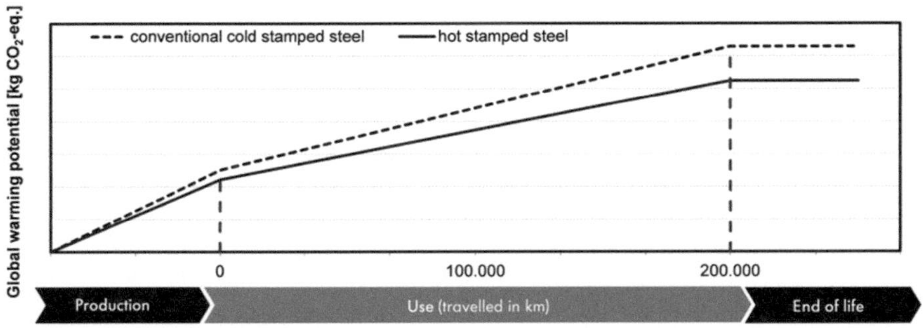

Fig. 23.3 Carbon footprint comparison: cold stamped steel vs. hot stamped steel

One good example is the analysis of hot stamped steels in comparison to conventional steel. Hot stamped steels are low-alloy steels with a special aluminum-silicon coating that is heated to 900 °C before the forming process. While the steel is formed, it is hardened by cooling it down abruptly. Therefore the forming process is clearly more energy-intensive than the conventional cold stamping process. The advantage of hot stamping is the much higher strength of the steel part. This property enables thinner and lighter steels that still have the same or even better crash performance than conventional steels. All in all, for an exemplary part, this results in a weight reduction of 20 % and a corresponding lower demand for raw materials. After considering the entire life cycle and the realized weight reduction, the hot stamped steel is advantageous in comparison to the conventional steels as shown in Fig. 23.3. Due to lower material demand and the resulting reduction for raw material extraction and steel production, in this case the lightweight alternative is at an advance even before the first meters are driven with the car. With each driven meter the lightweight effect can unfold on top.

Therefore hot stamped steel is a good example for lightweight strategy which offers environmental advantage from the first mile on.

4 Analysis and Derivation of Measurable Technical Targets

The statement of more or less abstract results from different environmental impact categories as a result of a LCA is barely promising. To enable change in development and to influence new developments, a derivation of measurable technical targets and indicators is needed. In addition, these targets must be based on crucial success factors for LCE that considerably influence the vehicle's environmental impact. With an ideal conception, the targets can be used on a universal scale and are not bound to specific concepts or assumptions.

Only technical targets in the language of the recipient can be used to support the developers and decision makers in charge. Thus, the target will not be in direct context of environmental impact categories like global warming potential (kg

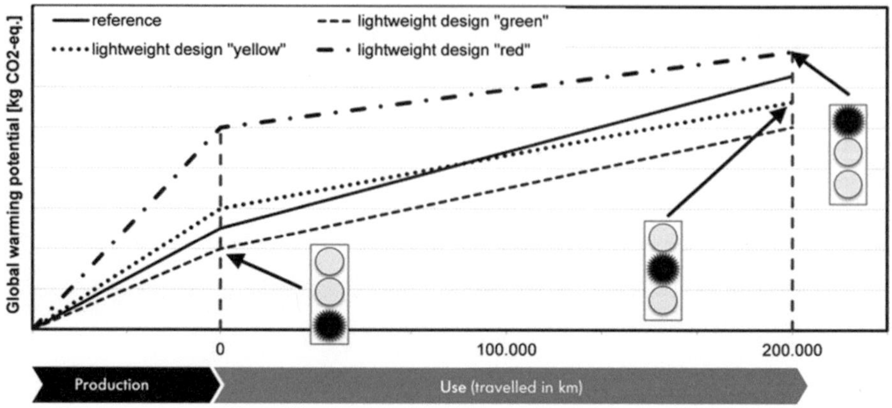

Fig. 23.4 Derivation of traffic lights from the life cycle illustration

CO_2-equivalent) or photochemical ozone creation potential (kg ethene-equivalent), but in the context of vehicle engineering like fuel consumption (liter per 100 km), or mass reduction with lightweight design (necessary weight reduction in kg or %). In addition to such use-phase-oriented targets it is useful to express targets in the context of production engineering, too. Targets with a high influence factor can be material efficiency or share of secondary material from closed-loop recycling.

A very intuitive form for the communication of targets can be the visualization as traffic lights. The traffic lights are derived from the LCA-common life cycle illustration to rate lightweight designs as shown in Fig. 23.4.

The green traffic light is reserved for the best case – an alternative design performing better than the reference right from production (as in the hot stamping example). If the advantage is realized during use, for example by reduced fuel consumption, the alternative will be awarded the yellow traffic light. In this case there will be a break-even at a certain kilometrage. However, if an alternative cannot show its advantage over the use of the product phase, the red traffic light appears.

As this illustration is still linked to a specific part or situation, the illustration has to be generalized. How a general approach looks like will be shown with the help of the press hardening steel example.

4.1 *Calculating Targets for Hot Stamped Steel*

Lightweight design may be the ideal case to show the need of a general target and illustration. The decisive factor to assess a design in advance is the weight reduction in comparison to a reference. Usually this factor is not available until the design is actually construed and the design process finished. To support the design from the beginning, a target value for a necessary weight reduction is required. How to quickly calculate a target for lightweight design (w_{rel}) is shown in Eq. (23.1):

$$w_{rel} = \frac{EIP_a - EIP_r}{EIP_a + EIRV_{LC}} \tag{23.1}$$

Thereby the environmental impact for the production of 1 kg of a reference material (EIP_r) and 1 kg of an alternative material (EIP_a) plus an environmental impact reduction value over life cycle ($EIRV_{LC}$) is needed. This value reflects the reduction of an environmental impact per km (e.g., g CO_2/km) over an assumed running distance in km for a weight reduction of 1 kg.

The calculated target shows the relative weight reduction needed to perform better than the reference after the use phase, parallel to the transition of the yellow to the red range shown in Fig. 23.5. With the ratio of EIP_r to EIP_a the value can also be calculated for the transition from production to use (green to yellow range). These calculations are conducted on the example of hot formed steel in comparison to cold formed steel in crash applications for global warming potential.

With a typical weight reduction of around 20 % it is obvious that the technology in this case is not critical and the hot formed part is in advance already after production. Furthermore, developers and decision makers can easily see that a weight reduction of less than 5 % is critical. In this case the savings in material and fuel do not compensate the higher burdens of the hot forming process. A weight reduction between roughly 5 % and 10 % is sufficient to obtain an advantage over lifecycle but depends on the usage of the vehicle. It may be appropriate to analyze this case in detail.

If a light weight alternative's typical weight reduction is outside the desired ranges, measures have to be considered to lower the environmental impact. Those measures can be rated again with the illustration in Fig. 23.6.

A measure in the use phase, e.g., a powertrain adaption, and the resulting higher fuel reduction, will cut the red range towards a lower necessary weight reduction. The use of more secondary materials as a measure for lowering the impact in production would even expand the green range and thus enable the use of a lightweight design even more. That is how the best and most efficient measures can be identified and compared in a very neat way.

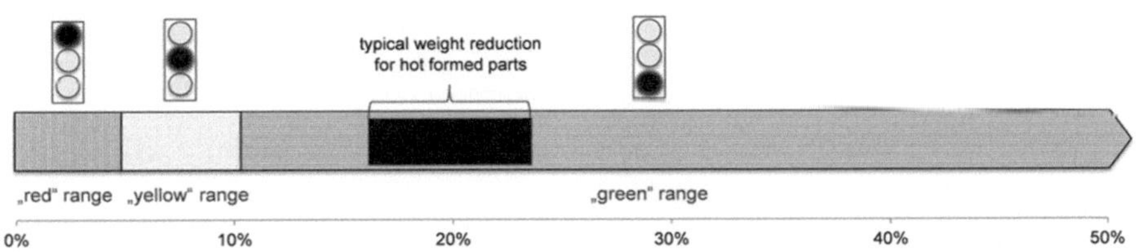

Fig. 23.5 Necessary weight reduction for a hot formed part in comparison to a cold formed part

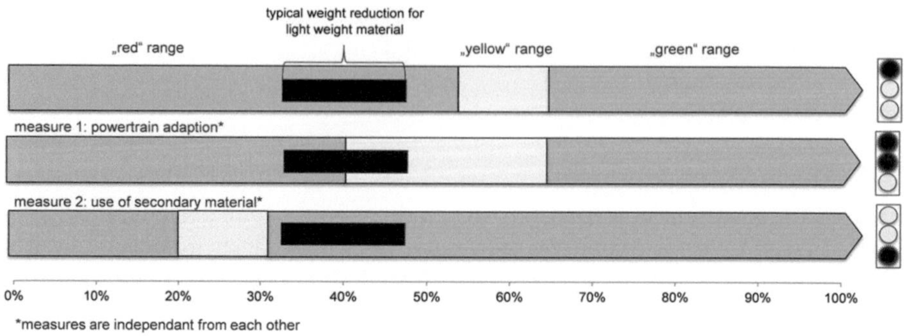

Fig. 23.6 Comparison of measure with target value in sight

5 Conclusion

The developed life cycle engineering approach is a helpful environmental management tool to support design for environment. The progress in this methodology is the ability to calculate measurable targets – such as weight reduction, fuel reduction on a vehicle level, or the amount of secondary material – on the basis of LCA results. With the help of an example in lightweight design, the successful implementation and application of LCA derived technical targets in a company was shown. It is important to note that LCA is not used here for comparing the environmental performance between competing materials or technologies. Instead, life cycle engineering shows the technical roadmap of measures that must be taken in order to assure environmental progress over the entire life cycle. In this way, life cycle engineering mainly focuses on the management and controlling of actions to improve the environmental performance of products. Environmental performance improvement is key for a life cycle management initiative in a company.

References

ISO (2006a) ISO 14040 Environmental management – life cycle assessment – principles and framework. International Organisation, Geneva

ISO (2006b) ISO 14044 Environmental management – life cycle assessment – Requirements and guidelines. International Organisation, Geneva

Krinke S, Koffler C, Deinzer G, Heil U (2010) An integrated life cycle approach to lightweight automotive design. ATZ Worldwide 112(6):36–42

Remmen A, Jensen AA, Frydendal J (2007) Life cycle management. A business guide to sustainability. UNEP DTI

Rohde-Brandenburger K (2014) Was bringen 100 kg Gewichtsreduzierung im Verbrauch? –eine physikalische Berechnung'. In: Siebenpfeiffer W (ed) Leichtbau-Technologien im Automobilbau. Springer Fachmedien, Wiesbaden, pp 179–189

Schweimer GW, Levin M (2000) LCI Golf A4. Retrieved from http://www.volkswagenag.com/content/vwcorp/info_center/en/publications/2007/01/Golf_A4__Life_Cycle_Inventory.bin.html/binarystorageitem/file/golfa4_english.pdf

Schweimer GW, Schuckert M (1996) Sachbilanz eines Golf: Ganzheitliche Betrachtungen im Automobilbau – Rohstoffe – Produktion – Nutzung – Verwertung. VDI-Berichte 1307, Düsseldorf, pp 235–255

UNEP/SETAC (2009) Life cycle management – how business uses it to decrease footprint, create opportunities and make value chains more sustainable. UNEP Paris

Volkswagen AG (2013) Group sustainability report 2013. Wolfsburg. Retrieved from http://sustainabilityreport2013.volkswagenag.com/sites/default/files/pdf/en/Volkswagen_SustainabilityReport_2013.pdf

Warsen J, Krinke S (2012) The life cycle approach at Volkswagen. ATZ Worldwide 114(7–8):4–9

Chapter 24
Managing Life Cycle Sustainability Aspects in the Automotive Industry

Marzia Traverso, Patrick Kim, Stefan Brattig, and Volkmar Wagner

Abstract An important target on the way to a sustainable development is ensuring a more sustainable production and a reduced consumption of energy and material resources. The automotive industry is considered one of the most strategic contributors to reach this objective. On the one hand, the automotive sector represents a meaningful pillar for the economic stability and social welfare of several countries; on the other hand, with its huge supply chain, it is often a cause of negative environmental as well as social impacts. In the past, the main focus was the reduction of environmental impacts in the use phase first, and then along the entire life cycle. More recently, the attention has moved to include the social performance according to the three pillars concept of sustainability. If the target of achieving a more sustainable development is clear, the methodology and tools that should be used to assess the sustainability performance of a vehicle along its life cycle still needs to be settled. This chapter presents and discusses the benefits and challenges of methods and tools that have been used in the automotive industry.

Keywords Life cycle assessment • Life cycle costing • Life cycle management • Life cycle sustainability assessment • Social hotspot database • Social life cycle assessment • Social sustainability • Sustainability • UNEP/SETAC life cycle initiative

1 Introduction

Since the publication of the Brundtland report (WCED 1987), a new concept of economic development has been defined. According to it, a real long-term economic growth cannot be achieved without the insurance of social equality and environmental protection. An overview of intervention areas towards sustainable development were identified and reported by United Nations Secretary-General's High-level Panel on Global Sustainability in the report 'Resilient People, Resilient Planet: A

M. Traverso (✉) • P. Kim • S. Brattig • V. Wagner
Product Sustainability, BMW Group, Knorrstraße 147, 80788 Munich, Germany
e-mail: marzia.traverso@bmw.de

G. Sonnemann, M. Margni (eds.), *Life Cycle Management*,
LCA Compendium – The Complete World of Life Cycle Assessment,
DOI 10.1007/978-94-017-7221-1_24

future worth choosing' (UN 2012). The Panel's list of sustainability actions to be implemented consists of: eradicating poverty; reducing inequality and making growth inclusive; making production and consumption more sustainable, combating climate change and respecting a range of other planetary boundaries; enabling consumers to make sustainable choices and to advance responsible behaviour individually and collectively; managing resources and enabling a twenty-first century green revolution in the fields of agriculture, energy and technology. Among these challenges, the automotive industry can play a meaningful role along its whole supply chain in the achievement of more sustainable production and consumption. In the past, the main focus was placed on the environmental performance of the vehicle with particular attention to the use phase. In recent years, thanks to the media and non-governmental organizations, the interest in the social and working conditions of employees and the social impact on the local communities has been increasing.

According to several scientific references the assessment of sustainability performance of a product can be obtained with the life cycle sustainability assessment (LCSA) as defined in (Finkbeiner et al. 2010; Klöpffer 2008; UNEP/SETAC 2012):

$$\mathrm{LCSA} = \mathrm{LCA} + \mathrm{LCC} + \mathrm{SLCA} \tag{24.1}$$

This formal equation means that to assess the sustainability performance of the product, we need to consider jointly the life cycle assessment (LCA) (ISO 2006), the life cycle costing (LCC) (Schmidt 2003) and the social life cycle assessment (SLCA) (UNEP/SETAC 2009) within the same or equivalent system boundary and related to the same functional unit of the considered product.

If the formal definition of this method has found the agreement of the scientific community, its implementation still presents a lot of challenges. These challenges are particularly significant in the automotive sector where the product has a long and complex life cycle and most of the experience has been matured for the assessment of the environmental performances.

If we are looking at the life cycle of a conventional vehicle and its relative environmental impacts, the use phase still constitutes the major portion (Fig. 24.1). It plays a meaningful role in the energy consumption and it has a direct economic impact on the customer. With the past increase of crude oil prices, and the introduction of emission limits for the vehicles, the interest for less consuming and more efficient cars has led most automotive companies to reduce the cars' weight by choosing lighter materials and to improve their efficiency in the use phase.

An example of a strategy to reach this target is the Efficient Dynamics strategy developed and adopted by BMW Group. This includes improving the aerodynamics of the vehicles, choosing lighter materials for reducing the tailpipe emissions and fuel consumption, and introducing functions such as auto start/stop, air vent control and brake energy regeneration (BMW Group 2010). Focusing the attention only on the use phase has resulted in the reduction of the environmental impact this step of the life cycle, but it does not necessarily improve the overall environmental performance of the car. Indeed, when we introduce more light-materials, components

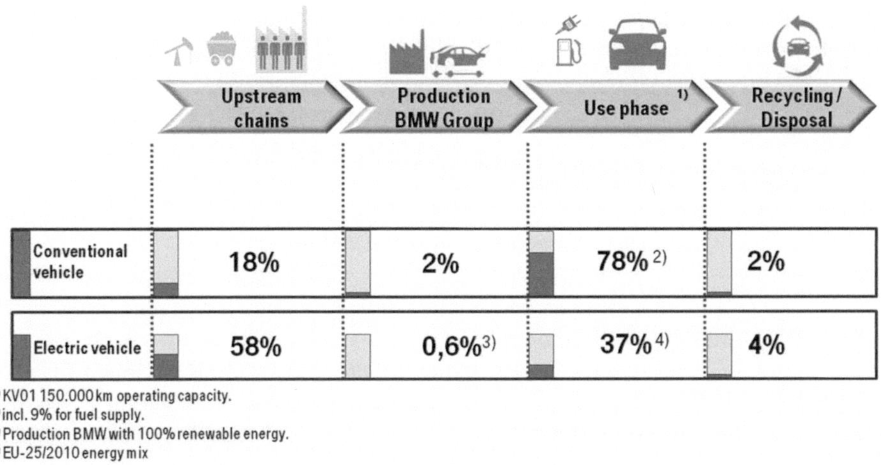

Fig. 24.1 Global warming potential [CO_2e] results along the vehicle life cycle at BMW Group

we often move the environmental impact from the use phase to the manufacturing one. That is the main reason why BMW Group develops its vehicles based on a life cycle philosophy.

If we consider a more comprehensive sustainability performance of a product then the focus on only the environmental impacts, or even more so on just one environmental impact category, e.g. global warming potential (GWP) (IPCC 2001) is not enough. Focusing the attention only on the reduction of CO_2e leads to the usage of more renewable raw materials in a product, but this can cause a negative impact on biodiversity as well as generate competition with the food sector in developing countries if not managed properly. Moreover, the agriculture sector is considered a compartment with high social risks; in fact working conditions, wages, and human rights are often social issues of this sector and must be evaluated.

A picture of the available and currently used techniques to assess environmental, economic and social sustainability performances of a product are presented in this chapter with their benefits and limits.

2 Assessing Sustainability Performance Within a Vehicle Life Cycle

2.1 Life Cycle Assessment of a Car

Life cycle assessment (LCA) is recognized as a scientifically valid method for assessing the environmental performance of a product and it is the only life cycle technique which has been standardized so far (ISO 2006b: ISO 14040).

The LCA according to the ISO 14040/44 (ISO 2006a: ISO 14040, ISO 2006b: ISO 14044) is currently used at BMW Group as a supporting decision-making tool in the development process of the car to orient designers and engineers towards developing a car with a better environmental performance. The LCA is also used as monitoring tool to check the product concept and its relative environmental impacts in each phase of the development process.

Reduction targets for each environmental impact category can be established according to the company and national strategy, and the relative improvements of a car compared with its previous model can consequently be assessed and confirmed. The main focus of the automotive sector is still on global warming potential (GWP) (IPCC 2001) for the reduction of greenhouse gases along the entire life cycle of the vehicle, but the other impact categories are monitored as well. Realizing a car with a better LCA balance than its previous model is only possible by acting on the use phase as well as in the manufacturing one. Examples of measures used at BMW Group to reach a higher environmental performance are: use of more secondary source of materials such as metals and thermoplastics, promotion of less energy-intensive materials, components produced with renewable energy sources, and improvement of the recyclability of the car's components at the end of life (EoL).

According to the ISO 14040/44 (2006) you can carry out an LCA on a car and obtain different results. This happens because the ISO 14040 and ISO 14044 (ISO 2006a: ISO 14040, ISO 2006b: ISO 14044) leave a lot of freedom and flexibility in the implementation of the method; several parameters, such as system boundaries, allocation criteria, data sources and database used can be established differently case by case and can strongly affect the results. The important element for making the LCA study valid is its transparency and the reproducibility of its implementation. Most LCA studies at BMW Group are used to support internal decision-making processes and they contain confidential information that cannot be easily published. In order to conform to the ISO 14044 regarding the reproducibility of the implementation, and to be able to use part of the LCA results in the company's communication, it is necessary to ensure that the entire process, the data, the results and the related communication material are validated and proofed by a third party certification office. This procedure is commonly used in the German automotive sector, and the proliferation of different "certificates" has increased in the last few years. BMW Group reports its LCA results in the Sustainability Value Report (BMW Group 2013), and with the Environmental Certification validated by TÜV SÜD (BMW Group 2014).

Considering this, what are the further levers to improve the environmental performance of a vehicle?

If it is true that a reasonable improvement has been reached and is still ongoing in the use phase, it is also true that more efforts have to be made to improve it by considering the manufacturing phase. The manufacturing phase of a car includes all phases from extraction of raw materials throughout the components' production to the assembly and manufacture of the car itself. The last part is normally directly under the control of the OEM (Original Equipment Manufacturer), whereas the rest is carried out by the upstream supply chain. The necessity to find new levers and new potentials in the supply chain has led to a push towards a more direct cooperation with our suppliers for a more sustainable production. This is clearly a big challenge in the automotive sector where the supply chain is long and complex and involves a large number of actors. The approach used in this case can be very different company to company and from car to car. A proliferation of activities in sectors strongly related with the automotive industry has started to improve the environmental and/or sustainability performances and to harmonize the methodologies used. An example of these initiatives is the Aluminium Stewardship Initiative.

The Aluminium Stewardship Initiative (ASI) was launched in 2012 to foster greater sustainability and transparency throughout the aluminium industry. It involves several key industry players in the aluminium supply chain such as BMW Group, Rio Tinto, Novelis, to name just a few (Aluminium Stewardship Initiative 2014). The main target of this initiative is to develop an Aluminium Stewardship Initiative Standard to address the environmental, social and governance aspects of the aluminium value chain. The ASI aims, by mobilising a broad base of stakeholders, to establish and promote responsible environmental, social and governance practices across the aluminium value chain. Throughout this standard it will be possible to further support the closed-loop recycling process in the aluminium production system and the inventory data retrieval for a transparent LCA.

At BMW, the LCA is also often used as an assessment method to compare different mobility concepts: conventional and electro-mobility, public transportation, car sharing. It is used to understand the potential impact of the development of one of these concepts on a large scale or to compare concepts among one another towards a more sustainable choice.

2.2 *Social and Economic Performances*

According to the definition of sustainability, economic and social factors should also be considered to establish the comprehensive sustainability performance of a product.

The economic performance in terms of the business case of a project/product is always considered, but it does not necessarily follow the framework introduced by the LCC (Swarr et al. 2011). LCC is defined as a compilation and assessment of all costs related to a product for a specific stakeholder over its entire life cycle, from production to use, maintenance and disposal (Swarr et al. 2011; Hunkeler et al. 2008). The life cycle costing implementation and results thus strongly depend on the chosen stakeholder. Indeed different data should be collected if we want to consider the costs paid and revenues obtained by the manufacturing company or when they are related to the consumer.

The LCC can also be used as a standalone method to assess economic and environmental performance of a product. In this case a monetization of the environmental externalities and economic costs is necessary. The monetization allows one to translate the relative environmental impacts (externalities) into costs, for example the carbon taxes. In general, the life cycle costing is more used in terms of cost engineering and it is performed by the industry with a different perspective than the LCA to calculate the business case of the project/product.

The social impacts are usually considered in the industry at corporate level; this means that social impacts are assessed per company and/or per each site of the company and usually the evaluation is extended only up to the first tier supplier.

This approach does not allow the integration of the results of the corporate social performance in the evaluation of a more sustainable component or material. To integrate the social considerations in the product development process, data on the social impact of a product life cycle should be collected.

According to the definition of UNEP/SETAC Life Cycle Initiative, the SLCA is defined as the assessment of social performance of a product along its life cycle and by considering at least five stakeholder groups: workers, customers, local community, society, and value chain actors not including consumers (UNEP/SETAC 2009). The SLCA is the youngest technique among the life cycle thinking approaches, and even if an UNEP/SETAC guideline and the relative Methodological Sheets (UNEP/SETAC 2013; Benoît-Norris et al. 2011b) to support LCA users in the implementation have been published, the scientific community as well as the industry has not come to a complete agreement on several crucial points. Challenges of SLCA are: selection and definition of a valid and commonly accepted set of indicators, availability of inventory data at product level, definition of pathways from inventory to the impact assessment, and so on.

A big challenge is the necessity to collect primary data from company and local communities. The only example of database developed according to the UNEP/SETAC guideline is the Social Hotspot Database (SHDB) (Benoît-Norris et al. 2011a). This database uses a Top-Down Input/Output approach to derive the social hotspot risk at country and sector level. The SHDB project offers an online database that allows users to browse data on social risks by sector, country, or risk theme. There are 227 countries and 57 economic sectors to choose from. The data comprehensively addresses social issues on human rights, working conditions, community impacts and governance issues, via a set of nearly 150 risk indicators grouped within 22 themes. Examples of sectors covered by the database are transport, transport equipment, motor vehicles and parts, and metal. It is clear that the SHDB can be used to make a screening on the potential risks of these sectors by country, but further efforts have to be paid for a primary data collection.

If on one hand, the scientific community is still debating the necessity and scientific validity of the social assessment at product level, on the other hand the media's, NGOs', and customers' concern about the consequences of buying a product that is produced under bad social conditions is increasing.

No specific initiative has been made at product level from the automotive sector. Most of the activities that have been developed are to improve the sustainability and social performance of the supply chain, by identifying social hotspots. An example of this initiative was the development of a common sustainability questionnaire among the OEMs of the automotive sector to screen the sustainability performance of the suppliers.

The most recent industrial social assessment initiatives that have been developed at product level are: the Roundtable of Product Social Metrics, Social Metrics for chemical sector by WBCSD, and case studies conducted on the SLCA of electronics products by the Sustainability Consortium. There are no specific initiatives on the SLCA from the automotive sector.

BMW Group is one of the founding members of the Roundtable of Product Social Metrics. In 2013, for the first time, a group of companies decided to sit down together and to define, based on the literature, the ILO Conventions and their strategy in order to develop a common methodology to assess social performance of a product along its life cycle.

The project is now running the third phase and the main results of the previous phases are included in the *Handbook of Product Social Impact Assessment* (Fontes et al. 2014). Six case studies were realized, mainly among the Roundtable members, to implement and validate the methodology presented in the handbook. This initiative, that also involved BMW Group, shows that the SLCA is a feasible and practicable methodology. The main challenge in its implementation is the necessity to directly involve the entire supply chain to collect primary social inventory data. Because these data are very sensitive, a strong cooperation and linking interests between OEM (requesting the primary data) and each 1-tier supplier (data giver) are necessary.

This challenge also existed for the environmental LCA 20 or 30 years ago, when software and databases such as GaBi® (PE International 2010) and SimaPro® (PRé Consultants 2013) were not available and performing an LCA of a product cost significant effort in terms of time and money just to collect primary data.

3 Results and Discussion

The long walk to a more sustainable production and consumption is a trip that needs to be taken by the actors of different sectors in a participatory and harmonized manner. The automotive industry can play a strategic role in creating this partnership and cooperation and it will affect all sectors connected with the automotive one, such as metals or electronics. Many challenges still have to be faced in moving towards a comprehensive life cycle sustainability assessment of a vehicle and some of them are methodological. A relative harmonized approach is used to assess the environmental performance of a product along its life cycle according to ISO 14040/44. BMW uses the LCA not only for assessing its vehicles, but as a supporting tool to address choices and decisions along the entire development process, from product concept creation to the vehicle's start-of-production. Measurable targets for the full life cycle are identified and established in the earliest stages of the development process. These are monitored and documented during vehicle development and the results independently certified as requested by the ISO 14040 standard. Therefore the LCA is mainly used as an internal supporting decision-making tool towards a more sustainable production.

The SLCA has taken the first step towards a more harmonized approach with the *Handbook of Product Social Impact Assessment*, even if the scientific community has not yet found a methodological solution for all implementation steps. Indeed, according to the LCA approach, more efforts have to be made to develop characterization factors which allow the translation of the inventory indicators (such as Living Wage and Working Hours) into the impact categories (Human Well-being, or Human Dignity).

References

Aluminium Stewardship Initiative (2014) http://aluminium-stewardship.org/. Accessed 12 Feb 2015

Benoît-Norris C, Aulisio D, Norris GA, Hallisey-Kepka C, Overakker S, Vickery Niederman G (2011a) A social hotspot database for acquiring greater visibility in product supply chains: overview and application to orange juice. In: Finkbeiner M (ed) Towards life cycle sustainability management. Springer, 53–62. doi:10.1007/978-94-007-1899-9_6

Benoît-Norris C, Vickery-Niederman G, Valdivia S, Franze J, Traverso M, Ciroth A, Mazijn B (2011b) Introducing the UNEP/SETAC methodological sheets for subcategories of social LCA. The International Journal of Life Cycle Assessment. Springer, Berlin/Heidelberg, pp 1–9, 31 May 2011. ISSN: 0948–3349

BMW Group (2010) BMW efficient dynamics. Joy powered by tomorrow's technologies – today as a standard. http://www.bmw.com/com/en/insights/technology/efficient_dynamics/phase_2/_shared/catalogue/pdf/efficient_dynamics_catalogue.pdf. Accessed 22 Feb 2015

BMW Group (2013) Sustainable value report 2013. http://www.bmwgroup.com/com/en/responsibility/sustainability/sustainability-management/index.html. Accessed 22 Feb 2015

BMW Group (2014) Environmental certification BMW i3. http://www.bmwgroup.com/com/en/_common/_pdf/Enviromental_Certification_i3.pdf. Accessed 22 Feb 2015

Finkbeiner M, Schau E, Lehmann A, Traverso MT (2010) Life cycle sustainability assessment. Sustainability 2(10):3309–3322. http://www.mdpi.com/2071-1050/2/10/3309/

Fontes J, Bolhuis A, Bogaers K, Saling P, van Gelder R, Traverso M, Das Gupta J, Bosch H, Morris D, Woodyard D, Bell L, van der Merwe R, Laubscher M, Jacobs M, Challis D, Alvarado C, Duclaux C, Slaoui Y, Culley H, Zink S (2014) Handbook of product social impact assessment Version 2.0 PRé sustainability

Hunkeler D, Lichtenvort K, Rebitzer G (eds) (2008) Environmental life cycle costing. SETAC, Pensacola, FL (US) in collaboration with CRC Press, Boca Raton

IPCC (2001) Climate change 2000, impacts, adaptation and vulnerability, contribution of working group II to the third assessment report of the intergovernmental panel on climate change. Cambridge University Press, Cambridge

ISO (2006a) ISO 14040 – Environmental management – life cycle assessment – principles and framework. Geneva

ISO (2006b) ISO 14044 – Environmental management – life cycle assessment – requirements and guidelines. Geneva

Klöpffer W (2008) Life cycle sustainability assessment of products. Int J Life Cycle Assess 13(2):89–95

PE International (2010) Gabi 4.3 LCA software. Leinfelden-Echterdingen. www.gabi-software.com/uk-ireland/index/

PRé Consultants (2013) Introduction to LCA with SimaPro 8. http://www.pre-sustainability.com/download/SimaPro8IntroductionToLCA.pdf. Accessed 22 Feb 2015

Schmidt WP (2003) Life cycle costing as part of design for environment environmental business cases. Int J Life Cycle Assess 8(3):167–174. doi:10.1007/BF02978464

Swarr T, Hunkeler D, Klöpffer W, Pesonen H-L, Ciroth A, Brent AC, Pagan R (2011) Environmental life cycle costing: a code of practice. SETAC Press, Pensacola

UN (2012) United Nations secretary-general's high-level panel on global sustainability. Resilient people, resilient planet: a future worth choosing. United Nations, New York

UNEP/SETAC (2009) Guidelines for social life cycle assessment of products. 103 p

UNEP/SETAC (2012) Towards life cycle sustainability products. Edition UNEP/SETAC LCI

UNEP/SETAC (2013) The methodological sheets for social life cycle assessment – a practical support to implement SLCA of Product. http://www.lifecycleinitiative.org/wp-content/uploads/2013/11/S-LCA_methodological_sheets_11.11.13.pdf. Accessed 22 Feb 2015

WCED (1987) Our common future. Oxford University Press, New York

Chapter 25
Life Cycle Management as a Way to Operationalize the Creating Shared Value Concept in the Food and Beverage Industry: A Case Study

Angela Adams, Urs Schenker, and Yves Loerincik

Abstract Nestlé is using life cycle management approaches to operationalize the Creating Shared Value concept that is strongly promoted by the company. Following the success of the implementation of PIQET, an LCA software for packaging designers to integrate environmental criteria in their decision-making, Nestlé has decided to develop a specific tool moving from packaging to product ecodesign tool. This tool, EcodEX, is linked to the recipe system of the company to facilitate the realization and improve the quality of LCA. The number of users has increased to reach 700 people today. However, there have been many challenges to overcome, such as: the availability and the management of inventory data, following-up the latest methodology, as well as the training and education of users.

Keywords Creating Shared Value • Data development • Data management • Ecodesign • EcodEX • Education • Life cycle assessment • Life cycle management • Nestlé • PIQET • Training

1 Introduction

Systematically applying Creating Shared Value principles in practice and integrating them into the way a company is doing business is a huge challenge, and in some cases, even a paradigm change. Life cycle management is an important and powerful approach that Nestlé uses to operationalize creating shared value.

A. Adams (✉) • Y. Loerincik
Quantis-International, EPFL Innovation Park, Bât D, CH-1015 Lausanne, Switzerland
e-mail: angela.adams@quantis-intl.com

U. Schenker
Nestlé Research Center, Vers-chez-les-Blanc, CH-1000 Lausanne 26, Switzerland

G. Sonnemann, M. Margni (eds.), *Life Cycle Management*,
LCA Compendium – The Complete World of Life Cycle Assessment,
DOI 10.1007/978-94-017-7221-1_25

Nestlé adopted a simplified ecodesign approach early in the design phase since many years through the implementation of PIQET (Karli et al. 2010), a tool that optimizes the environmental performance of packaging. Since then, packaging designers using life cycle assessment to support their design choices is significantly increased.

Several studies (Humbert et al. 2009; Kim et al. 2013) have shown that in the food and beverage industry the product itself generally dominates the impacts of any life cycle assessment (LCA) and that, comparatively, packaging has a limited impact on the environment. The resolution to expand the ecodesign approach to the products was therefore an obvious decision for Nestlé. To do so, a new simplified ecodesign tool, EcodEX (Schenker et al. 2014), has been adopted to evaluate the current and new products, taking into account the whole life cycle, including consumer behavior.

2 Implementing EcodEX: A Product Design Tool at Nestlé

2.1 *Initial Brief*

Prior to the choice to develop a specific tool for product ecodesign, a detailed requirement analysis was completed indicating that the ideal tool has to:

- Be user-friendly: because the users are not LCA experts and results must be quickly generated at a moment's notice, the tool has to be easy and quickly ready for use.
- Be linked to Nestlé's recipe management IT system in order to ensure high quality ingredient data, processing data and eventually other types of data, without spending a lot of time on data collection.
- Give reliable results: although the tool is a simplified tool, the results must be as robust as possible to support relevant decisions.

The direct benefits of implementing EcodEX include products with better environmental performances as well as conscious decision-making where a design alternative with a better environmental performance is selected (e.g., for reasons other than environmental reasons, such as legal compliance, consumer preference or quality).

2.2 *A Growing Number of Users*

EcodEX (Selerant Corporation 2015) has been regularly used by the Nestlé R&D community for approximately 1 year now. To date, approximately 700 users have analyzed over 900 LCA scenarios. In addition, approximately 1,5000 packaging scenarios which were generated using PIQET over the last 6 years are available. The results of these scenarios have influenced the development process of new products in several ways:

A. If the environmental performance of the new design is of less quality than before:
   - Product designers may work on an alternative design that improves the environmental performance of the product. In the past, in the absence of EcodEX, designers were not always aware that their designs had negative impacts on the environment.
   - The product development may be interrupted or stopped altogether because environmental performance is of key importance to the business unit and market. Without the information provided by EcodEX, decision-makers are not in the position to take timely decisions in the earlier stages of the development process. In cases where LCA studies were performed by an external consultant, the conclusions could not be implemented in the design process.
   - A decision may be made where the project goes ahead despite the adverse environmental effects because of other criteria such as consumer preference, nutrition, quality, or legal requirements. In this case, EcodEX provided all information available for the decision making process and may lead to follow-up projects in order to improve.
B. If the environmental performance of the new design improves, the results can be shared internally to gain support for the new product development. This may result in a faster adoption of the product, or the use of the product in other markets. Many EcodEX users consider the tool to be very useful to address upcoming needs for consumers.

The main objective of the ecodesign process is to ensure that the environmental performance of Nestlé's products is taken into account in internal decision-making. An added benefit is that the process identifies external communication opportunities. If in line with the brand positioning, such external communications can be achieved through comprehensive life cycle assessments with external consultants, because EcodEX does not comply directly with the requirements of the ISO 14040/14044 standards (ISO 14040 2006) for external communication of comparative assertions, given third party critical reviews are not part of the system. This may be a future development.

2.2.1 The Tivall Example

Tivall is a Nestlé brand that offers healthy and nutritious meat substitute products. *Tivall* products have been specifically developed to be nutritionally balanced and to be a complete source of protein fortified with nutrients that may otherwise be lacking in a meat-free diet. The *Tivall* range is mainly based on soya and wheat proteins.

Nestlé has used environmental LCAs to compare the environmental performance of a *Tivall* vegetarian burger (Hälsans Kök) with a conventional meat burger. The results show that the vegetarian *Tivall* burgers have a considerably improved environmental performance for all indicators compared to conventional meat burgers. The greenhouse gas emissions, for instance, are approximately 80 % lower. The results show that the vast majority of environmental impacts of the beef burger are related to beef production. For the vegetarian burger, distribution also plays a relatively important role (in addition to the ingredients), because they are distributed while frozen.

The example illustrates that healthy and tasty vegetarian dishes can play an important role in improving the environmental performance of nutrition. Given the evidence for this and other similar studies, Nestlé has launched several R&D projects to further investigate new sources of protein with improved environmental performance.

2.2.2 Increasing Internal Knowhow

Another key benefit of EcodEX is that product designers' understanding of sustainability is greatly increased as they have been trained to do the environmental assessments themselves, rather than picking up the information from LCA reports or being informed about the results and recommendations. At Nestlé, it can very clearly be seen that a systematic and holistic value chain approach is used. This takes into consideration responsible sourcing, ingredients, processing, distribution, consumer use processes, and food waste across the entire cycle, rather than only packaging and transportation distances as was the main focus at the beginning of the sustainability journey of Nestlé.

Life cycle management at Nestlé played an important role in expanding the scope of the eco-design from packaging to the full product life cycle and in the implementation and deployment of EcodEX tool across whole Nestlé. The training has been systematically organized and has a strong influence.

2.3 Challenges

The development and implementation of a tool like EcodEX faces many challenges. They can be classified into four main areas: methodology, data, technical, and training. With the exception of technical challenges that will not be addressed in this chapter, hereby we briefly describe the remaining challenges that are very much related to the nature of LCA.

2.3.1 Methodology

The importance of life cycle assessment for external communication is likely to increase in the future. While EcodEX is intended for internal decision-making, consumer communication teams may want to use the results for external communication as well. Therefore, it is important that the methodology in the tool is as closely aligned with externally-developed methodologies as possible. This is not a straightforward task because external methodologies are subject to change and not all requirements are known in sufficient detail. The ISO 14040 sets the frame, and the ENVIFOOD protocol (Food SCP RT 2013) gives additional specifications for the food and beverage sector. With the on-going Product Environmental Footprint (PEF) initiative at the European level (European Commission 2013), a series of Product Category Rules (PCR) will be considered in the tool development. A technical constraint for the tool is therefore able to follow the rapid evolution of the methodology.

2.3.2 Data

There are about 10,000 of ingredient specifications in the recipe management system. These specifications had to be matched with a reduced number (couple hundred) available life cycle inventory (LCI) datasets. The need for representative data is critical. Until very recent, the access to good quality LCI profiles was an issue in the food and beverage sector. To address this challenge, Nestlé partnered with external stakeholders to initiate the World Food LCA Database, a multi-company and authority-initiative, led by the consulting firm Quantis, and the research center Agroscope, to mutually develop quality food data (Peano et al. 2012). This project will lead to the development of hundreds of specific LCI profiles in the field of food and beverage, feeding EcodEX, and increasing the accuracy of LCAs. Other database projects such as Agri-footprint have also produced datasets, leaving us with the challenge of data harmonization (methodology, format).

Both the evolution of the methodology and the generation or use of new data leads to two main challenges. The first, which has been briefly discussed above, is to ensure that the tool is capable of following the evolution of methodologies and updating data. The second challenge is to make sure that the process to update and deploy these changes is compatible with the way the tool is used. In particular, an update could have an influence on a decision that is currently being taken or has just been taken by a designer. This therefore implies having a tool to manage data, including quality guidelines or a validation process, for instance, as well as a process to test, update, inform, and support users. Nestlé has developed its own data management guidelines and has a data quality manager.

2.3.3 Training and Support

Last but not least, users have to be extensively trained to gain the necessary understanding on sustainability and life cycle assessment, which inevitably requires time and significant resources. Most users have a background in nutrition, food science, process engineering, or packaging technology, which provides them valuable insight in their focus area. In order to be able to make meaningful decisions on the environmental impacts of the products they develop, they need to expand their skill sets to include this topic. It cannot be expected that the quality of the assessments is comparable to that of a trained LCA specialist at the beginning. Nestlé has trained the designers with the LCP board game, a "serious" game to learn about the concept of life cycle assessment that has proven to be very powerful (Fig. 25.1). But an initial training course is not sufficient. Practice and frequent use of the tool, along with

Fig. 25.1 The LCP board game is the ideal tool to educate employees to life cycle thinking and life cycle assessment

ongoing training through webinars (conference calls with desktop-sharing) and a review of the studies performed ensure that EcodEX users will perform high quality assessments over time.

Another challenge that had to be addressed is the reliability of a simplified eco-design tool and level of confidence on its outcomes. To verify this, Nestlé decided to evaluate the outcomes of EcodEX against stand-alone LCA case studies.

3 Conclusion

Nestlé's objective was to deploy an ecodesign approach for the development of new products. So far it is a success considering the numerous studies that have been performed and the way the tool has influenced designers' decisions accounting for environmental criteria. A technical solution alone, i.e. a tool, is, however, not sufficient to guarantee the success of such a sustainability project. Nestlé's eco-design initiative requires a transversal approach involving diverse teams. Life cycle management ensuring collaboration between different departments within the company and with stakeholders outside the organization is key to implement and sustain such an initiative. Sustainability will certainly be a key differentiator in the coming years, and it is very likely that today's active and engaged companies will have the best tools to be the leaders of tomorrow.

Continuous improvement of Nestlé's environmental performance is challenging and requires a holistic, interdisciplinary, and collaborative approach. A tool like EcodEX combined with appropriate internal business processes provides a very strong base to achieve this.

References

European Commission (2013) RECOMMENDATION (2013/179/EU) on the use of common methods to measure and communicate the life cycle environmental performance of products and organisations. Annex II (Product Environmental Footprint (PEF) Guide))

Food SCP RT (2013) ENVIFOOD protocol, environmental assessment of food and drink protocol. European Food Sustainable Consumption and Production Round Table (SCP RT), Working group 1, Brussels

Humbert S, Loerincik Y, Rossi V, Margni M, Jolliet O (2009) Life cycle assessment of spray dried soluble coffee and comparison with alternatives (drip filter and capsule espresso). J Clean Prod 17(15):1351–1358, Environmental Impacts of Food Production and Consumption, A research report completed for the Department for Environment, Food and Rural Affairs by Manchester Business School

ISO 14040 (2006) Environmental management – life cycle assessment – principles and framework. Geneva

Jean-Michel Moutot, Ganaël Bascoul, Marketig et développement durable: Stratégie de la valeur étendure, Edition Dunod

Karli L, Horne R, Carre A (2010) PIQET: the design and development of an online 'streamlined' LCA tool for sustainable packaging design decision support. Int J Life Cycle Assess 15(6):608–620

Kim D, Thoma G, Nutter D, Milani F, Ulrich R, Norris G (2013) Life cycle assessment of cheese and whey production in the USA. Int J Life Cycle Assess 18(5):1019–1035

Peano L, De Schryver A, Humbert S, Loerincik Y, Gaillard G, Lansche J, Nemecek T (2012) The world food LCA database project: towards more accurate food datasets. In: Proceedings 2nd LCA conference, Nov 2012, France

Schenker U, Espinoza-Orias N, Popovic D (2014) EcodEX: a simplified ecodesign tool to improve the environmental performance of product development in the food industry. In: Proceedings of the 9th international conference on life cycle assessment in the agri-food sector, San Francisco, Oct 2014

Selerant Corporation (2015) EcodEX, Ecodesign Software. http://www.selerant.com/main/en-us/solutions/ecodesign.aspx

Index

G. Sonnemann, M. Margni (eds.), *Life Cycle Management*,
LCA Compendium – The Complete World of Life Cycle Assessment,
DOI 10.1007/978-94-017-7221-1

MIX
Papier aus verantwortungsvollen Quellen
Paper from responsible sources
FSC® C105338

If you have any concerns about our products,
you can contact us on
ProductSafety@springernature.com

In case Publisher is established outside the EU,
the EU authorized representative is:
Springer Nature Customer Service Center GmbH
Europaplatz 3, 69115 Heidelberg, Germany

Printed by Libri Plureos GmbH
in Hamburg, Germany